Recent Advances in the Design of Structures with Passive Energy Dissipation Systems

Recent Advances in the Design of Structures with Passive Energy Dissipation Systems

Special Issue Editors

Giuseppe Ricciardi

Dario De Domenico

Ruifu Zhang

MDPI • Basel • Beijing • Wuhan • Barcelona • Belgrade

Special Issue Editors
Giuseppe Ricciardi
Department of Engineering,
University of Messina
Italy

Dario De Domenico
Department of Engineering,
University of Messina
Italy

Ruifu Zhang
Department of Disaster
Mitigation for Structures,
Tongji University
China

Editorial Office
MDPI
St. Alban-Anlage 66
4052 Basel, Switzerland

This is a reprint of articles from the Special Issue published online in the open access journal *Applied Sciences* (ISSN 2076-3417) from 2019 to 2020 (available at: https://www.mdpi.com/journal/applsci/special_issues/Structures_Passive_Energy_Dissipation).

For citation purposes, cite each article independently as indicated on the article page online and as indicated below:

LastName, A.A.; LastName, B.B.; LastName, C.C. Article Title. *Journal Name* **Year**, *Article Number*, Page Range.

ISBN 978-3-03936-060-4 (Pbk)
ISBN 978-3-03936-061-1 (PDF)

Contents

About the Special Issue Editors

Giuseppe Ricciardi has been full professor at the Department of Engineering of the University of Messina, Italy, since 2015. Previously, he was a researcher (1995–2000) and an associate professor (2000–2015) at the same university. He was also a visiting professor at the College of Engineering of the Florida Atlantic University, USA, in 2000. He has scientific responsibility for the structural section of the CERISI laboratory (earthquake engineering testing), as well as for other national and international projects in earthquake engineering. He also holds two patents for scientific inventions in the field of earthquake engineering.

Dario De Domenico is a researcher at the University of Messina, Italy. After graduating in civil engineering (2010), he obtained his PhD in materials and structural engineering at the University of Reggio Calabria (2014). After his doctoral studies, he spent one year in industry as a research project engineer (2015–2016). He has also been a visiting researcher at both the Deptartment of Civil and Structural Engineering, University of Sheffield, UK, and at the Laboratory of Mechanics and Materials, Aristotle University of Thessaloniki, Greece. His research interests include earthquake engineering, innovative structural control techniques, and mechanics of materials, especially reinforced concrete. He also holds two patents for scientific inventions in the field of earthquake engineering.

Ruifu Zhang graduated from Tongji University with a PhD in civil engineering. After graduation, he was a post-doctoral researcher at UC Berkeley and Tohoku University. He gained extensive industry experience prior to joining the Tongji University in 2015, at which he is currently an associate professor in the College of Civil Engineering. His main research interests are structure seismic safety and intelligent disaster prevention structure with a special focus on inerter systems.

applied sciences

MDPI

Editorial

Editorial for "Recent Advances in the Design of Structures with Passive Energy Dissipation Systems"

Dario De Domenico [1,*], Giuseppe Ricciardi [1] and Ruifu Zhang [2]

1. Department of Engineering, University of Messina, 98166 Messina, Italy; gricciardi@unime.it
2. Department of Disaster Mitigation for Structures, Tongji University, Shanghai 200092, China; zhangruifu@tongji.edu.cn

* Correspondence: dario.dedomenico@unime.it; Tel.: +39-0906765921

Received: 7 April 2020; Accepted: 16 April 2020; Published: 19 April 2020

Keywords: seismic isolation; energy dissipation devices; tuned mass damper; negative stiffness device; inerter system; damped structures

1. Editorial

Civil engineering structures and infrastructures are inherently vulnerable to exceptional loads related to natural disasters, primarily earthquakes, tsunamis, strong winds, and floods. Consequently, one of the major challenges in the structural engineering field for the last decades has been to conceptualize, develop, and implement effective protective systems to mitigate such vulnerability, and to improve structural robustness and resilience. Base isolation and passive energy dissipation systems have demonstrated their effectiveness in coping with different kinds of environmental forces, including earthquakes and winds, as documented in theoretical and numerical studies and shaking table tests, as well as evidence from how they actually behaved during real catastrophic events. These structural protective systems traditionally include elastomeric bearings, lead rubber bearings, sliding friction pendulum, and various kinds of dampers, such as metallic, viscous, viscoelastic, friction dampers, tuned mass dampers and tuned liquid dampers.

The working mechanism underlying the aforementioned technologies is well known, and basic methods for their rational design and implementation are well established. Notwithstanding, there is an ever-growing interest in developing novel analytical and/or numerical tools to design structures equipped with optimally configured devices. Indeed, the design of such devices benefits from the current state-of-the-art algorithms and solvers for their optimization, which are constantly evolving. Other recent advances in this field concerns the development of cutting-edge models to reliably capture a series of complex nonlinear phenomena characterizing the hysteresis of such devices, calibrated based on experimental findings. On the other hand, the family of devices and dissipative elements for structural control keeps broadening due to the increasing performance demands of structures and due to new progress achieved in material science and mechanical engineering. In this context, recent advances include new strategies to develop the concept of energy dissipation into innovative devices, including negative stiffness devices, inerter-based systems, low-cost replaceable systems and dampers with a phased behavior. Although the development of new technologies generally follows established practice and underlies basic working principles, existing design methods for conventional devices are not always straightforwardly applicable to these new devices. Thus, there is an urgent need for revisiting design methodologies for such emerging technologies. Other significant contributions concern the development of hybrid protective systems based on energy dissipation devices that are conventional in their working mechanism, but that are combined together in a non-conventional arrangement so that their dynamic behavior is more effective than existing technologies.

Following these research motivations, this Special Issue collects 13 papers focused on structural protective systems applied to structures and infrastructures, including both traditional and innovative devices, using conventional and advanced design methodologies. In the Editors' opinion, each article contains undisputable scientific novelties from various perspectives (analytical, numerical, experimental, conceptual, implementation issues), proposes benchmark or emblematic engineering projects, and represents a major contribution in the design of structures with passive energy dissipation systems. The Editors hope that this article collection can somehow contribute, even if modestly, to the continuous research for more effective mitigation of the risks that natural disasters pose to humankind.

Some papers in the Special Issue concern the development of inerter-based vibration control strategies, and their deployment in civil engineering structures and infrastructures. The inerter modifies the inertial properties of the structures while adding negligible physical mass, and it may be a cost-effective solution for both seismic [2,4] and wind [1,3] vibration control. Some other papers concern the development of efficient design methods that exploit state-of-the-art algorithms and solvers for the optimization of the devices [11,13]. One paper concerns the challenging task of designing lead rubber bearings considering different performance requirements [10]. The performance of some variants and improvements of the classical concept of tuned mass damper is also analyzed in various applications and configurations [6,8,12]. Novel dissipation devices are finally developed and analyzed from an experimental and a numerical point of view [5,7,9]. A brief description of each article is given below.

In paper [1], Wang et al. present free vibration analysis of a taut-cable with two inertial mass dampers (IMDs) either symmetrically placed along the cable, or installed at the same end of the cable. The results in terms of the supplemental modal damping ratio provided by the IMDs for the two installation configurations and for different values of the damping coefficient are obtained by complex modal analysis, and are critically discussed. The novel contributions of this paper are twofold: (1) the authors demonstrate that IMDs have a superior control performance over traditional viscous dampers (VDs); (2) they also notice that a single IMD may be incapable of providing supplemental modal damping in a super-long cable, especially for the multimode cable vibration mitigation. A wide parametric study is presented to investigate the effect of damper positions and damper properties on the control performance of the cable in practical applications.

In paper [2], Xie et al. propose a novel inerter-based vibration control system called cable-bracing inerter system (CBIS), in which tension-only cables are interposed in between the inerter device and the main frame for translation-to-rotation conversion. This paper has the merit of contributing to widening the knowledge on the possible implementation technologies of the inerter, besides the rack-and-pinion mechanism, ball-screw device, fluid inerter, and inerter with clutch. Although in this paper the analysis is limited to a single-degree-of-freedom (SDOF) system equipped with the proposed CBIS, the authors demonstrate that this configuration can effectively be used for rapid seismic retrofit of structures, benefitting from ease of installation, deformation relaxation at the connecting joints, and an adaptive layout for nonconsecutive-story deployment. The proposed system reveals a mass amplification effect and a non-contacting damping mechanism. Through a parametric study, the influence of dimensionless parameters, such as inertance-mass ratio, stiffness ratio and additional damping ratio on vibration mitigation are studied in terms of displacement response and force output. A performance-oriented multi-objective design framework is also established in order to identify the parameters of the CBIS that can satisfy the target vibration mitigation.

In paper [3], Wang et al. investigate the wind-induced response control of high-rise buildings through inerter-based vibration absorbers, including the tuned mass damper inerter (TMDI) and the tuned inerter damper (TID). The analysis is performed on a real 340 m tall building analyzed as case study. A realistic wind-excitation model is adopted, based on experimental measurements from wind tunnel tests obtained for a scaled prototype of the benchmark building, which accounts for the actual cross-section of the structure and the existing surrounding conditions. The results are analyzed in terms of wind-induced displacement and acceleration response. Performance-based optimization of the TMDI and the TID is carried out to find a good trade-off between displacement and acceleration-response mitigation, with the installation floor being an explicit design variable, in addition to frequency and damping ratio, and considering different wind directions. The authors demonstrate that the optimally designed TMDI/TID can achieve better wind-induced vibration mitigation than the conventional tuned mass damper (TMD), while allocating lower or null attached mass, especially in terms of acceleration response.

In paper [4], Zhao et al. propose a displacement-dependent damping inerter system (DDIS) for seismic response control. This configuration is alternative to conventional configurations of commonly used inerter systems utilizing a velocity-dependent damping. The proposed configuration implies a displacement-dependent element (DDE) in parallel to the inerter device and in series with a tuning spring, which is found to generate a larger control force in the early stage of excitation in comparison to a viscous-damping inerter system (VDIS). The DDE is governed by a bilinear elastoplastic constitutive behavior. Although in this paper the analysis is limited to a SDOF system equipped with the proposed DDIS, the authors demonstrate the influence of various DDIS-parameters through a wide parametric study based on stochastic dynamic analysis. The stochastic linearization method is used to handle the nonlinear terms, and three performance indicators related to the displacement, acceleration and filtered energy response are analyzed. Then, the seismic response is evaluated in the time domain, taking the non-linearity into account and considering both artificial and natural records. It is found that the interaction between inerter, spring and the DDE is particularly effective for the structural control. The inerter amplifies the deformation of the DDE in the DDIS by over 60%; thus, the DDIS is characterized by a higher energy dissipation capability, namely as damping enhancement effect. Because of the damping and mass-enhancement mechanism, the proposed DDIS considerably reduces the structural displacement and acceleration, and is more effective than a VDIS especially the early stage of the seismic response.

In paper [5], Zheng et al. investigate, from an experimental and numerical point of view, a novel curved steel plate damper to improve the seismic performance of structures. Analytical formulae to determine the key design parameters of the damper, namely elastic stiffness, yield strength, and yield displacement, are derived. Experimental tests are carried out on four prototypes of metallic dampers, characterized by different geometric properties, so as to identify the most effective combination of parameters in terms of stability of hysteresis and energy dissipation performance. Finite element simulations are also performed to simulate the loading process of the specimens, to investigate the strain and stress distributions and to validate the design formulae proposed in this research work.

In paper [6], Meng et al. propose a two-degree of freedom tuned mass damper (2DOFs TMD) for vibration mitigation of a suspension bridge. The simultaneous action of the two TMDs makes it possible to control both bending and torsional modes of the bridge deck. Parameters of the proposed 2DOFs TMD are optimized through a control problem, with decentralized static output feedback for minimizing the response of the bridge deck, and a graphical approach is introduced to arrange flexible beams properly according to the exact constraints. It is found that the synthetic approach, based on both the graphical approach and parameterized compliance, is an effective way to design the TMD with the expected DOFs, in order to accomplish expected modes. Moreover, experimental findings on a small scale prototype demonstrate the ability of the TMDs of suppressing several vibration modes under laboratory conditions.

In paper [7], Mena et al. develop a new low-cost energy dissipation system for application to precast concrete structures. This solution is particularly appealing for residential structures in developing countries, in which precast footings, precast structural walls, and precast concrete slabs are present. The system is based on a new connection between the precast foundation and the precast structural wall, through a series of threaded steel bars that undergo plastic deformation during a seismic event. The advantages of the new system are experimentally evaluated via pushover tests performed on a single connection, on a structural frame, and on a real-scale three-story precast concrete building. Based on the obtained experimental results, the proposed device proves to be an effective strategy to increase the ductility and to mitigate the structural damage in the structural members, as the energy dissipation is mostly concentrated in the low-cost energy dissipation device. It is concluded that the proposed energy dissipation device makes it possible to reach the performance level of "immediate occupancy", according to the American standards ACI374.2R-13.

In paper [8], Stanikzai et al. investigate the seismic response of different structural control systems, including traditional base isolated buildings and three hybrid control solutions combining the base isolation (BI) with: (a) a single TMD at the top of the building; (b) multiple tuned mass dampers (MTMDs); (c) distributed tuned mass dampers (d-TMDs). The structural control performance of the various vibration absorbers is studied considering two buildings (5-story and 10-story), and including a set of 40 earthquake ground motions, with different scale factors to capture different intensity levels. An incremental dynamic analysis (IDA) with increasing peak ground accelerations (PGAs) is performed to develop simplified fragility curves for the maximum target isolator displacement. In line with other literature studies, the combination of BI and TMD leads to a significant reduction of the isolation bearing displacements, along with a reduction of the top floor acceleration and base shear. Additionally, it is found that the MTMDs placed at the top floor and d-MTMDs on different floors of the buildings are more efficient in reducing the probability of failure of the BI building when compared to a single TMD solution.

In paper [9], Zhu et al. propose a so-called horizontal-connection and energy-dissipation structure (HES), which could be employed for horizontal connection of prefabricated shear wall structural system. This system consists of an external replaceable energy dissipation (ED) zone, mainly for energy dissipation, and an internal stiffness lifting (SL) zone for enhancing the load-bearing capacity. The ED zone may be easily replaced after damage at the end of the seismic event, while the SL zone can increase the load-carrying capacity. Through the combination of the two zones, the load-displacement curves of the HES exhibits a "double-step" behavior, which is desired to meet performance requirements at different levels of the earthquake excitation. The system is investigated through detailed finite element simulations aimed at investigating the influence of the design parameters of the connections, such as aspect ratio, shape of the plate in the ED zone and displacement threshold in the SL zone. A customized hysteretic behavior is obtained, and a phased energy dissipation performance can be particularly useful for improving the seismic performance of prefabricated shear wall structures against large and super-large earthquakes.

In paper [10], Ju et al. study the vibration mitigation effects of base isolation realized with lead rubber bearings (LRBs) in high-tech factories. The authors consider a wide spectrum of external excitations in terms of disturbing frequencies, namely seismic, wind and moving crane loads. They also develop a three-dimensional finite element model, including soil-structure interaction effects. The authors critically discuss the obtained results in view of the achievement of different performance requirements under different types of external excitation. In particular, large initial stiffness is useful to reduce micro-vibrations due to moving crane loads and wind loads, as well as during small or moderate earthquakes, while small final LRB stiffness is necessary to reduce the seismic displacements during strong earthquakes. It is found that the effectiveness of the seismic isolation is excellent for earthquakes with short dominant periods but it decreases with increasing the dominant periods. Since micro vibration is a major concern for high-tech factories, an appropriate design of LRBs should entail a large initial stiffness and a small ratio of the final stiffness over the initial stiffness.

In paper [11], Naranjo-Pérez et al. develop a motion-based design method under uncertainty conditions for the vibration mitigation of stay cables under wind-induced vibrations. A robust design of the devices is carried out based on a constrained multi-objective optimization problem, wherein the a multi-objective function is defined in terms of characteristic parameters of the damping devices, and an inequality constraint is additionally included to guarantee an acceptable probability of failure of the structural system. Following the United States Federal Highway Administration guidelines, the design criterion is governed by the compliance of the vibration serviceability limit state, quantitatively indicated by a reliability index being greater than a threshold value. The performance of the proposed design method is numerically validated considering the longest stay cable of the Alamillo bridge (Spain) equipped with different passive damping devices, namely viscous, elastomeric and friction dampers.

The proposed motion-based design method turns out to be more effective than a conventional method, by reducing the size and the budget of the devices, which facilitates its feasibility of implementation.

In paper [12], Rupakhety et al. explore the effectiveness of shared tuned mass damper (STMD) in reducing seismic pounding of adjacent buildings. The authors carefully revisit the idea of STMD reported in a paper from the literature [*Earthq. Eng. Struct. Dyn.* **2001**, *30*, 1185–1201]. In particular, they noted that, strictly speaking, such solution does not act like a shared tuned mass damper. Optimal parameters of the STMD are evaluated by minimizing the cost function using a genetic algorithm. Two optimal (tuning) parameters are found: the first solution corresponds to the device being tuned to one of the two buildings, thus being a classical TMD, and not a shared TMD; the second solution corresponds to a very stiff system, in which the TMD mass hardly moves, thus it is equivalent to a viscous connection between the two adjacent buildings. In the second solution, the TMD mass introduces no benefit while, counterproductively, it adds unnecessary load to the structure. Any reduction in response resulting from the STMD is due to the viscous coupling of the two buildings, rather than to the tuned vibration of the STMD mass. Based on the authors' study, the STMD strategy is not effective. This conclusion is obtained based on results from a large set of real earthquake ground motions, including 462 ground motion records from 110 earthquakes recorded in Europe and the Middle East.

In paper [13], Palacios-Quiñonero et al. develop an optimal passive actuation scheme of multibuilding systems composed of both interstory and interbuilding linear viscous dampers. Unlike other literature studies that are limited to one or at the most two adjacent buildings, the paper addresses a set of five identical planar frames. Optimization is carried out using a hybrid discrete-continuous formulation, based on H∞ objective function combined with genetic algorithm approaches and parallel computing techniques. The optimal position and the optimal damping coefficient of the devices are determined through the developed design procedure. Three different classes (or configurations) of distributed damping systems are analyzed, with the frames being subjected to the El Centro ground motion. The resulting seismic performance is analyzed in terms of the peak interstory drift response of the various buildings and story-accelerations peak values, with an eye for the pounding risk. The proposed design methodology proves to be very effective from a computational point of view, and promising for application to large-scale multibuilding systems.

Author Contributions: Writing—Original Draft Preparation D.D.D.; Writing—Review & Editing, G.R. and R.Z. All authors have read and agreed to the published version of the manuscript.

Funding: This research received no external funding.

Acknowledgments: We would like to thank all authors for their valuable contributions, the many dedicated referees for their useful guidance to improve the papers, the Editorial Team of Applied Sciences and, above all, Sharon Wang for the professional assistance, support and kindness demonstrated during the management of this Special Issue.

Conflicts of Interest: The authors declare no conflict of interest.

References

1. Wang, Z.; Yue, F.; Gao, H. Free Vibration of a Taut Cable with Two Discrete Inertial Mass Dampers. *Appl. Sci.* **2019**, *9*, 3919. [CrossRef]
2. Xie, L.; Ban, X.; Xue, S.; Ikago, K.; Kang, J.; Tang, H. Theoretical Study on a Cable-Bracing Inerter System for Seismic Mitigation. *Appl. Sci.* **2019**, *9*, 4096. [CrossRef]
3. Wang, Q.; Qiao, H.; De Domenico, D.; Zhu, Z.; Xie, Z. Wind-Induced Response Control of High-Rise Buildings Using Inerter-Based Vibration Absorbers. *Appl. Sci.* **2019**, *9*, 5045. [CrossRef]
4. Zhao, Z.; Zhang, R.; Jiang, Y.; De Domenico, D.; Pan, C. Displacement-Dependent Damping Inerter System for Seismic Response Control. *Appl. Sci.* **2020**, *10*, 257. [CrossRef]
5. Zheng, J.; Zhang, C.; Li, A. Experimental Investigation on the Mechanical Properties of Curved Metallic Plate Dampers. *Appl. Sci.* **2020**, *10*, 269. [CrossRef]

6. Meng, F.; Wan, J.; Xia, Y.; Ma, Y.; Yu, J. A Multi-Degree of Freedom Tuned Mass Damper Design for Vibration Mitigation of a Suspension Bridge. *Appl. Sci.* **2020**, *10*, 457. [CrossRef]
7. Mena, Á.; Franco, J.; Miguel, D.; Mínguez, J.; Jiménez, A.C.; González, D.C.; Vicente, M.Á. Experimental Campaign of a Low-Cost and Replaceable System for Passive Energy Dissipation in Precast Concrete Structures. *Appl. Sci.* **2020**, *10*, 1213. [CrossRef]
8. Stanikzai, M.H.; Elias, S.; Rupakhety, R. Seismic Response Mitigation of Base-Isolated Buildings. *Appl. Sci.* **2020**, *10*, 1230. [CrossRef]
9. Zhu, L.; Kong, L.; Zhang, C. Numerical Study on Hysteretic Behaviour of Horizontal-Connection and Energy-Dissipation Structures Developed for Prefabricated Shear Walls. *Appl. Sci.* **2020**, *10*, 1240. [CrossRef]
10. Ju, S.H.; Yuantien, C.C.; Hsieh, W.K. Study of Lead Rubber Bearings for Vibration Reduction in High-Tech Factories. *Appl. Sci.* **2020**, *10*, 1502. [CrossRef]
11. Naranjo-Pérez, J.; Jiménez-Alonso, J.F.; Díaz, I.M.; Quaranta, G.; Sáez, A. Motion-Based Design of Passive Damping Systems to Reduce Wind-Induced Vibrations of Stay Cables under Uncertainty Conditions. *Appl. Sci.* **2020**, *10*, 1740. [CrossRef]
12. Rupakhety, R.; Elias, S.; Olafsson, S. Shared Tuned Mass Dampers for Mitigation of Seismic Pounding. *Appl. Sci.* **2020**, *10*, 1918. [CrossRef]
13. Palacios-Quiñonero, F.; Rubió-Massegú, J.; Rossel, J.M.; Karimi, H.R. Distributed Passive Actuation Schemes for Seismic Protection of Multibuilding Systems. *Appl. Sci.* **2020**, *10*, 2383. [CrossRef]

applied sciences

MDPI

Article

Free Vibration of a Taut Cable with Two Discrete Inertial Mass Dampers

Zhihao Wang [1,*], Fangfang Yue [1] and Hui Gao [1,2]

1 International Joint Research Lab for Eco-Building Materials and Engineering of Henan Province, North China University of Water Resources and Electric Power, Zhengzhou 450045, China; ffyue1993@126.com (F.Y.); hgao1993@126.com (H.G.)

2 Key Laboratory of Concrete and Prestressed Concrete Structure of Ministry of Education, Southeast University, Nanjing 210096, China

* Correspondence: wangzhihao@ncwu.edu.cn; Tel.: +86-150-9340-8299

Received: 15 August 2019; Accepted: 12 September 2019; Published: 18 September 2019

Abstract: Recently, inertial mass dampers (IMDs) have shown superior control performance over traditional viscous dampers (VDs) in vibration control of stay cables. However, a single IMD may be incapable of providing sufficient supplemental modal damping to a super-long cable, especially for the multimode cable vibration mitigation. Inspired by the potential advantages of attaching two discrete VDs at different locations of the cable, arranging two external discrete IMDs, either at the opposite ends or the same end of the cable is proposed to further improve vibration mitigation performance of the cable in this study. Complex modal analysis based on the taut-string model was employed and extended to allow for the existence of two external discrete IMDs, resulting in a transcendental equation for complex wavenumbers. Both asymptotic and numerical solutions for the case of two opposite IMDs or the case of two IMDs at the same end of the cable were obtained. Subsequently, the applicability of asymptotic solutions was then evaluated. Finally, parametric studies were performed to investigate the effects of damper positions and damper properties on the control performance of a cable with two discrete IMDs. Results showed that two opposite IMDs can generally provide superior control performance to the cable over a single IMD or two IMDs at the same end. It was also observed that attaching two IMDs at the same end of the cable had the potential to achieve significant damping improvement when the inertial mass of the IMDs is appropriate, which seems to be more promising than two opposite IMDs for practical application.

Keywords: stay cable; vibration control; hybrid control; inertial mass damper; viscous damper

---

1. Introduction

With the flourishing development of materials and construction technologies, civil engineering structures are becoming larger, lighter, and more flexible, especially for long-span bridges. Cable-stayed is a common option for bridges in the medium to long-span ranges due to its unique structural formation, economic advantage, and esthetic value [1]. However, as important load-bearing components of cable-stayed bridges, stay cables are highly susceptible to dynamic excitations due to their high flexibility and low intrinsic damping [2,3]. Frequent and excessive amplitude cable vibrations may lead to fatigue failure of cables. These problems may inevitably shorten the service life and cause the risk of losing public confidence in cable-stayed bridges. To guarantee structural safety, several solutions have been proposed to dampen cable vibrations, which include modifying aerodynamic surface of cables [4], connecting cables together via cross tie [5], and attaching external dampers on cables [6–9].

Though these practical measures have been well applied in the field, each has its own shortcomings. Changing the surface of the cable is difficult to implement for retrofit and may increase drag forces at high wind velocities [10]. Cross-ties are incapable of direct energy dissipation and make the aesthetics of cable-stayed bridges deteriorate [11]. Compared to the two methods above, attaching external dampers on the cable seems to be more promising. Nevertheless, the installation location of a passive viscous damper is typically restricted to within a few percentage points of the cable length from the cable anchorage [12]. As expected, passive viscous dampers cannot provide sufficient damping to eliminate vibrations for super-long cables, such as the Sutong Bridge, with cables nearly 600 m long. Moreover, the results based on both theoretical and experimental studies indicated that the existence of the cable sag [13,14], the cable flexural rigidity [15,16], the damper stiffness [17], and the damper support stiffness [18] or their coexistence [19–23] would have adverse impacts on the efficiency of passive viscous dampers.

An active damper can produce a force-deformation relationship with the negative-stiffness behavior that benefits damper efficiency when the linear quadratic regulator (LQR) algorithm is employed [24,25]. However, active dampers often require high power sources beyond practical limits and are thus rarely used for cable vibration mitigation in real bridges. Alternatively, semiactive dampers, which can produce similar hysteresis and achieve control performance comparable to that of active dampers, were proposed [26–29]. For instance, the semiactive control based on magnetorheological dampers has been successfully applied on the Dongting Lake Bridge [30], Binzhou Bridge [31], and Sutong Bridge [32]. Compared to active dampers, semiactive dampers require less power. Nevertheless, possible implementations of semiactive dampers on site still require an external stable power supply, a sensing system, and a controller, which seems to be complicated and costly. This fact has inspired researchers to introduce a negative stiffness mechanism into passive dampers to mitigate cable vibrations.

Recently, several representative passive dampers with negative stiffness mechanisms, including pre-spring negative stiffness dampers (pre-spring NSDs) [33,34] and magnetic negative stiffness dampers (magnetic NSDs) [35,36], have been successfully developed. Negative stiffness dampers have well demonstrated to be capable of providing superior damping over that of traditional passive viscous dampers [37–39]. However, extremely large passive negative stiffness may make the NSD lose its stability. Alternatively, an inerter has the potential to provide similar negative stiffness without a stability problem [40]. Many inerter-based absorber layouts have been proposed, and their control performance advantages have been proven for civil engineering structures [41–59]. As for the vibration suppression of cables, typical inertial mass dampers (IMDs) [60–65] and tuned inerter dampers [66,67] were well developed, and their significant improvement on the achievable modal damping ratio of the cable was verified via both theoretical and experimental investigations.

With the increasing cable length, it may be difficult to attain a desired level of supplemental modal damping with a single damper or a pair of dampers installed near the deck anchorage. Hence, some hybrid techniques have been further proposed. The idea of combining external dampers with cross-ties for cable vibration control was considered, which not only addresses the deficiencies of these two commonly used countermeasures but also still retains their respective merits [68–73]. A hybrid damper system, combining a viscous damper and a tuned mass damper, can overcome the shortcomings of single type of dampers and improve effectiveness and robustness in suppressing cable vibration [74]. In addition, application of two viscous dampers or two high-damping rubber dampers at different locations of a cable was proposed [75–77]. The results have shown that when two viscous dampers are installed at opposite ends of a cable, their damping effects are approximately the sum of the contributions from each damper [77]. However, when they are at the same cable end, the maximum modal damping ratio of the cable is determined by a single damper at the further distance, indicating no benefits over a single damper configuration [77].

Inspired by the potential advantages of attaching two external discrete viscous dampers (VDs) on a cable, this study aimed to evaluate the feasibility of a cable with two discrete IMDs, either on the opposite end or on the same end of the cable, to improve the vibration mitigation performance of the cable in each mode. Complex modal analysis based on the taut-string model was employed and extended to allow for the existence of two external discrete IMDs. The formulation for free vibration of a taut cable with two discrete IMDs was established, and corresponding complex wavenumber equations of free damped vibration were derived. The asymptotic and numerical solutions of the wavenumber equation were obtained, and the applicability of asymptotic solutions was then evaluated. Finally, parametric studies were performed to investigate the effects of damper positions and damper properties on the control performance of the cable with two discrete IMDs.

2. Formulation of the Cable–IMD System

A taut cable with two transversely attached inertial mass dampers is shown in Figure 1. The length, the mass per unit length, and the tension of the cable are L, m, and T, respectively. The coordinate system defines that the x-axis and the v-axis are along the cable chord and the transverse direction, respectively. $x^* = L - x$ represents the coordinate from the right end of the cable. Two discrete IMDs are respectively installed at distances x_1 and x_2 from the left end of the cable ($x_2 \geq x_1$). The distance between the right IMD and the right end is denoted as $x_2^* = L - x_2$. The damping coefficient and inertial mass of the jth IMD are denoted as c_j and b_j ($j = 1, 2$), respectively. The equation of motion of the cable–IMD system is given by:

$$T\frac{\partial^2 v}{\partial x^2} - m\frac{\partial^2 v}{\partial t^2} = \sum_{j=1}^{2} F_{\mathrm{IMD}j}(t)\delta(x - x_j), \tag{1}$$

where $v(x,t)$ is the cable transverse displacement and $\delta(\cdot)$ is delta function to specify the location of the damping force $F_{\mathrm{IMD}j}$ at $x = x_j$.

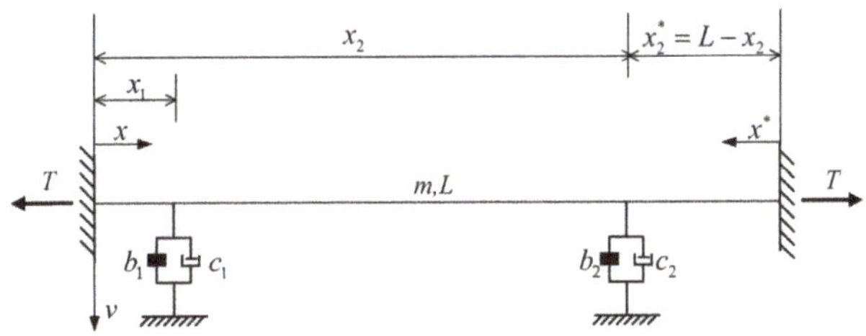

Figure 1. The taut cable with two discrete inertial mass dampers.

Under free vibration, applying separation of variables, the cable transverse displacement and the IMD force can be respectively expressed as:

$$v(x,t) = \widetilde{v}(x)e^{i\omega t}, F_{\mathrm{IMD}j}(t) = \widetilde{F}_{\mathrm{IMD}j}e^{i\omega t} \tag{2}$$

where $i^2 = -1$, ω is a complex natural frequency of the cable, and $\widetilde{v}(x)$ is the corresponding complex mode shape. To find $\widetilde{v}(x)$, the cable can be considered as a multispan structure connected at the damper locations [77]. Substituting Equation (2) into Equation (1), $\widetilde{v}(x)$ in each span needs to satisfy a homogeneous equation [77]:

$$\frac{d^2\widetilde{v}}{dx^2} + \beta^2\widetilde{v} = 0 \begin{cases} 0 \leq x \leq x_1 \\ x_1 \leq x \leq x2 \\ 0 \leq x^* \leq x_2^* \end{cases}, \tag{3}$$

where $\beta = \omega\sqrt{m/T}$ refers to the wavenumber.

Applying boundary conditions at cable ends, i.e., $\widetilde{v}(0) = \widetilde{v}(L) = 0$, and the transverse displacement compatibility conditions at damper locations, i.e., $\widetilde{v}(x_1) = \widetilde{v_1}$, $\widetilde{v}(x_2^*) = \widetilde{v_2}$, the general solution of Equation (3) can be further written in the form [77]:

$$\widetilde{v}(x) = \begin{cases} \widetilde{v_1}\frac{\sin\beta x}{\sin\beta x_1} & 0 \le x \le x_1 \\ \widetilde{v_1}\frac{\sin\beta(x_2-x)}{\sin\beta(x_2-x_1)} + \widetilde{v_2}\frac{\sin\beta(x-x_1)}{\sin\beta(x_2-x_1)} & x_1 \le x \le x2, \\ \widetilde{v_2}\frac{\sin\beta x^*}{\sin\beta x_2^*} & 0 \le x^* \le x_2^* \end{cases} \tag{4}$$

where $\widetilde{v}_j$ is the mode shape amplitude at the jth damper location.

At damper locations, there is:

$$T(\frac{dv}{dx}\Big|_{x_j^+} - \frac{dv}{dx}\Big|_{x_j^-}) = \widetilde{F}_{\mathrm{IMD}j}(t). \tag{5}$$

Substituting Equation (4) into Equation (5), it yields:

$$\begin{cases} \cot\beta x_1 + \cot(x_2 - x_1) - \frac{\widetilde{v_2}}{\widetilde{v_1}}\frac{1}{\sin\beta(x_2-x_1)} = -\frac{\widetilde{F}_{\mathrm{IMD1}}/T}{\beta\widetilde{v_1}} \\ -\frac{\widetilde{v_2}}{\widetilde{v_1}}\frac{1}{\sin\beta(x_2-x_1)} + \cot\beta(x_2 - x_1) + \cot\beta x_2^* = -\frac{\widetilde{F}_{\mathrm{IMD2}}/T}{\beta\widetilde{v_1}} \end{cases}. \tag{6}$$

Substituting $\widetilde{v_1}/\widetilde{v_2}$ from the second one into the first one of Equation (6) and rearranging, the characteristic equation of the wavenumber β is derived as:

$$\begin{aligned} &(\cot\beta x_1 + \frac{\widetilde{F}_{\mathrm{IMD1}}/T}{\beta\widetilde{v_1}})(\cot\beta x_2^* + \frac{\widetilde{F}_{\mathrm{IMD2}}/T}{\beta\widetilde{v_2}}) + \\ &(\cot\beta x_1 + \frac{\widetilde{F}_{\mathrm{IMD1}}/T}{\beta\widetilde{v_1}} + \cot\beta x_2^* + \frac{\widetilde{F}_{\mathrm{IMD2}}/T}{\beta\widetilde{v_2}})\cot\beta(x_2 - x_1) = 1 \end{aligned} \tag{7}$$

3. Two Opposite IMDs

3.1. The Wavenumber Equation

The damper force of the jth IMD can be expressed as [45]:

$$F_{\mathrm{IMD}j}(t) = b_j\frac{\partial^2 v(x_j,t)}{\partial t^2} + c_j\frac{\partial v(x_j,t)}{\partial t} \text{ or } \widetilde{F}_{\mathrm{IMD}j} = -b_j\omega^2\widetilde{v}_j + c_j\omega\widetilde{v}_j. \tag{8}$$

When two IMDs are installed at different ends of the cable, substituting Equation (8) into Equation (7) and using trigonometric relations, Equation (7) can be rearranged to the form relating to x_1 and x_2^* as:

$$\tan\beta L = \frac{A_1 + iB_1}{C_1 + iD_1}, \tag{9}$$

$$A_1 = -\chi_1\sin^2\beta x_1 - \chi_2\sin^2\beta x_2^* + (\chi_1\chi_2 - \eta_1\eta_2)\sin\beta x_1\sin\beta x_2^*\sin\beta(x_1 + x_2^*), \tag{9a}$$

$$B_1 = \eta_1\sin^2\beta x_1 + \eta_2\sin^2\beta x_2^* - (\chi_1\eta_2 + \chi_2\eta_1)\sin\beta x_1\sin\beta x_2^*\sin\beta(x_1 + x_2^*), \tag{9b}$$

$$\begin{aligned} C_1 = {} & 1 - \chi_1\sin\beta x_1\cos\beta x_1 - \chi_2\sin\beta x_2^*\cos\beta x_2^* + \\ & (\chi_1\chi_2 - \eta_1\eta_2)\sin\beta x_1\sin\beta x_2^*\cos\beta(x_1 + x_2^*) \end{aligned} \tag{9c}$$

$$\begin{aligned} D_1 = {} & \eta_1\sin\beta x_1\cos\beta x_1 + \eta_2\sin\beta x_2^*\cos\beta x_2^* - \\ & (\chi_1\eta_2 + \chi_2\eta_1)\sin\beta x_1\sin\beta x_2^*\cos\beta(x_1 + x_2^*) \end{aligned} \tag{9d}$$

where $\eta_j = \frac{c_j}{\sqrt{mT}}$ and $\chi_j = \frac{b_j\omega}{\sqrt{mT}}$ represent the dimensionless damping coefficient and the dimensionless inertial mass of the jth IMD, respectively.

The form of Equation (9) is suitable for solutions, either for asymptotic solutions or numerical solutions by iteration.

3.2. Asymptotic Solution

The following assumptions are introduced [12,17,77]: (1) The locations of IMDs are very close to the ends, i.e., $x_1, x_2^* \ll L$; (2) the wavenumber β_n of each mode n ($n = 1, 2, \ldots$) has a small perturbations $\Delta\beta_n = \beta_n - \beta_n^0$ from the undamped value $\beta_n^0 = n\pi/L$. The assumptions above lead to the following approximations:

$$\tan(\beta_n L) \cong \beta_n L - n\pi \ \sin(\beta_n x_1) \cong \beta_n^0 x_1 \ \sin(\beta_n x_2^*) \cong \beta_n^0 x_2^* \ \cos(\beta_n x_1) \cong \cos(\beta_n x_2^*) = 1. \tag{10}$$

The asymptotic formula for the wavenumber β_n takes the form:

$$\beta_n L \cong n\pi + \beta_n^0 \frac{E_1 + iF_1}{G_1 + iH_1}, \tag{11}$$

$$E_1 = -\overline{b_1} x_1 - \overline{b_2} x_2^* + (\overline{b_1 b_2} - \overline{c_1 c_2})(x_1 + x_2^*), \tag{11a}$$

$$F_1 = \overline{c_1} x_1 + \overline{c_2} x_2^* - (\overline{b_1 c_2} + \overline{b_2 c_1})(x_1 + x_2^*), \tag{11b}$$

$$G_1 = 1 - \overline{b_1} - \overline{b_2} + (\overline{b_1 b_2} - \overline{c_1 c_2})(x_1 + x_2^*), \tag{11c}$$

$$H_1 = \overline{c_1} + \overline{c_2} - (\overline{b_1 c_2} + \overline{b_2 c_1})(x_1 + x_2^*), \tag{11d}$$

where $\overline{c_1} = \eta_1 \beta_n^0 x_1$ and $\overline{c_2} = \eta_1 \beta_n^0 x_2^*$ represent dimensionless damping coefficient groups, while $\overline{b_1} = \chi_1 \beta_n^0 x_1$ and $\overline{b_2} = \chi_2 \beta_n^0 x_2^*$ represent dimensionless inertial mass groups.

The complex eigen-frequency corresponding to the wavenumber β_n is denoted as ω_n. The nth supplemental modal damping radio of a cable ξ_n can be obtained by [17]:

$$\xi_n = \frac{\mathrm{Im}[\omega_n]}{|\omega_n|} = \frac{\mathrm{Im}[\beta_n]}{|\beta_n|} \cong \frac{\mathrm{Im}[\Delta\beta_n]}{|\beta_n^0|}. \tag{12}$$

Substituting Equation (11) into Equation (12), the asymptotic supplemental modal damping ratio of the cable is finally derived as:

$$\xi_n \cong \frac{\overline{c_1}}{(1 - \overline{b_1})^2 + (\overline{c_1})^2} \frac{x_1}{L} + \frac{\overline{c_2}}{(1 - \overline{b_2})^2 + (\overline{c_2})^2} \frac{x_2}{L}. \tag{13}$$

If two identical IMDs are symmetrically installed at the cable for practical implementation, some simplifications in the notation can be introduced, i.e., $x = x_1 = L - x_2 = x_2^*$, $b_1 = b_2 = b$, $\chi_1 = \chi_2 = \chi$, $\overline{b_1} = \overline{b_2} = \overline{b}$, $c_1 = c_2 = c$, $\eta_1 = \eta_2 = \eta$, and $\overline{c_1} = \overline{c_2} = \overline{c}$. Equation (13) can be further simplified as:

$$\xi_n = \frac{2\overline{c}}{(1 - \overline{b})^2 + (\overline{c})^2} \frac{x_1}{L}. \tag{14}$$

3.3. Numerical Solution

The numerical solution of the wavenumber β_n to Equation (9) is obtained by the fixed point iteration [12,14], which starts from the undamped wavenumber β_n^0. Substituting β_n^0 into the right side of the Equation (9), the resulting value β_n^1 is obtained. Similarly, with current estimate β_n^k, $k = 1, 2, 3 \ldots$, a new estimate β_n^{k+1} will be derived. The iterative scheme is given by:

$$\beta_n^{k+1} L = n\pi + \arctan \frac{\overline{A}_1 + i\overline{B}_1}{\overline{C}_1 + i\overline{D}_1}, \tag{15}$$

$$\overline{A}_1 = -\chi_1 \sin^2 \beta_n^k x_1 - \chi_2 \sin^2 \beta_n^k x_2^* + (\chi_1 \chi_2 - \eta_1 \eta_2) \sin \beta_n^k x_1 \sin \beta_n^k x_2^* \sin \beta_n^k (x_1 + x_2^*), \tag{15a}$$

$$\overline{B}_1 = \eta_1 \sin^2 \beta_n^k x_1 + \eta_2 \sin^2 \beta_n^k x_2^* - (\chi_1 \eta_2 + \chi_2 \eta_1) \sin \beta_n^k x_1 \sin \beta_n^k x_2^* \sin \beta_n^k (x_1 + x_2^*), \tag{15b}$$

$$\begin{aligned} \overline{C}_1 = 1 - \chi_1 \sin \beta_n^k x_1 \cos \beta_n^k x_1 - \chi_2 \sin \beta_n^k x_2^* \cos \beta_n^k x_2^* + \\ (\chi_1 \chi_2 - \eta_1 \eta_2) \sin \beta_n^k x_1 \sin \beta_n^k x_2^* \cos \beta_n^k (x_1 + x_2^*) \end{aligned} \tag{15c}$$

$$\begin{aligned} \overline{D}_1 = \eta_1 \sin \beta_n^k x_1 \cos \beta_n^k x_1 + \eta_2 \sin \beta_n^k x_2^* \cos \beta_n^k x_2^* - \\ (\chi_1 \eta_2 + \chi_2 \eta_1) \sin \beta_n^k x_1 \sin \beta_n^k x_2^* \cos \beta_n^k (x_1 + x_2^*) \end{aligned} \tag{15d}$$

Finally, the supplemental modal damping ratio of a cable with two opposite IMDs can be calculated by Equation (12) after solving the wavenumber β_n.

3.4. Comparison of Asymptotic and Numerical Solutions

Figure 2 shows the comparison of asymptotic and numerical complex wavenumbers of a cable with two symmetric identical IMDs for various inertial masses. Two IMDs are assumed to be respectively installed at distances x_1 of 1%L and x_2 of 99%L from the left end of the cable, i.e., $x = x_1 = x_2^* = L - x_2 = 1\%L$. When inertial masses remain constant and damping coefficients of two IMDs increase from zero to infinity, the loci, which nearly trace a semicircular contour, start from the undamped wavenumber and finally attach to the real axis. According to Equation (12), the damping properties result from the imaginary part of the wavenumber. Maximum supplemental modal damping can be obtained at the top point of the semicircle [17]. The diameter of the loci is quite small but increases with the increase of the inertial masses of IMDs, indicating that two symmetric identical IMDs have slight influences on the damped frequency of the cable and can achieve higher supplemental modal damping ratios than traditional VDs. By comparing asymptotic and numerical complex wavenumbers, it is seen that two results coincide well with each other for small or moderate inertial mass ($\chi \leq 0.6/(n\pi x/L)$) adopted in the IMDs. Nevertheless, the results deviate significantly from each other when the big inertial mass ($\chi = 0.9/(n\pi x/L)$) shown in Figure 2d is adopted. Hence, solutions via numerical iteration are used to accurately predict the maximum supplemental modal damping ratio of the cable and corresponding optimal damper size of the IMD in the following discussions.

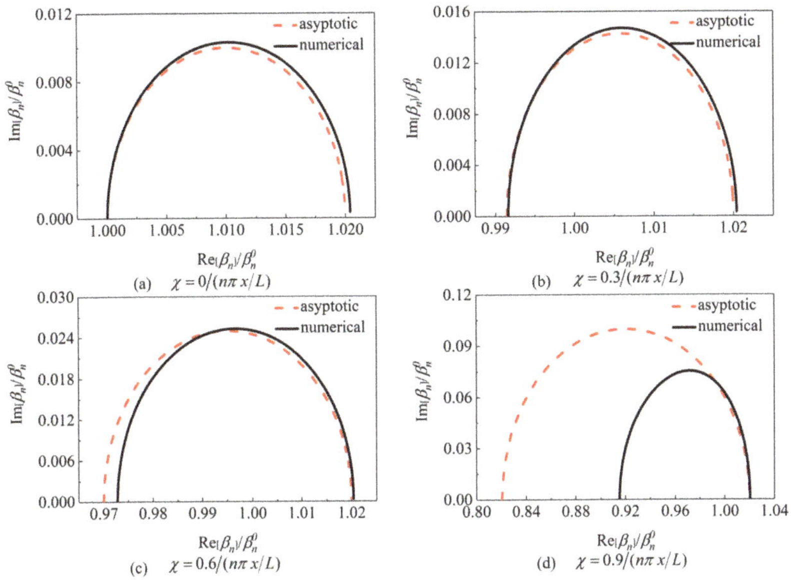

Figure 2. Comparison of asymptotic and numerical complex wavenumbers of a cable with two symmetric identical inertial mass dampers (IMDs) ($x = x_1 = x_2^* = 1\%L$).

3.5. Parametric Studies

Figure 3 presents the supplemental modal damping ratio of a cable with two symmetric identical VDs versus damping coefficients. For the convenience of comparisons, the results of a cable with a single VD are also shown. It is observed that symmetrically attaching two VDs on the opposite end of a cable is favorable to increasing the maximum supplemental damping ratio of the cable, and its maximum supplemental modal damping ratio is asymptotically the sum of contributions from each VD separately. The findings above are quite consistent with those reported in previous studies [76,77].

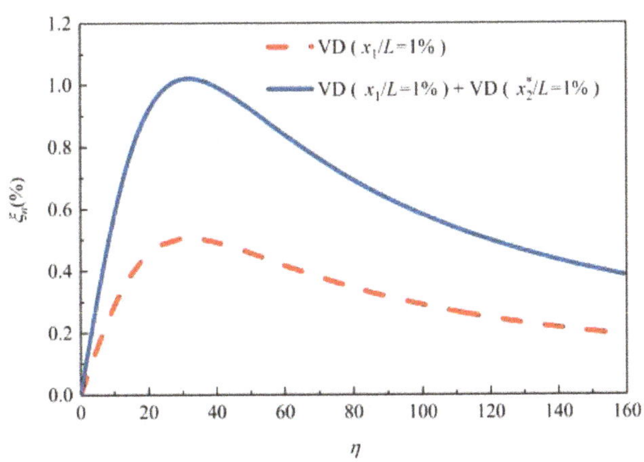

Figure 3. The supplemental modal damping ratio curve of a cable with a single viscous damper (VD) or two symmetric identical VDs ($x = x_1 = x_2^* = 1\%L$).

Figure 4 presents the supplemental modal damping ratio of a cable with two symmetric identical IMDs or a single IMD versus damping coefficients for various inertial masses. It is clear that two opposite IMDs can provide superior control performance to the cable over a single IMD. Figure 5 directly compares the maximum supplemental modal damping ratio of a cable equipped with a single IMD or two symmetric identical IMDs. For two symmetric identical IMDs with small or medium inertial mass, similarly to the case of two symmetric identical VDs, the maximum achievable supplemental damping ratio is approximately doubled with that provided by a single IMD. It indicates that two opposite IMDs are almost independent from each other. This finding may explain why the optimal damping coefficients of the IMDs for both the single configuration and two-symmetric configuration are similar to each other in magnitude, as shown in Figure 4. Moreover, the maximum achievable supplemental modal damping ratio of a cable provided by two symmetric identical IMDs is larger than that provided by a single IMD or two symmetric identical VDs.

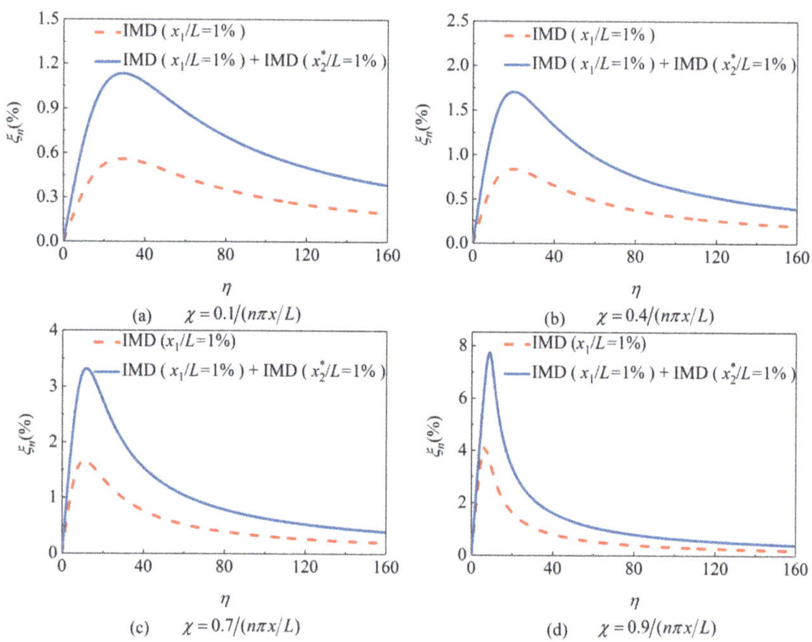

Figure 4. The supplemental modal damping ratio curve of a cable with a single IMD or two symmetric identical IMDs ($x = x_1 = x_2^* = 1\%L$).

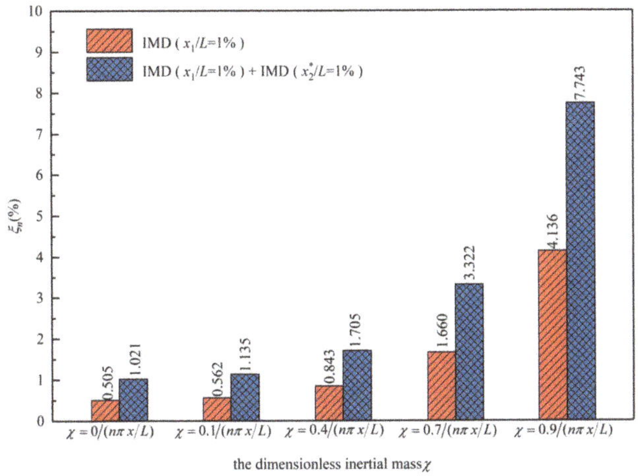

Figure 5. The maximum achievable supplemental modal damping ratio of a cable equipped with a single IMD or two symmetric identical IMDs ($x = x_1 = x_2^* = 1\%L$).

4. Two IMDs at the Same End

4.1. The Wavenumber Equation

When two IMDs are installed at the same end of the cable, substituting Equation (8) into Equation (7) and rearranging terms gives the following expression relating to x_1 and x_2:

$$\tan\beta L = \frac{A_2 + iB_2}{C_2 + iD_2}, \tag{16}$$

$$A_2 = -\chi_1 \sin^2\beta x_1 - \chi_2 \sin^2\beta x_2 + (\chi_1\chi_2 - \eta_1\eta_2)\sin\beta x_1 \sin\beta x_2 \sin\beta(x_2 - x_1), \tag{16a}$$

$$B_2 = \eta_1 \sin^2\beta x_1 + \eta_2 \sin^2\beta x_2 - (\chi_1\eta_2 + \chi_2\eta_1)\sin\beta x_1 \sin\beta x_2 \sin\beta(x_2 - x_1), \tag{16b}$$

$$\begin{aligned} C_2 = 1 - \chi_1 \sin\beta x_1 \cos\beta x_1 - \chi_2 \sin\beta x_2 \cos\beta x_2 + \\ (\chi_1\chi_2 - \eta_1\eta_2)\sin\beta x_1 \cos\beta x_2 \sin\beta(x_2 - x_1) \end{aligned} \tag{16c}$$

$$\begin{aligned} D_2 = \eta_1 \sin\beta x_1 \cos\beta x_1 + \eta_2 \sin\beta x_2 \cos\beta x_2 - \\ (\chi_1\eta_2 + \chi_2\eta_1)\sin\beta x_1 \cos\beta x_2 \sin\beta(x_2 - x_1) \end{aligned} \tag{16d}$$

4.2. Asymptotic Solution

Similarly to the case of two opposite IMDs, assuming two IMDs locations $x_1, x_2 \ll L$ and the wave number β_n of the damped cable to be small perturbations from β_n^0, Equation (16) can be simplified as:

$$\beta_n L \cong n\pi + \beta_n^0 \frac{E_2 + iF_2}{G_2 + iH_2}, \tag{17}$$

$$E_2 = -\overline{b_1}x_1 - \overline{b_2}x_2 + (\overline{b_1b_2} - \overline{c_1c_2})(x_2 - x_1), \tag{17a}$$

$$F_2 = \overline{c_1}x_1 + \overline{c_2}x_2 - (\overline{b_1c_2} + \overline{b_2c_1})(x_2 - x_1), \tag{17b}$$

$$G_2 = 1 - \overline{b_1} - \overline{b_2} + (\overline{b_1b_2} - \overline{c_1c_2})(1 - x_1/x_2), \tag{17c}$$

$$H_2 = \overline{c_1} + \overline{c_2} - (\overline{b_1c_2} + \overline{b_2c_1})(1 - x_1/x_2). \tag{17d}$$

From Equations (12) and (17), the asymptotic modal damping ratio of a cable with two IMDs at the same end can be obtained as:

$$\xi_n \cong \frac{\mathrm{Im}[\Delta\beta_n]}{|\beta_n^0|} = \frac{F_2G_2 - E_2H_2}{(G_2)^2 + (H_2)^2}. \tag{18}$$

4.3. Numerical Solution

Similarly to the case of two opposite IMDs, Equation (16) can be solved for the wavenumber using the fixed-point iteration. Starting from the undamped wavenumber β_n^0, the iterative scheme is given by:

$$\beta_n^{k+1}L = n\pi + \arctan\frac{\overline{A}_2 + i\overline{B}_2}{\overline{C}_2 + i\overline{D}_2}, \tag{19}$$

$$\overline{A}_2 = -\chi_1 \sin^2\beta_n^k x_1 - \chi_2 \sin^2\beta_n^k x_2 + (\chi_1\chi_2 - \eta_1\eta_2)\sin\beta_n^k x_1 \sin\beta_n^k x_2 \sin\beta_n^k(x_2 - x_1), \tag{19a}$$

$$\overline{B}_2 = \eta_1 \sin^2\beta_n^k x_1 + \eta_2 \sin^2\beta_n^k x_2 - (\chi_1\eta_2 + \chi_2\eta_1)\sin\beta_n^k x_1 \sin\beta_n^k x_2 \sin\beta_n^k(x_2 - x_1), \tag{19b}$$

$$\begin{gathered}\overline{C}_2 = 1 - \chi_1 \sin\beta_n^k x_1 \cos\beta_n^k x_1 - \chi_2 \sin\beta_n^k x_2 \cos\beta_n^k x_2 + \\ (\chi_1\chi_2 - \eta_1\eta_2)\sin\beta_n^k x_1 \cos\beta_n^k x_2 \sin\beta_n^k(x_2 - x_1)\end{gathered} \tag{19c}$$

$$\begin{gathered}\overline{D}_2 = \eta_1 \sin\beta_n^k x_1 \cos\beta_n^k x_1 + \eta_2 \sin\beta_n^k x_2 \cos\beta_n^k x_2 - \\ (\chi_1\eta_2 + \chi_2\eta_1)\sin\beta_n^k x_1 \cos\beta_n^k x_2 \sin\beta_n^k(x_2 - x_1)\end{gathered} \tag{19d}$$

After solving numerically for the wavenumber, the supplemental modal damping ratio of a cable with two IMDs at the same end can be determined from Equation (12).

4.4. Comparison of Asymptotic and Numerical Solutions

Figure 6 shows the comparison of asymptotic and numerical complex wavenumbers of a cable with two IMDs at the same end for various inertial masses, where two IMDs are installed at distances x_1 of 1%L and x_2 of 2%L from the left end of the cable, i.e., $x_1 = 1\%L, x_2 = 2\%L$. Seeing that two IMDs are usually identical, some simplifications in the notation are introduced, i.e., $b_1 = b_2 = b$, $\chi_1 = \chi_2 = \chi$, $c_1 = c_2 = c$, and $\eta_1 = \eta_2 = \eta$. Similar to the case of two opposite IMDs, the loci start from the undamped wavenumber along a semicircular contour and finally attach to the real axis when damping coefficients of the IMDs increase from zero to infinity, and the effect of two IMDs installed at the same end of the cable on the cable frequency is also not significant. For a cable with two IMDs at the same end, although the asymptotic complex wavenumber agrees well with the numerical solution when the small inertial mass ($\chi \leq 0.3/(n\pi x_2/L)$) is used, it will lose accuracy when moderate or large inertial mass ($0.6/(n\pi x_2/L) \leq \chi \leq 0.9/(n\pi x_2/L)$) is adopted. Compared to the case of two opposite IMDs, prediction accuracies of the asymptotic solution are found to be relatively poor when two IMDs are installed at the same end of the cable. Hence, numerical results are used for the following parametric study.

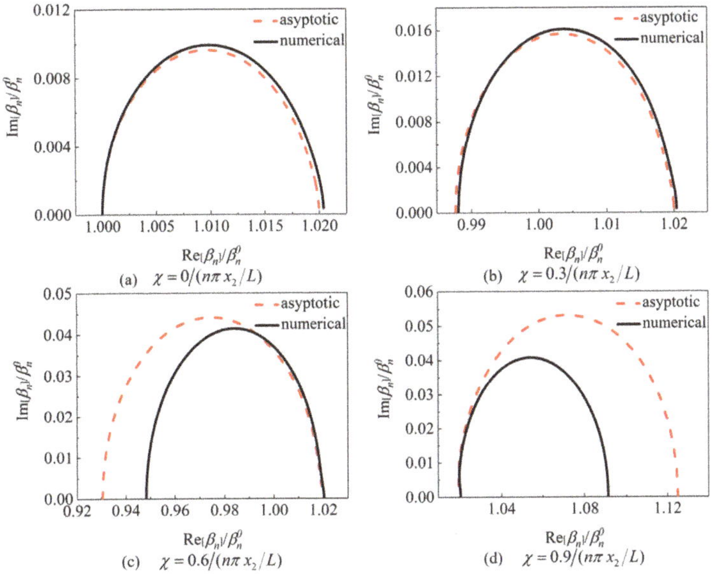

Figure 6. Comparison of asymptotic and numerical complex wavenumbers of a cable with two identical IMDs at the same end ($x_1 = 1\%L, x_2 = 2\%L$).

4.5. Parametric Studies

Figure 7 presents the supplemental modal damping ratio of a cable with two identical VDs at the same end or a single VD versus damping coefficients. It is observed that attaching two VDs at the same ends of the cable may help to reduce the damper size but cannot increase the maximum supplemental modal damping ratio. Moreover, its maximum modal damping ratio is slightly smaller than that provided by a single VD at the further distance. These observations are in agreement with previous findings [76,77].

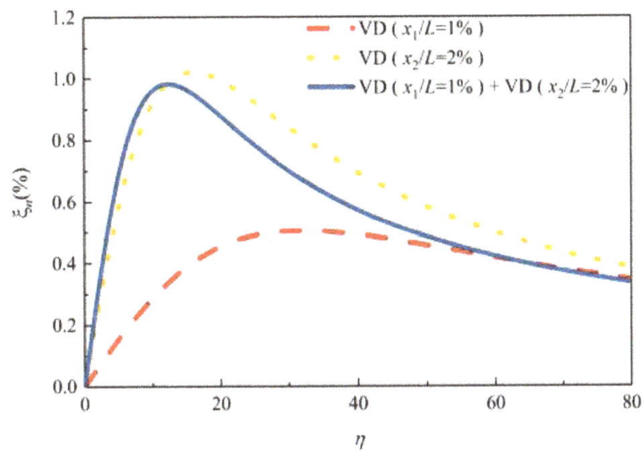

Figure 7. The modal damping ratios curves of a cable equipped with a single VD or two identical VDs at the same end ($x_1 = 1\%L, x_2 = 2\%L$).

Figure 8 presents the supplemental modal damping ratio of a cable with two identical IMDs at the same end or a single IMD versus damping coefficients of the IMD. If two IMDs with relatively small or big inertial masses ($\chi \leq 0.1/(n\pi x_2/L)$ or $\chi = 0.9/(n\pi x_2/L)$) are installed at the same end of the cable, similarly to the case of two VDs at the same end of a cable, there is no advantage of increasing the maximum modal damping ratio over that of a single IMD. However, if moderate inertial mass ($0.4/(n\pi x_2/L) \leq \chi \leq 0.7/(n\pi x_2/L)$) of the IMD is used, it is interesting to observe that two IMDs at the same end can lead to smaller optimum damping coefficients and larger maximum supplemental modal damping ratios than that of a single IMD at a bigger distance.

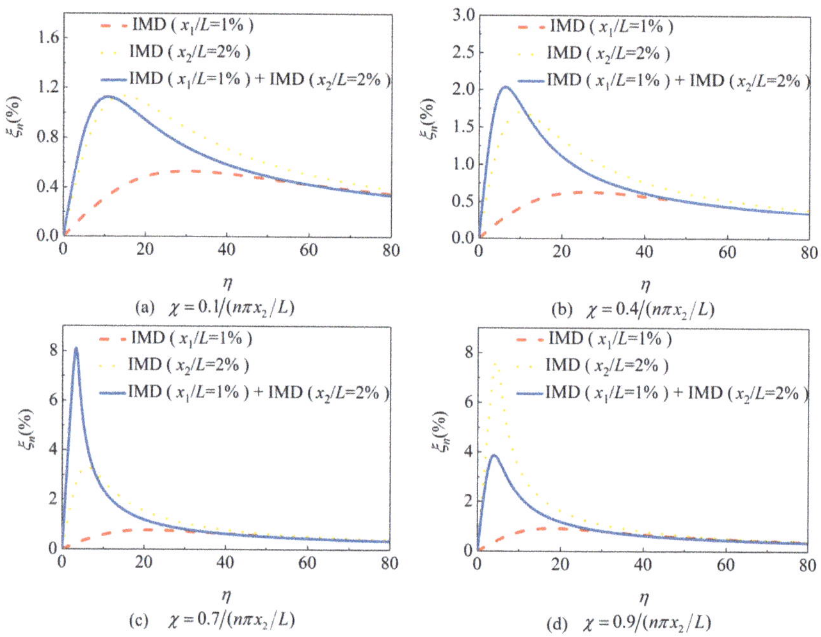

Figure 8. The modal damping ratios curves of a cable equipped with a single IMD or two identical IMDs at the same end ($x_1 = 1\%L, x_2 = 2\%L$).

The maximum achievable supplemental modal damping ratios of a cable equipped with a single IMD and two IMDs at the same end are directly compared in Figure 9. It is worth noting that the maximum supplemental modal damping ratio provided by two IMDs is higher than the sum of contributions from each IMD when inertial mass $\chi = 0.7/(n\pi x_2/L)$ is used. Though the strategy of two opposite IMDs has demonstrated that it can provide superior control performance, installing a damper at cable-tower anchorage is difficult and inconvenient. Thus, attaching two IMDs with appropriate inertial mass installed at the same end of the cable seems to be more promising for practical application.

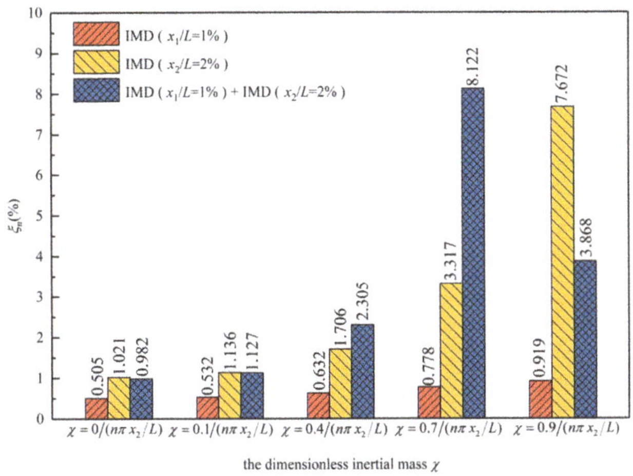

Figure 9. The maximum achievable supplemental modal damping ratio of a cable equipped with a single IMD or two identical IMDs at the same end ($x_1 = 1\%L, x_2 = 2\%L$).

5. Conclusions

In this paper, the combined damping effect of two discrete IMDs on a stay cable, either on the opposite end or the same end, was theoretically investigated in comparison with a single IMD, especially for the single-mode cable vibration control. Results showed that the maximum supplemental modal damping ratio of a cable provided by two opposite IMDs with small or moderate inertial mass is approximately the sum of contributions from each IMD. However, damping performances of the cable with two opposite IMDs will be reduced when the IMDs adopt relatively large inertial mass, in which the superposition effect of each IMD gets weak. As for a cable with two IMDs at the same end, the maximum modal damping ratio of the cable is smaller than that of a single IMD at the further distance when the IMDs adopt relatively small or large inertial mass. Fortunately, when the inertial mass of the IMD is appropriate, attaching two IMDs at the same end of the cable is able to obtain a larger maximum modal damping ratio than that of a single IMD at a bigger distance, which is even more than the sum of contributions from each IMD. Generally, attaching two opposite IMDs on a cable has shown better control performance than two IMDs at the same end. However, installing a damper at cable-tower anchorage is difficult and inconvenient. As an alternative, attaching two IMDs with appropriate inertial mass at the same end of a cable seems to be more promising for practical application. However, it is still necessary to explore the performance of two IMDs for the multimode cable vibration control, especially for super-long cables, which will be our consideration for further study.

Author Contributions: Z.W., F.Y., and H.G. conceived the idea of this research; Z.W. and F.Y. wrote the paper; H.G. reviewed and revised the paper.

Funding: The research was financially supported by the National Natural Science Foundation of China (Grant No. 51878274).

Conflicts of Interest: The authors declare no conflict of interest.

References

1. Mao, J.X.; Wang, H.; Feng, D.M.; Tao, T.Y.; Zheng, W.Z. Investigation of dynamic properties of long-span cable-stayed bridges based on one-year monitoring data under normal operating condition. *Struct. Control Health Monit.* **2018**, *25*, e2146. [CrossRef]

2. Zhou, H.J.; Xu, Y.L. Wind-rain-induced vibration and control of stay cables in a cable-stayed bridge. *Struct. Control Health Monit.* **2007**, *14*, 1013–1033. [CrossRef]
3. Wang, Z.H.; Chen, Z.Q.; Gao, H.; Wang, H. Development of a self-powered magnetorheological damper system for cable vibration control. *Appl. Sci.* **2018**, *8*, 118. [CrossRef]
4. Kleissl, K.; Georgakis, C.T. Comparison of the aerodynamics of bridge cables with helical fillets and a pattern-indented surface. *J. Wind Eng. Ind. Aerod.* **2012**, *104*, 166–175. [CrossRef]
5. He, X.H.; Cai, C.; Wang, Z.J.; Jing, H.Q.; Chen, Z.Q. Experimental verification of the effectiveness of elastic cross cross-ties in suppressing wake wake-induced vibrations of staggered stay cables. *Eng. Struct.* **2018**, *167*, 151–165. [CrossRef]
6. Pacheco, B.M.; Fujino, Y.; Sulekh, A. Estimation curve for modal damping in stay cables with viscous damper. *J. Struct. Eng.* **1993**, *119*, 1961–1979. [CrossRef]
7. Xu, Y.L.; Zhou, H.J. Damping cable vibration for a cable-stayed bridge using adjustable fluid dampers. *J. Sound Vib.* **2007**, *306*, 349–360. [CrossRef]
8. Mekki, O.B.; Auricchio, F. Performance evaluation of shape-memory-alloy superelastic behavior to control a stay cable in cable-stayed bridges. *Int. J. Non-Lin. Mech.* **2011**, *46*, 470–477. [CrossRef]
9. Soltane, S.; Montassar, S.; Ben Mekki, O.; El Fatmi, R. A hysteretic Bingham model for MR dampers to control cable vibrations. *J. Mech. Mater. Struct.* **2015**, *10*, 195–206. [CrossRef]
10. Miyata, T.; Yamada, H.; Hojo, T. Experimental study on aerodynamic characteristics of cables with patterned surface. *J. Struct. Eng.* **1994**, *40*, 1065–1076.
11. Caracoglia, L.; Jones, N.P. In-plane dynamic behavior of cable networks. Part 2: Prototype prediction and validation. *J. Sound Vib.* **2005**, *279*, 993–1014. [CrossRef]
12. Krenk, S. Vibration of a taut cable with an external damper. *J. Appl. Mech.* **2000**, *67*, 772–776. [CrossRef]
13. Xu, Y.L.; Yu, Z. Vibration of inclined sag cables with oil dampers in cable stayed bridges. *J. Bridge Eng.* **1998**, *3*, 194–203. [CrossRef]
14. Krenk, S.; Nielsen, S.R.K. Vibrations of a shallow cable with a viscous damper. *Proc. R. Soc. Lond. A* **2002**, *458*, 339–357. [CrossRef]
15. Hoang, N.; Fujino, Y. Analytical study on bending effects in a stay cable with a damper. *J. Eng. Mech.* **2007**, *133*, 1241–1246. [CrossRef]
16. Main, J.A.; Jones, N.P. Vibration of tensioned beams with intermediate damper. II: Damper near a support. *J. Eng. Mech.* **2007**, *133*, 379–388. [CrossRef]
17. Krenk, S.; Høgsberg, J.R. Damping of cables by a transverse force. *J. Eng. Mech.* **2005**, *131*, 340–348. [CrossRef]
18. Huang, Z.; Jones, N.P. Damping of taut-cable systems: Effects of linear elastic spring support. *J. Eng. Mech.* **2011**, *137*, 512–518. [CrossRef]
19. Tabatabai, H.; Mehrabi, A.B. Design of mechanical viscous dampers for stay cables. *J. Bridge Eng.* **2000**, *5*, 114–123. [CrossRef]
20. Fujino, Y.; Hoang, N. Design formulas for damping of a stay cable with a damper. *J. Struct. Eng.* **2008**, *134*, 269–278. [CrossRef]
21. Cheng, S.H.; Darivandi, N.; Ghrib, F. The design of an optimal viscous damper for a bridge stay cable using energy-based approach. *J. Sound Vib.* **2010**, *329*, 4689–4704. [CrossRef]
22. Fournier, J.A.; Cheng, S.H. Impact of damper stiffness and damper support stiffness on the efficiency of a linear viscous damper in controlling stay cable vibrations. *J. Bridge Eng.* **2014**, *19*, 04013022. [CrossRef]
23. Javanbakht, M.; Cheng, S.H.; Ghrib, F. Control-oriented model for the dynamic response of a damped cable. *J. Sound Vib.* **2019**, *442*, 249–267. [CrossRef]
24. Li, H.; Liu, M.; Ou, J.P. Negative stiffness characteristics of active and semi-active control systems for stay cables. *Struct. Control Health Monit.* **2008**, *15*, 120–142. [CrossRef]
25. Iemura, H.; Pradono, M.H. Advances in the development of pseudo-negative-stiffness dampers for seismic response control. *Struct. Control Health Monit.* **2009**, *16*, 784–799. [CrossRef]
26. Johnson, E.A.; Christenson, R.E.; Spencer, B.F. Semiactive damping of cables with sag. *Comput. Aided Civ. Inf.* **2003**, *18*, 132–146. [CrossRef]
27. Duan, Y.F.; Ni, Y.Q.; Ko, J.M. State-derivative feedback control of cable vibration using semiactive magnetorheological dampers. *Comput. Aided Civ. Inf.* **2005**, *20*, 431–449. [CrossRef]
28. Duan, Y.F.; Ni, Y.Q.; Zhang, H.M.; Spencer, B.F.; Ko, J.M.; Fang, Y. Design formulas for vibration control of taut cables using passive MR dampers. *Smart Struct. Syst.* **2019**, *23*, 521–536.

29. Huang, H.W.; Liu, T.T.; Sun, L.M. Multi-mode cable vibration control using MR damper based on nonlinear modeling. *Smart Struct. Syst.* **2019**, *23*, 565–577.
30. Chen, Z.Q.; Wang, X.Y.; Ko, J.M.; Ni, Y.Q.; Spencer, B.F.; Yang, G. MR damping system for mitigating wind-rain induced vibration on Dongting Lake Cable-Stayed Bridge. *Wind Struct.* **2004**, *7*, 293–304. [CrossRef]
31. Li, H.; Liu, M.; Li, J.H.; Guan, X.C.; Ou, J.P. Vibration control of stay cables of Shandong Binzhou Yellow River Highway Bridge by using magnetorheological fluid dampers. *J. Bridge Eng.* **2007**, *12*, 401–409. [CrossRef]
32. Weber, F.; Distl, H. Amplitude and frequencyindependent cable damping of Sutong Bridge and Russky Bridge by magnetorheological dampers. *Struct. Control Health Monit.* **2015**, *22*, 237–254. [CrossRef]
33. Chen, L.; Sun, L.M.; Nagarajaiah, S. Cable with discrete negative stiffness device and viscous damper: Passive realization and general characteristics. *Smart Struct. Syst.* **2015**, *15*, 627–643. [CrossRef]
34. Zhou, P.; Li, H. Modeling and control performance of a negative stiffness damper for suppressing stay cable vibrations. *Struct. Control Health Monit.* **2016**, *23*, 764–782. [CrossRef]
35. Shi, X.; Zhu, S.Y. Magnetic negative stiffness dampers. *Smart Mater. Struct.* **2015**, *24*, 072002. [CrossRef]
36. Shi, X.; Zhu, S.Y.; Li, J.Y.; Spencer, B.F. Dynamic behavior of stay cables with passive negative stiffness dampers. *Smart Mater. Struct.* **2016**, *25*, 75044. [CrossRef]
37. Shi, X.; Zhu, S.Y.; Nagarajaiah, S. Performance comparison between passive negative stiffness damper and active control in cable vibration mitigation. *J. Bridge Eng.* **2017**, *22*, 04017054. [CrossRef]
38. Shi, X.; Zhu, S.Y.; Spencer, B.F. Experimental study on passive negative stiffness damper for cable vibration mitigation. *J. Eng. Mech.* **2017**, *143*, 04017070. [CrossRef]
39. Javanbakht, M.; Cheng, S.H.; Ghrib, F. Refined damper design formula for a cable equipped with a positive or negative stiffness damper. *Struct. Control Health. Monit.* **2018**, *25*, e2236. [CrossRef]
40. Smith, M.C. Synthesis of mechanical networks: The inerter. *IEEE Ttans. Automat. Control* **2002**, *47*, 1648–1662. [CrossRef]
41. Ikago, K.; Saito, K.; Inoue, N. Seismic control of single-degree-of-freedom structure using tuned viscous mass damper. *Earthq. Eng. Struct. Dyn.* **2012**, *41*, 453–474. [CrossRef]
42. Takewaki, I.; Murakami, S.; Yoshitomi, S.; Tsuji, M. Fundamental mechanism of earthquake response reduction in building structures with inertial dampers. *Struct. Control Health Monit.* **2012**, *19*, 590–608. [CrossRef]
43. Lazar, I.F.; Neild, S.A.; Wagg, D.J. Using an inerter-based device for structural vibration suppression. *Earthq. Eng. Struct. Dyn.* **2014**, *43*, 1129–1147. [CrossRef]
44. Nakamura, Y.; Fukukita, A.; Tamura, K.; Matsuoka, T.; Hiramoto, K.; Sunakoda, K. Seismic response control using electro-magnetic inertial mass damper. *Earthq. Eng. Struct. Dyn.* **2014**, *43*, 507–527. [CrossRef]
45. Makris, N.; Kampas, G. Seismic protection of structures with supplemental rotational inertia. *J. Eng. Mech.* **2016**, *142*, 04016089. [CrossRef]
46. De Domenico, D.; Ricciardi, G. An enhanced base isolation system equipped with optimal tuned mass damper inerter (TMDI). *Earthq. Eng. Struct. Dyn.* **2018**, *47*, 1169–1192. [CrossRef]
47. De Domenico, D.; Deastra, P.; Ricciardi, G.; Sims, N.D.; Wagg, D.J. Novel fluid inerter based tuned mass dampers for optimised structural control of base-isolated buildings. *J. Frankl. Inst.* **2019**, *356*, 7626–7649. [CrossRef]
48. De Domenico, D.; Ricciardi, G. Improving the dynamic performance of base-isolated structures via tuned mass damper and inerter devices: A comparative study. *Struct. Control Health Monit.* **2018**, *25*, e2234. [CrossRef]
49. De Domenico, D.; Ricciardi, G. Optimal design and seismic performance of tuned mass damper inerter (TMDI) for structures with nonlinear base isolation systems. *Earthq. Eng. Struct. Dyn.* **2018**, *47*, 2539–2560.
50. Javidialesaadi, A.; Wierschem, N.E. Three-element vibration absorber–inerter for passive control of single-degree-of-freedom structures. *J. Vib. Acoust.* **2018**, *140*, 061007. [CrossRef]
51. Pan, C.; Zhang, R.F. Design of structure with inerter system based on stochastic response mitigation ratio. *Struct. Control Health Monit.* **2018**, *25*, e2169. [CrossRef]
52. Xu, K.; Bi, K.M.; Han, Q.; Li, X.P.; Du, X.L. Using tuned mass damper inerter to mitigate vortex-induced vibration of long-span bridges: Analytical study. *Eng. Struct.* **2019**, *182*, 101–111. [CrossRef]
53. Huang, Z.W.; Hua, X.G.; Chen, Z.Q.; Niu, H.W. Performance evaluation of inerter-based damping devices for structural vibration control of stay cables. *Smart Struct. Syst.* **2019**, *23*, 615–626.

54. Luo, J.; Macdonald, J.H.; Jiang, J.Z. Identification of optimum cable vibration absorbers using fixed-sized-inerter layouts. *Mech. Mach. Theory* **2019**, *140*, 292–304. [CrossRef]
55. Luo, J.N.; Jiang, J.Z.; Macdonald, J.H.G. Cable vibration suppression with inerter-based absorbers. *J. Eng. Mech.* **2019**, *145*, 04018134. [CrossRef]
56. Sun, H.X.; Zuo, L.; Wang, X.Y.; Peng, J.; Wang, W.X. Exact H_2 optimal solutions to inerter-based isolation systems for building structures. *Struct. Control Health Monit.* **2019**, *26*, e2357. [CrossRef]
57. Wang, Z.H.; Gao, H.; Wang, H.; Chen, Z.Q. Development of stiffness-adjustable tuned mass dampers for frequency retuning. *Adv. Struct. Eng.* **2019**, *22*, 473–485. [CrossRef]
58. Zhang, R.F.; Zhao, Z.P.; Dai, K.S. Seismic response mitigation of a wind turbine tower using a tuned parallel inerter mass system. *Eng. Struct.* **2019**, *180*, 29–39. [CrossRef]
59. Zhu, H.P.; Li, Y.M.; Shen, W.A.; Zhu, S.Y. Mechanical and energy-harvesting model for electro-magnetic inertial mass dampers. *Mech. Syst. Signal Process.* **2019**, *120*, 203–220. [CrossRef]
60. Lu, L.; Duan, Y.F.; Spencer, B.F.; Lu, X.L.; Zhou, Y. Inertial mass damper for mitigating cable vibration. *Struct. Control Health Monit.* **2017**, *24*, e1986. [CrossRef]
61. Cu, V.H.; Han, B.; Pham, D.H.; Yan, W.T. Free vibration and damping of a taut cable with an attached viscous mass damper. *KSCE J. Civ. Eng.* **2018**, *22*, 1792–1802. [CrossRef]
62. Shi, X.; Zhu, S.Y. Dynamic characteristics of stay cables with inerter dampers. *J. Sound Vib.* **2018**, *423*, 287–305. [CrossRef]
63. Wang, Z.H.; Gao, H.; Xu, Y.W.; Chen, Z.Q.; Wang, H. Impact of cable sag on the efficiency of inertial mass damper in controlling stay cable vibrations. *Smart Struct. Syst.* **2019**, *24*, 83–94.
64. Wang, Z.H.; Yue, F.F.; Wang, H.; Gao, H.; Fan, B.Q. Refined study on free vibration of a cable with an inertial mass damper. *Appl. Sci.* **2019**, *9*, 2271. [CrossRef]
65. Wang, Z.H.; Xu, Y.W.; Gao, H.; Chen, Z.Q.; Xu, K.; Zhao, S.B. Vibration control of a stay cable with a rotary electromagnetic inertial mass damper. *Smart Struct. Syst.* **2019**, *23*, 627–639.
66. Lazar, I.F.; Neild, S.A.; Wagg, D.J. Vibration suppression of cables using tuned inerter dampers. *Eng. Struct.* **2016**, *122*, 62–71. [CrossRef]
67. Sun, L.M.; Hong, D.X.; Chen, L. Cables interconnected with tuned inerter damper for vibration mitigation. *Eng. Struct.* **2017**, *151*, 57–67. [CrossRef]
68. Caracoglia, L.; Jones, N.P. Passive hybrid technique for the vibration mitigation of systems of interconnected stays. *J. Sound Vib.* **2007**, *307*, 849–864. [CrossRef]
69. Caracoglia, L.; Zuo, D. Effectiveness of cable networks of various configurations in suppressing stay-cable vibration. *Eng. Struct.* **2009**, *31*, 2851–2864. [CrossRef]
70. Zhou, H.J.; Sun, L.M.; Xing, F. Free vibration of taut cable with a damper and a spring. *Struct. Control Health Monit.* **2014**, *21*, 996–1014. [CrossRef]
71. Zhou, H.J.; Yang, X.; Sun, L.M.; Xing, F. Free vibrations of a two-cable network with near-support dampers and a cross-link. *Struct. Control Health Monit.* **2015**, *22*, 1173–1192. [CrossRef]
72. Ahmad, J.; Cheng, S.; Ghrib, F. Efficiency of an external damper in two-cable hybrid systems. *J. Bridge Eng.* **2017**, *23*, 04017138. [CrossRef]
73. Ahmad, J.; Cheng, S.H.; Ghrib, F. Combined effect of external damper and cross-tie on the modal response of hybrid two-cable networks. *J. Sound Vib.* **2018**, *417*, 132–148. [CrossRef]
74. Cu, V.H.; Han, B. A stay cable with viscous damper and tuned mass damper. *Aust. J. Struct. Eng.* **2015**, *16*, 316–323. [CrossRef]
75. Xu, Y.L.; Yu, Z. Mitigation of three-dimensional vibration of inclined sag cable using discrete oil dampers—II. Application. *J. Sound Vib.* **1998**, *214*, 675–693. [CrossRef]
76. Caracoglia, L.; Jones, N.P. Damping of taut-cable systems: Two dampers on a single stay. *J. Eng. Mech.* **2007**, *133*, 1050–1060. [CrossRef]
77. Hoang, N.; Fujino, Y. Combined damping effect of two dampers on a stay cable. *J. Bridge Eng.* **2008**, *13*, 299–303. [CrossRef]

applied sciences

MDPI

Article

Theoretical Study on a Cable-Bracing Inerter System for Seismic Mitigation

Liyu Xie [1], Xinlei Ban [1], Songtao Xue [1,2], Kohju Ikago [3], Jianfei Kang [1] and Hesheng Tang [1,4,*]

1. Department of Disaster Mitigation for Structures, Tongji University, Shanghai 200092, China; liyuxie@tongji.edu.cn (L.X.); 1610229@tongji.edu.cn (X.B.); xue@tongji.edu.cn (S.X.); 1832540@tongji.edu.cn (J.K.)
2. Department of Architecture, Tohoku Institute of Technology, Sendai 982-8577, Japan
3. International Research Institute of Disaster Science, Tohoku University, Sendai 980-0845, Japan; ikago@irides.tohoku.ac.jp
4. State Key Laboratory of Disaster Reduction in Civil Engineering, Tongji University, Shanghai 200092, China

* Correspondence: thstj@tongji.edu.cn; Tel.: +86-21-6598-2390

Received: 1 August 2019; Accepted: 24 September 2019; Published: 1 October 2019

Abstract: In this paper, cables are proposed to connect the inerter and main frame for translation-to-rotation conversion, i.e., the cable-bracing inerter system (CBIS), with a magnified mass and enhanced damping effect. This novel configuration has the benefits of deformation relaxation at the connecting joints, easy installation, and an adaptive layout for nonconsecutive-story deployment. Dynamic motion equations were established for a single degree-of-freedom (SDOF) model equipped with a CBIS. The influence of dimensionless parameters, such as inertance-mass ratio, stiffness ratio and additional damping ratio on vibration mitigation were studied in terms of displacement response and force output. A single objective and multiple objective optimal design method were developed for a CBIS-equipped structure based on a performance-oriented design framework. Finally, the mitigation effect was illustrated and verified by a numerical simulation in a time-domain. The results showed that a CBIS is an effective structural response mitigation device used to mitigate the response of structural systems under earthquake excitation. Using the proposed optimization method, CBIS parameters can be effectively designed to satisfy the target vibration control level.

Keywords: passive vibration control; inerter system; cable bracing; parametric study; optimal design

1. Introduction

To suppress the structural vibration induced by earthquakes or winds, various vibration control devices have been developed and widely applied [1–3]. Among them, the tuned mass damper (TMD) [4] has the simplest design and a concise vibration control mechanism, which consists of three classical mechanical components, namely mass elements, springs, and dampers. With the addition of a lumped mass in a TMD, the fundamental frequency of the main frame is tuned away from the dominating frequency range of the excitation. Part of the input energy is stored by the lumped mass in the form of kinetic energy, eventually dissipated by the dampers [5,6]. The spring and damper are typical two-terminal elements in the structure, and their output restoring force and damping force depends on the relative displacement and relative velocity between two terminals, respectively. On the other hand, the lumped mass suspended in the TMD [7] is a one-terminal element, and its inertia force exerted on the bearing frame is the product of its absolute acceleration and mass. To control seismic response effectively, the weight of TMDs contributes a significant portion of the entire structure, which demands additional bearing capacity of the main structure. For example, a 660-ton TMD was installed at the top of the Taipei 101 Building in Taiwan, China, whose weight is 0.4 percent of the primary structure, taking up nearly two stories of space for installation. The requests for extra space and the weight burden of the TMDs bring practical problems in real applications. Due to space constraints, an active mass damper [8] was installed in a tall TV tower in Nanjing, China to reduce its wind-induced response to replace the original plan of TMDs. An active mass damper has a smaller mass than a TMD, but it requires higher investment and maintenance costs.

To minimize the weight and dimension of a damper, an innovative inerter [9–14] for civil structures has been developed. For the same performance target, the required physical mass of the inerter is much smaller than that of the conventional TMD [12,15,16]. For the apparent mass to be much greater than its actual mass, a displacement amplification mechanism must be used, such as the rotation mechanism. The concept of inerter, a two-terminal inertial element, was initially introduced by Smith [17] in the early 2000s. The inertial force produced by an inerter is proportional to the relative acceleration between two terminals, which allows the inerter to utilize the acceleration difference between the adjacent floors for vibration mitigation. Another distinguishing feature is the mass amplification effect of inerters [9,18]. The inertance with the apparent mass can be several hundred times greater than its physical mass. An inerter behaves as a tuning element for absorbing vibration energy much like a lumped mass in a TMD does. The topology of an inerter system consists of three basic mechanical elements—the inerter, damping, and spring elements. The damping efficiency in the inerter system can be significantly enhanced by using the rotational amplifier compared to the traditional viscous dampers [17–19].

In civil engineering, a similar concept as inerters had been practiced independently in 1999, Arakaki et al. [20] used the ball screw mechanism to amplify the efficient output force of a viscous damper for suppressing vibrations induced by earthquakes. This was the first application of an inerter-based damper in civil engineering. Since then, various inerter-based devices have been developed, including tuned viscous mass dampers (TVMD) [9,21], tuned mass-damper–inerter systems (TMDI) [12], tuned inerter dampers (TIDs) [22], and so on. These inerter-based devices use rack pinion [22–24], ball screw [9,25], hydraulic [26–28] and electromagnetic [29–31] mechanisms to convert translational movement into rotational movement.

To establish an efficient and practical design method for structures with inerter systems, some theoretical analyses were carried out in the present study. Ikago et al. [9] derived a simple formula for optimal design of TVMD based on the fixed-point theory, which can be used as a design method in practice. Taking the inherent damping ability of a single degree-of-freedom (SDOF) structure into consideration, Pan et al. [32,33] proposed a demand-based optimal design method for a parallel-layout inerter system to satisfy performance demands with minimum control costs. Zhang et al. [34] investigated the impact of the mechanical layout of inerter systems on seismic-response mitigation of liquid-storage tanks, and Chen et al. [35] explored the influence of soil–structure interaction on structures equipped with an inerter system. After these designing measures were undertaken in the engineering applications, some novel inerter-based devices were developed.

Researchers had different approaches to equipping structures with these inerter-based devices. Hwang et al. [36] presented a ball–screw inerter system connected with a toggle brace to magnify the relative displacement between adjacent floors and showed that their system could be utilized effectively in structures even when the drift was small. Makris et al. [23] presented a rack–pinion–flywheel system supported by an infinitely stiff chevron frame and demonstrated that this system was particularly effective in suppressing the peak displacement of structures over long periods of time. Sugimura et al. [31] installed the TVMD system in a building in Tohoku, Japan. This building has seismic response control systems to upgrade the seismic safety of structures and facilities in the Tohoku building. It has traditional viscous dampers supporting the lower floors and the TVMD system supporting the upper floors. The TVMD was fixed between the adjacent floors using a support member with a relevant stiffness like steel. These braces can transmit bending moments, shear, and torque, and are sensitive to displacement at boundaries, which may induce the non-negligible moment and deformation at the TVMD terminals. Ball joints were used to release the deformation which could have induced the unwanted internal moment and torque in the brace. Cable bracing is the alternative method to ball joints for connecting inerters, since cables can only bear the axial tension force and release deformation other than the axial direction.

Tension-only cables are an important element in the seismic control systems and are mainly used to transmit control forces and to direct deformation from the main structure to energy dissipating devices. Samuel [37] uses cables to prevent progressive collapse of buildings. Kim et al. [38] proposed a rotational friction damper connected to tension-only braces to enhance the seismic resisting capacity of existing structures. Kurata et al. [39] developed a bracing system consisting of cables and a central energy dissipator. The tension-only cable design can increase the speed of construction by adopting simple connections with rapid installation features.

In this paper, we propose to use a pair of tension-only cables to transfer the story drift to the rotating flywheels, i.e., the cable-bracing inerter system (CBIS). Section 2 will introduce the concept of CBIS and establish the motion governing equation for a CBIS-equipped SDOF system excited by the ground motion. The frequency response functions of displacement and force output are derived for characteristic study. In Section 3, a parametric analysis is conducted to study the effects of CBIS parameters on structural seismic mitigation. In addition, a performance-based multi-objective H_2 norm optimum design method is proposed to design the CBIS. Design cases are carried out to illustrate the effects of CBIS and the effectiveness of the proposed design method. Section 4 draws the conclusions. These theoretical studies will lay the foundation for future experimental study.

2. Theoretical Analysis of a Cable-Bracing Inerter System

2.1. CBIS Concept

Cable bracing is the proposed mechanism of translation-to-rotation conversion for an inerter connected to a structure. Figure 1 shows an SDOF structure with a CBIS, which consists of a pair of bracing cables, a pair of conductor plates (flywheels), and a shaft. A pair of cables is pre-tensioned connecting the structural frame and the shaft diagonally. Both ends of the shaft are supported by a pair of shaft bearings mounted on the side plates fixed on the ground floor, making sure that only the shaft rotates. When inter-story drift occurs in structures, one of the cables will shorten and drive the shaft into rotation. The low-speed translational movement of the structure can be converted into a high-speed rotational motion of the conductor plates by cable bracing.

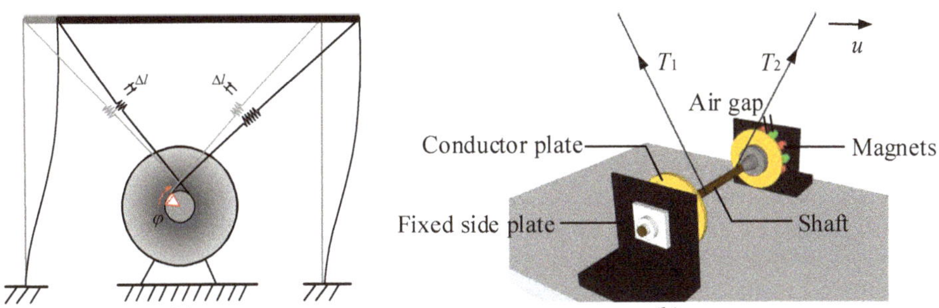

Figure 1. A single-degree-of-freedom (SDOF) structure with a CBIS: (**a**) structure with a CBIS and (**b**) detail of a CBIS. Note that the conductor plates serve as flywheels.

Conductor plates are fixed on the shaft and rotate together, meanwhile, several magnets with alternating magnetic polarization are allocated on the fixed side plate to generate the electromagnetic field. The rotational conductor plates and shaft function as an inerter. The fixed side plate and conductor plate form one eddy current damper when the rotation conductor plate is cutting through the electromagnetic field, thereby dissipating the vibration energy in the form of heat. As a result, the novel cable-bracing inerter system presented herein can obtain its inertance and enhance the energy dissipation capacity via the additional damping provided by the eddy current damping. Compared to the classic ball-screw mechanism of an inerter, CBIS is cost-effective and very simple.

2.2. Inerter Element

An inerter element is a two-terminal element. The output force is proportional to the relative acceleration between two terminals and can be expressed as:

$$p = m_{\mathrm{d}}(a_2 - a_1), \tag{1}$$

where p is the output force of the inerter element, and m_{d} is the inertance; moreover, a_1 and a_2 are the accelerations at the two terminals of the inerter element, as depicted in Figure 2.

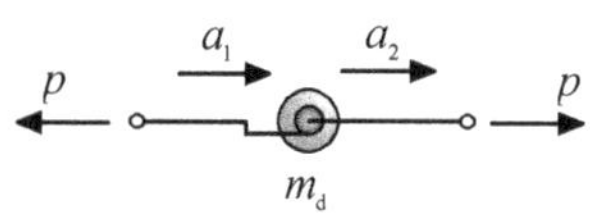

Figure 2. Mechanical model of an inerter element.

2.3. Layout of SDOF System with CBIS

The mechanical system of a CBIS consists of an additional damping element, an inerter element and a spring element. The damping element is set in parallel with the inerter element. The spring element is then connected with the paralleled inerter and damping element in a series. Figure 3 shows the layout and mechanical model of an SDOF structure equipped with a CBIS when the structure is deformed in a horizontal direction.

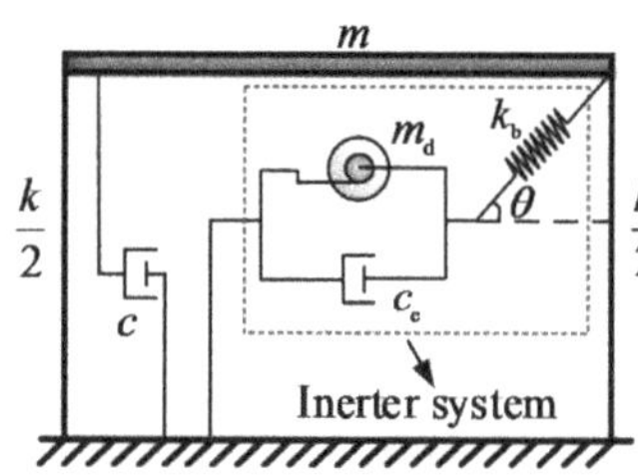

Figure 3. Layout of an SDOF system with a CBIS.

In Figure 3, m, c, and k are the mass, damping coefficient, and stiffness of the SDOF system, the primary structure, respectively. θ is the inclined angle of the diagonal cable; m_{d} is the inertance of the CBIS; $c_{\mathrm{e}} = c_{\mathrm{d}} \cos^2 \theta$ is the equivalent damping coefficient considering the inclined angle of the cables, where c_{d} is the damping coefficient of the damping element, and k_{b} is the stiffness of the supporting spring element. The output force of this inerter system is the resultant force of the inerter element and the eddy current damping element.

2.4. Motion Governing Equation of SDOF System with CBIS

When the structure is in a balanced state, the prestressed tension forces in both cables are T_0. If the structure starts to leave the balance position by moving to the right as illustrated in Figure 4, the diagonal cable on the right side drives the inerter to rotate clockwise. At this time, the force increment in the right cable is ΔT, and the force becomes T_2 ($T_2 = T_0 + \Delta T$). The tension in the left cable decreases ΔT and becomes T_1 ($T_1 = T_0 - \Delta T$).

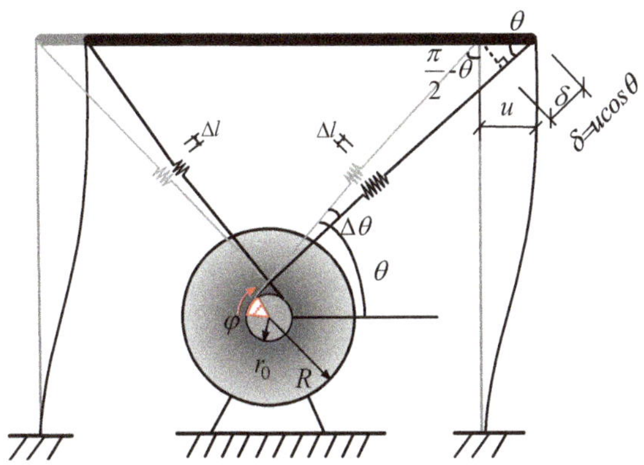

Figure 4. Transmission mechanism of cable-bracing inerter system.

When the structure has a positive deformation, it moves to the right with a relative displacement $u(t)$, and the shaft and the conductor plates rotate correspondingly. We assume there is no relative slippage between the cable and the shaft. Considering the axial stiffness of the one-sided cable $k_b{}^0$, the forces in the right cable T_2 and the left cable T_1 are as follows:

$$\begin{aligned} T_2 &= T_0 + k_b{}^0(u(t)\cos\theta - \varphi(t)r_0) \\ T_1 &= T_0 - k_b{}^0(u(t)\cos\theta - \varphi(t)r_0), \end{aligned} \tag{2}$$

where the rotational angle of the conductor plate is $\varphi(t)$. As shown in Figure 5, when the conductor plates rotate, the cable moves in its own axial direction. Thus, the angle difference $\Delta\theta$ from its balance position is trivial and can be ignored. $u(t)\cos\theta - \varphi(t)r_0$ is the axial elongation of the cable. During the operation of a CBIS, the tension force difference between two cables drives the shaft to rotate and is given as:

$$T_2 - T_1 = 2k_b{}^0(u(t)\cos\theta - \varphi(t)r_0). \tag{3}$$

The eddy currents cause a damping force that is proportional to the velocity of the conductive metal, which makes the eddy currents function like a viscous damper. According to the force equilibrium conditions, compatibility condition, and the layout of the system (as shown in Figure 3), the motion equation for this SDOF structure with a CBIS under earthquake excitations can be written as:

$$m\ddot{u}(t) + c\dot{u}(t) + ku(t) + k_b(u(t)\cos\theta - \varphi(t)r_0)\cos\theta = -ma_g(t), \tag{4}$$

where $u(t)$ is the relative displacement of the SDOF system, and the dots represent the derivative with respect to time t. $a_g(t)$ is the acceleration of the ground motion, k_b is the equivalent stiffness of two cables, and it is used to replace $2k_b{}^0$ in Equation (3). The motion equation for the CBIS is written as:

$$J\ddot{\varphi}(t) + c_d\dot{\varphi}(t)r_0{}^2 = k_b(u(t)\cos\theta - \varphi(t)r_0)r_0, \tag{5}$$

where J is the moment of inertia for the inerter, and r_0 is the radius of the shaft. The conductor plate serves as a flywheel whose moment of inertia can be calculated as:

$$J = m_I R^2/2, \tag{6}$$

where m_I is the physical mass of two conductor plates and the shaft, and R is the radius of gyration.

The governing motion equation of an SDOF structure can be described as:

$$\begin{bmatrix} m & 0 \\ 0 & J \end{bmatrix}\begin{Bmatrix} \ddot{u} \\ \ddot{\varphi} \end{Bmatrix}+\begin{bmatrix} c & 0 \\ 0 & c_{\mathrm{d}}r_0^2 \end{bmatrix}\begin{Bmatrix} \dot{u} \\ \dot{\varphi} \end{Bmatrix}+\begin{bmatrix} k+k_{\mathrm{b}}\cos^2\theta & -k_{\mathrm{b}}r_0\cos\theta \\ -k_{\mathrm{b}}r_0\cos\theta & k_{\mathrm{b}}r_0^2 \end{bmatrix}\begin{Bmatrix} u \\ \varphi \end{Bmatrix}=\begin{bmatrix} -ma_g \\ 0 \end{bmatrix}. \tag{7}$$

Therefore, Equation (7) can be expressed in matrix form as:

$$\mathbf{M}\ddot{\mathbf{X}}+\mathbf{C}\dot{\mathbf{X}}+\mathbf{K}\mathbf{X}=\mathbf{F}, \tag{8}$$

where **M**, **C**, **K** and **F** respectively represent the mass matrix, damping matrix, stiffness matrix and external excitation vector of the SDOF system with a CBIS. Equation (8) is converted into the state space form:

$$\mathbf{A}\dot{\overline{\mathbf{X}}}+\mathbf{B}\overline{\mathbf{X}}=\begin{Bmatrix} \mathbf{F} \\ \mathbf{0} \end{Bmatrix}, \tag{9}$$

where $\mathbf{0}=\{\mathbf{0},\mathbf{0}\}^T$, **A**, **B**, $\overline{\mathbf{X}}$ and $\dot{\overline{\mathbf{X}}}$ are determined as:

$$\mathbf{A}=\begin{bmatrix} c & 0 & m & 0 \\ 0 & c_{\mathrm{d}}r_0^2 & 0 & J \\ m & 0 & 0 & 0 \\ 0 & J & 0 & 0 \end{bmatrix}, \mathbf{B}=\begin{bmatrix} k+k_{\mathrm{b}}\cos^2\theta & -k_{\mathrm{b}}r_0\cos\theta & 0 & 0 \\ -k_{\mathrm{b}}r_0\cos\theta & k_{\mathrm{b}}r_0^2 & 0 & 0 \\ 0 & 0 & -m & 0 \\ 0 & 0 & 0 & -J \end{bmatrix}, \overline{\mathbf{X}}=\begin{Bmatrix} u \\ \varphi \\ \dot{u} \\ \dot{\varphi} \end{Bmatrix}, \dot{\overline{\mathbf{X}}}=\begin{Bmatrix} \dot{u} \\ \dot{\varphi} \\ \ddot{u} \\ \ddot{\varphi} \end{Bmatrix}. \tag{10}$$

Assume that the solution of Equation (9) has the form of:

$$\overline{\mathbf{X}}=\boldsymbol{\psi}\mathbf{e}^{\lambda t}, \dot{\overline{\mathbf{X}}}=\boldsymbol{\psi}\lambda\mathbf{e}^{\lambda t}, \tag{11}$$

where $\boldsymbol{\psi}$ is the eigenvector, substituting $\boldsymbol{\psi}=[\psi_1\ \psi_2]^{\mathrm{T}}$ into Equation (9), whereby we obtain:

$$(\mathbf{A}\lambda+\mathbf{B})\begin{Bmatrix} \boldsymbol{\psi} \\ \boldsymbol{\psi}\lambda \end{Bmatrix}=0. \tag{12}$$

The characteristic equation can be expressed as:

$$\det|\mathbf{A}\lambda+\mathbf{B}|=0. \tag{13}$$

The j th pair of eigenvalues are λ_{2j-1} and λ_{2j}, and the j th fundamental angular frequency ω_j can be obtained as:

$$\omega_j=|\lambda_{2j-1}|=|\lambda_{2j}|. \tag{14}$$

2.5. Frequency Response Function

In this section, the frequency response function of an SDOF structure with a CBIS under seismic excitation is obtained. To normalize the dynamic equation, the parameters can be defined as:

$$\omega_{\mathrm{s}}=\sqrt{\frac{k}{m}}, \tag{15}$$

$$\zeta=c/2m\omega_{\mathrm{s}}, \tag{16}$$

$$\xi=c_{\mathrm{d}}/2m\omega_{\mathrm{s}}, \tag{17}$$

where ω_s and ζ are the circular frequency and the inherent damping ratio of the original SDOF structure, respectively. ξ is the additional damping ratio provided by the CBIS. Through dimensionless processing, the following parameters can be defined for designing CBIS:

$$\mu = m_d/m, \tag{18}$$

$$\kappa = k_b \cos^2\theta/k, \tag{19}$$

where μ is inertance–mass ratio, a ratio of the inertance of CBIS to the mass of primary system. κ is the ratio of supporting spring stiffness in the horizontal direction to the primary stiffness k. Substituting the parameters in Equations (15)–(19) into Equation (9), the Laplace transformation of Equation (7) can be written as:

$$\begin{bmatrix} sU \\ s^2U \\ s\Phi \\ s^2\Phi \end{bmatrix} = \begin{bmatrix} 0 & 1 & 0 & 0 \\ -\omega_s{}^2-\kappa\omega_s{}^2 & -2\zeta\omega_s & \kappa\omega_s{}^2 r_0/\cos\theta & 0 \\ 0 & 0 & 0 & 1 \\ \kappa\omega_s{}^2 m r_0/J\cos\theta & 0 & -\kappa\omega_s{}^2 m r_0{}^2/J\cos^2\theta & -2\xi\omega_s r_0{}^2 m/J \end{bmatrix} \begin{bmatrix} U \\ sU \\ \Phi \\ s\Phi \end{bmatrix} - \begin{bmatrix} 0 \\ 1 \\ 0 \\ 0 \end{bmatrix} A_g(s), \tag{20}$$

where $s = i\omega$, and $A_g(s)$ is the Laplace transformation of $a_g(t)$. U, $\dot{U}$, Φ and $\dot{\Phi}$ are the Laplace transformations of u, $\dot{u}$, φ and $\dot{\varphi}$, respectively, and they can be solved from Equation (8):

$$\begin{bmatrix} U(s) \\ \dot{U}(s) \\ \Phi(s) \\ \dot{\Phi}(s) \end{bmatrix} = \begin{bmatrix} \frac{-(s^2\mu+2\xi\omega_s s\cos^2\theta+\kappa\omega_s{}^2)A_g(s)}{C(s,\xi,\zeta,\mu,\kappa,\omega_s)} \\ \frac{-s(s^2\mu+2\xi\omega_s s\cos^2\theta+\kappa\omega_s{}^2)A_g(s)}{C(s,\xi,\zeta,\mu,\kappa,\omega_s)} \\ \frac{-\kappa\omega_s{}^2\cos\theta A_g(s)}{D(s,\xi,\zeta,\mu,\kappa,\omega_s,r_0)} \\ \frac{-s\kappa\omega_s{}^2\cos\theta A_g(s)}{D(s,\xi,\zeta,\mu,\kappa,\omega_s,r_0)} \end{bmatrix}, \tag{21}$$

where

$$\begin{aligned} &C(s,\xi,\zeta,\mu,\kappa,\omega_s) = s^4\mu + 2s^3(\xi\cos^2\theta+\zeta\mu)\omega_s + s^2(\kappa+4\xi\zeta\cos^2\theta+\mu+\kappa\mu)\omega_s{}^2 \\ &+2s(\zeta\kappa+\xi\cos^2\theta+\kappa\xi\cos^2\theta)\omega_s{}^3+\kappa\omega_s{}^4 \\ &D(s,\xi,\zeta,\mu,\kappa,\omega_s,r_0) = s^4\mu r_0 + 2s^3(r_0\xi\cos^2\theta+r_0\zeta\mu)\omega_s + s^2(\kappa\mu r_0+4\xi\zeta\cos^2\theta r_0 \\ &+\mu r_0+\kappa r_0)\omega_s{}^2+2s(\xi\kappa r_0+\zeta\kappa r_0+\xi\cos^2\theta r_0)\omega_s{}^3+\kappa r_0\omega_s{}^4. \end{aligned} \tag{22}$$

The frequency–domain transfer function between $u(t)$ and input excitations can be easily obtained as:

$$H_U(s) = \frac{U(s)}{A_g(s)} = \frac{-(s^2\mu+2\xi\omega_s s\cos^2\theta+\kappa\omega_s{}^2)}{C(s,\xi,\zeta,\mu,\kappa,\omega_s)}. \tag{23}$$

The normalized force of the CBIS is defined as $F(t) = k_b(u(t)\cos\theta - \varphi(t)r_0)/m$, which is provided by the inertial mass element and the eddy current damping element. The frequency–domain transfer function between $F(t)$ and the input excitation is given as:

$$H_F(s) = \frac{F(s)}{A_g(s)} = \frac{-(2\kappa\xi\omega_s{}^3 s\cos^2\theta+\kappa\omega_s{}^2\mu s^2)}{A(s,\xi,\zeta,\mu,\kappa,\omega_s)\cos\theta}. \tag{24}$$

3. Parametric Study

In this section, parametric analyses will be performed to investigate the effects of CBIS parameters on structural seismic mitigation. The mitigation effect is represented in terms of the moduli of displacement and force frequency response function at the resonant frequency. Three arguments, namely the inertance–mass ratio μ, stiffness ratio κ, and additional damping ratio ξ, vary to study the control effect of a CBIS damper. The domain of these arguments is determined mathematically without considering practical aspects.

3.1. Analysis Index

The response of a structure at the resonant frequency is much greater than that at another frequency, so the peak value of the system's responses deserves much attention. To find out the maximum displacement response of the structure, the displacement amplification factor is referred to as the H_∞ norm and can be defined as:

$$H_\infty = \max\left\{\left|H_U(i\beta)\right|\omega_s^2\right\}, \tag{25}$$

where $\beta = \omega/\omega_s$, and it can be interpreted as the normalized frequency. This index is independent of the natural circular frequency ω_s of the original structure, and can be considered as a function, depending on the original structure's inherent damping ratio ζ and the CBIS parameters, μ, ξ, and κ. Henceforth, the intention of vibration control is to minimize the displacement amplification factor in terms of the H_∞ norm for a set of optimal parameters of the CBIS.

3.2. Parametric Analysis Results

Based on the analysis indexes mentioned above, a series of numerical cases were considered. In these cases, the inherent damping ratio for the main frame $\zeta = 0.02$ was assumed. Three inertance-mass ratios, namely, $\mu = 0.01$, 0.1 and 1.0, were used to study the vibration control effects by continuously varying κ and ξ within specified ranges. To describe the controlling index of the SDOF with a CBIS, contour plots were illustrated with ξ on the x-axis, κ on the y-axis, and $\max\left\{\left|H_U(i\beta)\right|\omega_s^2\right\}$ as the height, as shown in Figure 5.

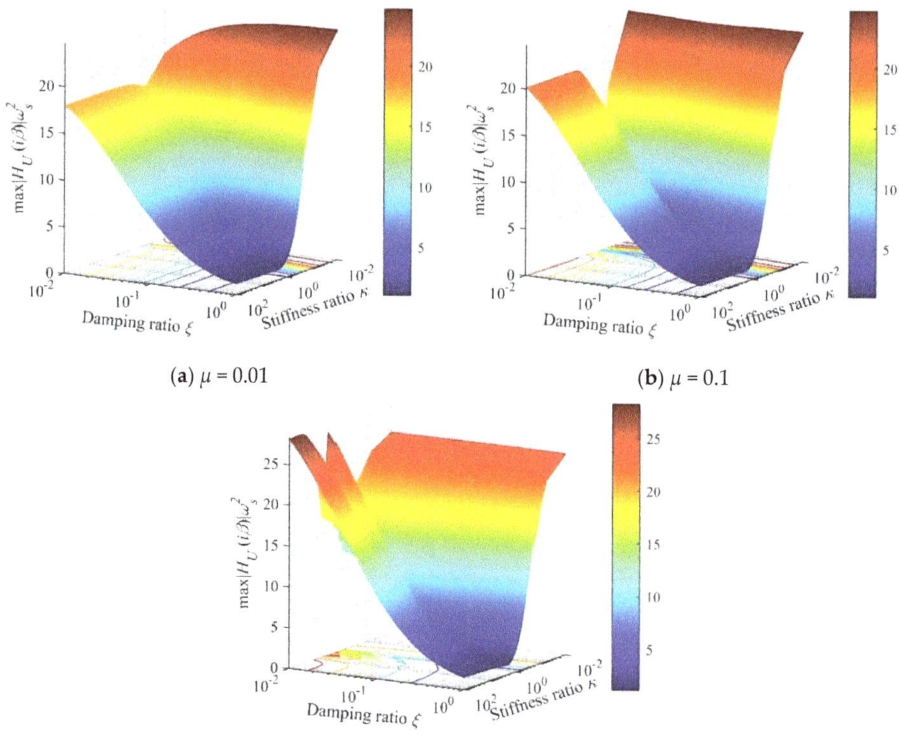

(**a**) $\mu = 0.01$ (**b**) $\mu = 0.1$

(**c**) $\mu = 1$

Figure 5. Displacement amplification factor of a CBIS-equipped SDOF structure with changes in $\kappa \in [0.01, 100]$ and $\xi \in [0.01, 1]$.

The displacement amplification factor of any point in the κ-ξ space is determined by the frequency response function of Equation (25). The magnitude is represented by the color intensity. The lowest displacement amplification factor is in the corner of the κ-ξ space, where the two parameters reach their upper bounds. As shown in Figure 5a–c, the desirable solution of the dynamic response ratio is determined by the feasible upper bounds of the CBIS parameters in the given ranges. Mathematically, the optimal configuration for vibration control requires the stiffness ratio and additional damping to be as large as possible. However, both are impossible to realize in actual engineering. It is necessary to introduce appropriate boundary conditions or constraints for practical optimization processes.

When the additional damping ratios are fixed, for example $\xi = 0.05, 0.1$, closed contour lines can be obtained (as shown in Figure 6). A very low point implies that the parameter set for optimal control can always be found in the inner part of every contour. This means that optimal solutions lie within the inner part of the parametric space. When the stiffness ratio closes to 1, and the inertance–mass ratio closes to 0.1, the displacement response reaches its lowest point. In this process, the optimal additional damping ratio ξ remains unknown. Therefore, the selection of a rational parameter set for the design of CBIS based on only the displacement response is difficult. However, for many situations, the optimization will involve a recursive process in which the optimal configuration keeps updating with a prescribed additional damping ratio ξ until the mitigation effect satisfies the objective.

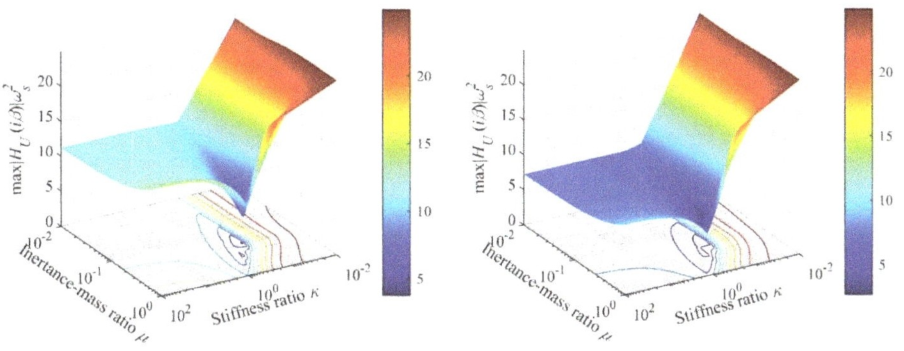

(a) $\xi = 0.05$ **(b)** $\xi = 0.1$

Figure 6. Displacement amplification factor of a CBIS-equipped SDOF structure with changes in $\kappa \in [0.01, 100]$ and $\mu \in [0.01, 1]$.

In Figure 6, the displacement responses reach their optimal points when the stiffness ratio closes to 1, and the inertance–mass ratio closes to $[0.1, 0.5]$,where the force response of the inerter system is relatively large (as shown in Figure 7). In the optimization design, it is unreasonable to consider only the structural displacement, or the force response provided by the inerter element. Therefore, both the displacement responses and the inerter element's force should be considered in the design of CBIS. The demand-oriented multi-objective optimum design method will be introduced in the next section, and the control force response of the CBIS will be brought into the optimization process as the secondary objective.

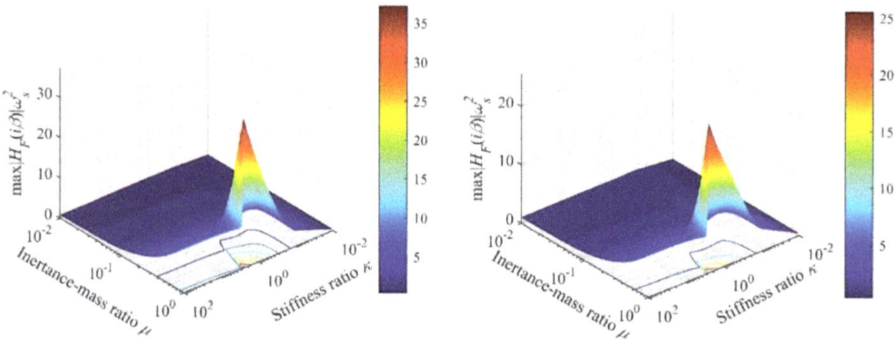

(**a**) $\xi = 0.05$ (**b**) $\xi = 0.1$

Figure 7. Force amplification factor of the inerter system.

3.3. Multi-Objective H_2 Norm Optimum Design

Three unknown parameters can be designed optimally according to performance demands. An optimization method is proposed to achieve the desired performance levels (structure's displacement u) with low control output force (CBIS's force f_d). In other words, the goal is to suppress the displacement of the structure as thoroughly as possible while minimizing the output force of the inerter system. Therefore, both the displacement response and the CBIS force should be considered in the design of this inerter system. Two dimensionless response variation ratios [32] are defined as: namely, the displacement reduction ratio, γ_U and the force ratio, γ_{F_d}, which can be expressed as:

$$\gamma_U(\zeta,\mu,\xi,\kappa) = \frac{\sigma_U}{\sigma_{U_0}} = \frac{\sqrt{\int_{-\infty}^{\infty} |H_U(i\omega)|^2 d\omega}}{\sqrt{\int_{-\infty}^{\infty} |H_{U_0}(i\omega)|^2 d\omega}}, \tag{26}$$

$$\gamma_{F_d}(\zeta,\mu,\xi,\kappa) = \frac{\sigma_{F_d}}{\sigma_{F_{d0}}} = \frac{\sqrt{\int_{-\infty}^{\infty} |H_{F_d}(i\omega)|^2 d\omega}}{\sqrt{\int_{-\infty}^{\infty} |H_{F_{d0}}(i\omega)|^2 d\omega}}. \tag{27}$$

In these expressions, γ_U is the ratio of the CBIS-equipped structure's displacement, compared with the primary structure, and γ_{F_d} is the dimensionless force of the CBIS. σ_U is the root mean square (RMS) of the output displacement response of the structure equipped with CBIS, and σ_{U_0} is the displacement response of the primary structure. σ_{F_d} is the force of the RMS response of the structure equipped with CBIS, and $\sigma_{F_{d0}}$ is the force RMS response of the original structure. $|H_{U_0}(i\omega)|$ and $|H_{F_{d0}}(i\omega)|$ are the displacement and damping force (caused by the inherent damping) transfer function moduli of the original structure, respectively. The optimization of CBIS can be expressed mathematically as:

$$\begin{aligned} &\underset{\mu,\xi,\kappa}{\text{minimize}}\left[\gamma_U(\mu,\xi,\kappa), \gamma_{F_d}(\mu,\xi,\kappa)\right], \\ &\text{subject to} \begin{cases} \mu_{\min} \le \mu \le \mu_{\max} \\ \xi_{\min} \le \xi \le \xi_{\max} \\ \kappa_{\min} \le \kappa \le \kappa_{\max} \end{cases}, \end{aligned} \tag{28}$$

where μ, ξ and κ are decision variables, and $\mu_{\min}$, $\xi_{\min}$, and $\kappa_{\min}$ are the lower bounds of μ, ξ and κ, while $\mu_{\max}$, $\xi_{\max}$, and $\kappa_{\max}$ are the upper bounds, respectively. Multi-objective optimization (MOO) is used to find the boundary of the feasible criterion space where all optimal points lie: namely, Pareto Front (shown in Figure 8). On this boundary, there are many points. Based on the demanding performance, a reasonable optimization parameter for $\gamma_U \leq 40\%$ and the corresponding γ_{F_d} can be found. Furthermore, a set of parameters was selected for time domain analysis, as shown in Table 1.

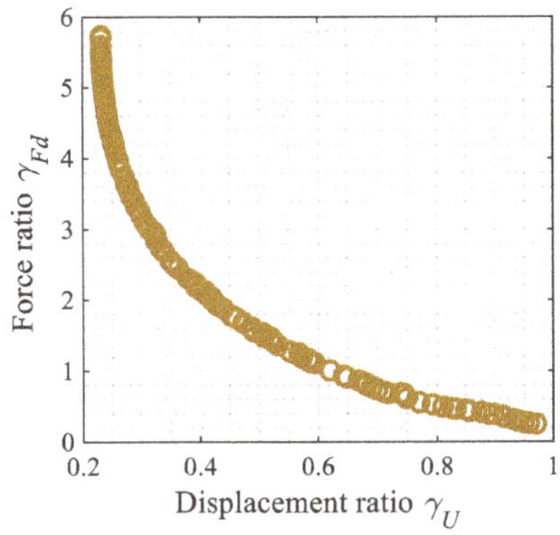

Figure 8. Pareto optimal front.

Table 1. Results of performance-based design optimization.

Optimized Design Parameters			**Displacement Response Ratio**	**Force Ratio**	**Energy Dissipation Efficiency**
Inertance-mass ratio	Damping ratio	Stiffness ratio	γ_U	γ_{F_d}	ψ
0.0198	0.1065	1.2728	0.40	2.1375	8.7517

One of the advantages of a CBIS over a viscous element is that it can enhance the energy dissipation efficiency. The energy dissipation enhancement mechanism is described by the factor ψ [32] (shown in Table 1), which equals the ratio of the response mitigation of an SDOF structure with CBIS to that of an SDOF structure with a viscous element having the same additional damping coefficient as the CBIS, that is:

$$\psi = \frac{\sigma_{U_0}(\zeta) - \sigma_U(\mu, \xi, \kappa)}{\sigma_{U_0}(\zeta) - \sigma_{U_0}(\zeta + \xi)}. \tag{29}$$

The degree of the energy dissipation enhancement of the CBIS can be adjusted by adding the following supplementary constraint condition to the optimization problem by:

$$\psi \geq \psi_0, \tag{30}$$

where ψ_0 is a constant during and the recommended range $1< \psi_0 \leq 2$, according to the many numerical case studies [32]. Here, ψ is 8.7517, which means that the inerter element has fully played its role, and, in the case of the same additional damping ratio, the energy dissipation efficiency of CBIS is 8.7517 times that of a purely viscous element. To illustrate the effects of CBIS on the seismic performance in the time domain, dynamic time–history analyses were conducted to further verify the design results under harmonic excitation and seismic excitations. Four seismic waves are used as external excitations—the El Centro record (1940, NS), the ground motion recorded at Tohoku University during the 2011 Tohoku earthquake (M = 9.0, PGA = 3.33 m/s^2), Kobe record (1995) and Chi–chi record (1999). The Tohoku wave occurred on 11 March 2011 and was part of the most powerful known earthquake in Japan. For the SDOF structure, the inherent damping ratio is $\zeta = 0.02$. The natural period on the rigid base is 1.00 s.

3.3.1. Harmonic Excitation

Using the performance-based optimization results in Table 1, the controlled and uncontrolled responses of the structure are compared under harmonic excitations. The natural frequency of the structure is 1 Hz and the frequency range of harmonic excitations is 0–3 Hz. Figure 9 shows displacement amplification factors of the uncontrolled and a CBIS-equipped SDOF structure. The peak value reduction effect is 71.75%. The response of a structure at the resonant frequency is significantly reduced by the CBIS.

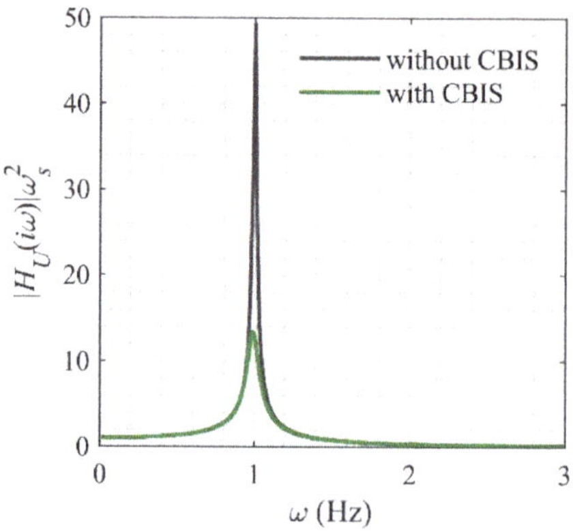

Figure 9. Displacement amplification factors.

3.3.2. Earthquake Excitation

Figures 10 and 11 show the SDOF structures' acceleration and displacement responses. The peak value and RMS value of the displacement and acceleration responses were chosen to evaluate the vibration mitigation performance of CBIS. These values are both important controlling indices in structural vibration control. The peak value reflects the dynamic response at a certain instant, whereas the RMS value indicates the vibration energy and reflects the responses over an entire period.

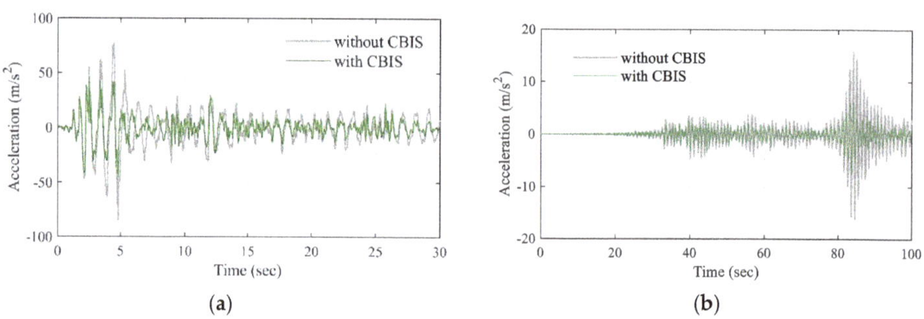

Figure 10. *Cont.*

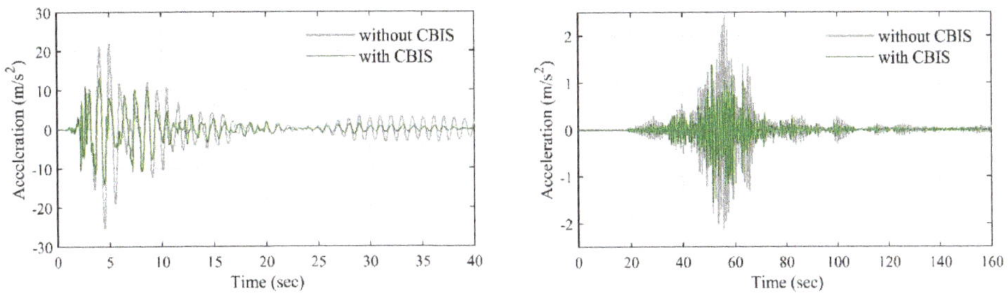

(c) (d)

Figure 10. Acceleration response time histories for two test frame roofs during (**a**) the El Centro wave, (**b**) the Tohoku wave recorded by Tohoku University, (**c**) the Kobe record and (**d**) the Chi–chi record.

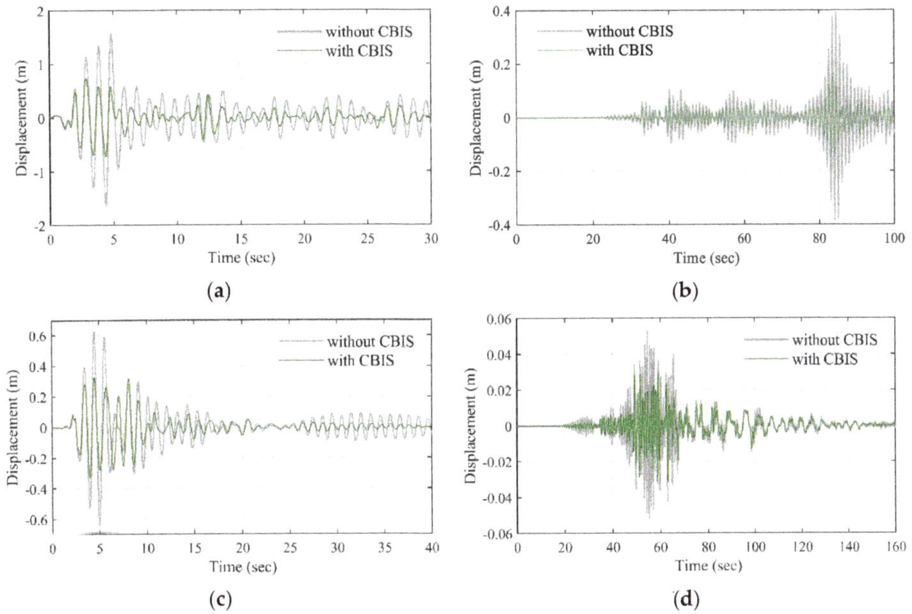

Figure 11. Displacement response time histories for two test frame roofs during (**a**) the El Centro wave, (**b**) the Tohoku wave recorded by Tohoku University, (**c**) the Kobe record and (**d**) the Chi–chi record.

The vibration reduction effect is defined as:

$$\text{Reduction effect} = \frac{\text{Response of uncontrolled structure} - \text{Response of controlled structure}}{\text{Response of uncontrolled structure}} \times 100\%. \quad (31)$$

The results are listed in Tables 2 and 3. The reduction effects were favorable under the El Centro wave, Tohoku wave, Kobe record and Chi–chi record. The best vibration control effects for the peak and RMS values of the acceleration responses were 52.08% and 45.71% (marked in bold), respectively. The values for the displacement responses were 55.56% and 52.50%, respectively.

Table 2. Acceleration responses at the roof of the test frame (m/s^2).

Seismic Input	El Centro Record		The Tohoku University Record		Kobe Record		Chi–Chi Record	
	Peak	RMS	Peak	RMS	Peak	RMS	Peak	RMS
Uncontrolled	8.77	1.74	16.59	1.85	25.47	4.75	2.44	0.35
Controlled	4.75	1.04	7.95	1.16	14.14	2.91	1.55	0.19
Reduction effects (%)	45.84	40.23	**52.08**	37.30	44.48	38.74	36.48	**45.71**

Table 3. Displacement responses at the roof of the test frame (m).

Seismic Input	El Centro Record		The Tohoku University Record		Kobe Record		Chi–Chi Record	
	Peak	RMS	Peak	RMS	Peak	RMS	Peak	RMS
Uncontrolled	0.168	0.040	0.396	0.046	0.646	0.129	0.053	0.009
Controlled	0.077	0.019	0.176	0.026	0.327	0.081	0.032	0.005
Reduction effects (%)	54.17	**52.50**	**55.56**	43.48	49.38	37.21	39.62	44.44

3.4. Ignoring the Flexibility of the Cable

To simplify the CBIS analytical model, the flexibility of the cable is neglected as shown in Figure 12.

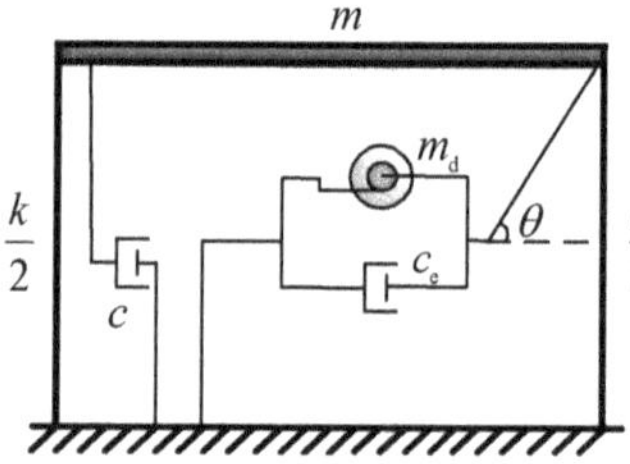

Figure 12. Simplified analytical model of the CBIS-equipped SDOF structure.

Consider this simplified CBIS-equipped SDOF structure as a model with no connection element flexibility. The relationship between the axial deformation of the cable and the rotational angle of the shaft is expressed as:

$$\varphi(t) = \frac{u(t)\cos\theta}{r_0}. \tag{32}$$

The equation of motion for an SDOF model with a CBIS is given by:

$$\left(m + \frac{J\cos^2\theta}{{r_0}^2}\right)\ddot{u}(t) + \left(c_0 + c_d\cos^2\theta\right)\dot{u}(t) + ku(t) = -ma_g(t). \tag{33}$$

In Equation (33), $\frac{J\cos^2\theta}{{r_0}^2}$ is m_d, which is namely the inertance of the inerter. From Equation (33), it can be understood that the utilization of CBIS induces the elongation of the natural period and increases the damping effect. The mass of the primary structure is affected, while the stiffness of the overall structure remains unaffected. By using the Laplace transformations, the transfer function can be easily obtained as:

$$H_U(s) = \frac{U(s)}{A_g(s)} = \frac{-1}{(1+\mu)s^2 + (2\zeta\omega_s + 2\xi\omega_s\cos^2\theta)s + \omega_s^2}. \tag{34}$$

We assume a single-floor frame structure, which can usually be treated as an SDOF structure. The key parameters of the structure and the inerter system are as follows: inherent damping ratio is $\zeta = 0.02$, the tilt angle of the cable is $\theta = \pi/4$, the inertance–mass ratio is $\mu = 0.1, 0.2$, and the damping ratio is ξ = 0.01, 0.05, 0.1. By using these parameters, the displacement amplification factor can be plotted as shown in Figure 13.

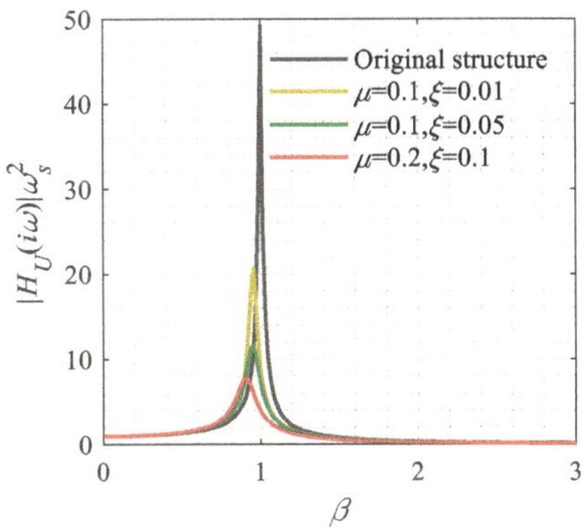

Figure 13. Displacement amplification factors of the SDOF structure with different inerter systems.

Figure 13 shows that inerter systems can significantly suppress the resonant response in a narrow band near the natural frequency of the primary structure but does not impact the other range of frequencies.

4. Conclusions

A novel inerter system for vibration control, which uses tension-only cables for translation-to-rotation conversion, was proposed in this paper. This device can be put into practical use with an inertia mass amplification element, enabling it to simultaneously achieve the displacement amplification effect. To study the performance of a CBIS on the seismic response mitigation of structures, the motion equations, both with and without flexibility of the cable, were derived and studied based on parametric analysis.

CBIS has the potential for seismic rapid retrofit of structures due to their easy installation and adaptive deployment. In this system, cables are used to convert translational deformation of the primary structure into the rotational motion of the fly wheels; thus, small actual mass can be amplified to large inertance by several hundred times. The proposed cable-bracing system can be adjusted for various frame configurations and design capacities. It can be installed in any direction and part of the structure as long as there exists relative deformation, not limited to horizontal vibration. This cable-bracing system uses simple connections with rapid and adjustable installation. It has the advantages of lower construction cost and easy replacement. The CBIS with a non-contacting damping mechanism shows excellent performance in the adjustable damping ratio by varying the air gap between the permanent magnet and the conductor.

The CBIS is an effective structural response mitigation device used to mitigate the response of structural systems under dynamic excitation. The peak and RMS responses of the SDOF structure were reduced after they were equipped with this system. To obtain a more rational parameter set for practical design, the demand-oriented multi-objective optimum design method is used to find the boundary of the feasible criterion space. Using the proposed method, the parameters of the CBIS can be effectively designed to satisfy the target vibration mitigation effects.

The vibration mitigation effect of an SDOF structure with the inerter system was analyzed in this paper. Future research will explore its application to multi-degree of freedom structures. Currently, a physical realization of the CBIS has been developed, and experimental verifications, including free vibration and shaking table tests, are underway.

Author Contributions: Conceptualization, L.X. and S.X.; methodology, X.B.; software, X.B., J.K.; validation, K.I.; formal analysis, L.X. and X.B.; data curation, J.K. and S.X.; writing—original draft preparation, X.B.; writing—review and editing, H.T.

Funding: This study was supported by the National Natural Science Foundation of China (Grant No. 51478356, No. 51778490), the Key Program for International S&T Cooperation Projects of China (Grant No. 2016YFE0127600), Open Research Fund Program of Guangdong Key Laboratory of Earthquake Engineering and Application Technology (Grant No. 2017B030314068), the Ministry of Science and Technology of China (Grant No. SLDRCE19-B-02), and the National Key R&D Program of China (Grant No. 2017YFC0703607).

Conflicts of Interest: The authors declare no conflict of interest.

References

1. Housner, G.W.; Bergman, L.A.; Caughey, T.K.; Chassiakos, A.G.; Claus, R.O.; Masri, S.F.; Skelton, R.E.; Soong, T.T.; Spencer, B.F.; Yao, J.T.P. Structural control: Past, present, and future. *J. Eng. Mech* **1997**, *123*, 897–971. [CrossRef]
2. Yao, J.T.P. Concept of structural control. *J. Struct. Div.* **1972**, *98*, 1567–1574.
3. Mahmoodi, P. Structural dampers. *J. Struct. Div.* **1969**, *95*, 1661–1672.
4. Frahm, H. Device for Damping Vibrations of Bodies. U.S. Patent 989,958, 18 April 1911.
5. Miyamoto, H.K.; Gilani, A.S.J.; Garza, J.; Stephen, A.M. Seismic Retrofit of a Landmark Structure Using a Mass Damper. In Proceedings of the Conference on Improving the Seismic Performance of Existing Buildings and Other Structures, San Francisco, CA, USA, 9–11 December 2009.
6. Tributsch, A.; Adam, C. Evaluation and analytical approximation of Tuned Mass Damper performance in an earthquake environment. *Smart Struct. Syst.* **2012**, *10*, 155–179. [CrossRef]
7. Sadek, F.; Mohraz, B.; Taylor, A.W.; Chung, R.M. A method of estimating the parameters of tuned mass dampers for seismic applications. *Earthq. Eng. Struct. Dyn.* **1997**, *26*, 617–635. [CrossRef]
8. Cao, H.; Reinhorn, A.M.; Soong, T.T. Design of an active mass damper for a tall TV tower in Nanjing, China. *Eng. Struct.* **1998**, *20*, 134–143. [CrossRef]
9. Ikago, K.; Saito, K.; Inoue, N. Seismic control of single-degree-of-freedom structure using tuned viscous mass damper. *Earthq. Eng. Struct. Dyn.* **2012**, *41*, 453–474. [CrossRef]
10. Ikago, K.; Sugimura, Y.; Saito, K.; Inoue, N. Simple Design Method for a Tuned Viscous Mass Damper Seismic Control System. In Proceedings of the 15th World Conference on Earthquake Engineering, Lisbon, Portugal, 24–28 September 2012.
11. Lazar, I.F.; Neild, S.A.; Wagg, D.J. Vibration suppression of cables using tuned inerter dampers. *Eng. Struct.* **2016**, *122*, 62–71. [CrossRef]
12. Marian, L.; Giaralis, A. Optimal design of a novel tuned mass-damper—Inerter (TMDI) passive vibration control configuration for stochastically support-excited structural systems. *Probab. Eng. Mech.* **2014**, *38*, 156–164. [CrossRef]
13. Saito, K.; Inoue, N. A study on optimum response control of passive control systems using viscous damper with inertial mass: Substituting equivalent nonlinear viscous elements for linear viscous elements in optimum control systems. *AIJ J. Technol. Des.* **2007**, *13*, 457–462. [CrossRef]
14. Saito, K.; Toyota, K.; Nagae, K.; Sugimura, Y.; Nakano, T.; Nakaminam, I.S.; Arima, F. Dynamic loading test and application to a high-rise building of viscous damping devices with amplification system. In Proceedings of the Third World Conference on Structural Control, Como, Italy, 11 April 2002.
15. Zhang, R.; Zhao, Z.; Dai, K. Seismic response mitigation of a wind turbine tower using a tuned parallel inerter mass system. *Eng. Struct.* **2019**, *180*, 29–39. [CrossRef]
16. Zhao, Z.; Zhang, R.; Lu, Z. A particle inerter system for structural seismic response mitigation. *J. Frankl. Inst.* **2019**, *193*, 110–120. [CrossRef]
17. Smith, M.C. Synthesis of mechanical networks: The inerter. *IEEE Trans. Automat. Control* **2002**, *47*, 1648–1662. [CrossRef]
18. Saito, K.; Sugimura, Y.; Nakaminami, S.; Kida, H.; Inoue, N. Vibration Tests of 1-Story Response Control System Using Inertial Mass and Optimized Softy Spring and Viscous Element. In Proceedings of the 14th World Conference on Earthquake Engineering, Beijing, China, 12–17 October 2008.
19. Chen, M.Z.; Hu, Y.; Huang, L.; Chen, G. Influence of inerter on natural frequencies of vibration systems. *J. Sound Vib.* **2014**, *333*, 1874–1887. [CrossRef]

20. Arakaki, T.; Kuroda, H.; Arima, F.; Inoue, Y.; Baba, K. Development of seismic devices applied to ball screw. Part 1: Basic performance test of RD-series. *AIJ J. Technol. Des.* **1999**, *5*, 239–244. [CrossRef]
21. Ikago, K.; Sugimura, Y.; Saito, K.; Inoue, N. Modal response characteristics of a multiple-degree-of-freedom structure incorporated with tuned viscous mass dampers. *J. Asian Architect. Build. Eng.* **2012**, *11*, 375–382. [CrossRef]
22. Lazar, I.F.; Neild, S.A.; Wagg, D.J. Using an inerter-based device for structural vibration suppression. *Earthq. Eng. Struct. Dyn.* **2014**, *43*, 1129–1147. [CrossRef]
23. Makris, N.; Kampas, G. Seismic protection of structures with supplemental rotational inertia. *J. Eng. Mech.* **2016**, *142*, 1–11. [CrossRef]
24. Papageorgiou, C.; Houghton, N.E.; Smith, M.C. Experimental testing and analysis of inerter devices. *J. Dyn. Syst. Meas. Control* **2009**, *131*, 101–116. [CrossRef]
25. Nakamura, Y.; Fukukita, A.; Tamura, K.; Yamazaki, I.; Matsuoka, T.; Hiramoto, K.; Sunakoda, K. Seismic response control using electromagnetic inertial mass dampers. *Earthq. Eng. Struct. Dyn.* **2014**, *43*, 507–527. [CrossRef]
26. Wang, F.C.; Hong, M.F.; Lin, T.C. Designing and testing a hydraulic inerter. *Proc. Inst. Mech. Eng. Part C J. Mech. Eng. Sci.* **2011**, *225*, 66–72. [CrossRef]
27. Kawamata, S. Liquid type mass damper with elongated discharge tube. U.S. Patent 4,872,649, 10 October 1989.
28. Smith, M.C.; Houghton, N.E.; Long, P.J.G.; Glover, A.R. Force-Controlling Hydraulic Device. U.S. Patent 8,881,876, 11 November 2014.
29. Høgsberg, J.; Brodersen, M.L.; Krenk, S. Resonant passive–active vibration absorber with integrated force feedback control. *Smart Mater. Struct.* **2016**, *25*, 047001. [CrossRef]
30. Wang, F.C.; Chen, Y.C.; Lee, C.H. Design and Optimization of Inerter Layouts for a Multi-Layers Building Model. In Proceedings of the 55th Annual Conference of the Society of Instrument and Control Engineers of Japan, IEEE, Tsukuba, Japan, 20–23 September 2016.
31. Sugimura, Y.; Goto, W.; Tanizawa, H.; Saito, K.; Ninomiya, T. Response Control Effect of Steel Building Structure Using Tuned Viscous Mass Damper. In Proceedings of the 15th World Conference on Earthquake Engineering, Lisbon, Portugal, 24–28 September 2012.
32. Pan, C.; Zhang, R.; Luo, H.; Li, C.; Shen, H. Demand-based optimal design of oscillator with parallel-layout viscous inerter damper. *Struct. Control Health Monit.* **2017**, *25*, e2051. [CrossRef]
33. Pan, C.; Zhang, R. Design of structure with inerter system based on stochastic response mitigation ratio. *Struct. Control Health Monit.* **2018**, *25*, e2169. [CrossRef]
34. Zhang, R.; Zhao, Z.; Pan, C. Influence of mechanical layout of inerter systems on seismic mitigation of storage tanks. *Soil Dyn. Earthq. Eng.* **2018**, *114*, 639–649. [CrossRef]
35. Chen, Q.; Zhao, Z.; Zhang, R.; Pan, C. Impact of soil–structure interaction on structures with inerter system. *J. Sound Vib.* **2018**, *433*, 1–15. [CrossRef]
36. Hwang, J.S.; Kim, J.; Kim, Y.M. Rotational inertia dampers with toggle bracing for vibration control of a building structure. *Eng. Struct.* **2007**, *29*, 1201–1208. [CrossRef]
37. Tan, S.; Astaneh-Asl, A. Use of steel cables to prevent progressive collapse of existing buildings. In Proceedings of the sixth conference on tall buildings in seismic regions, Los Angeles, CA, USA, 4 June 2003.
38. Kim, J.; Choi, H.; Min, K.W. Use of rotational friction dampers to enhance seismic and progressive collapse resisting capacity of structures. *Struct. Des. Tall Spec. Build.* **2011**, *20*, 515–537. [CrossRef]
39. Kurata, M.; Leon, R.T.; Desroches, R. Rapid seismic rehabilitation strategy: Concept and testing of cable bracing with couples resisting damper. *J. Struct. Eng.* **2012**, *138*, 354–362. [CrossRef]

applied sciences

MDPI

Article

Wind-Induced Response Control of High-Rise Buildings Using Inerter-Based Vibration Absorbers

Qinhua Wang [1,2], Haoshuai Qiao [1], Dario De Domenico [3,*], Zhiwen Zhu [1,2] and Zhuangning Xie [4]

[1] Department of Civil and Environmental Engineering, Shantou University, Shantou 515063, China; qinhuawang@stu.edu.cn (Q.W.); haoshuaiqiao@foxmail.com (H.Q.); zhuzw@stu.edu.cn (Z.Z.)
[2] Key Laboratory of Structure and Wind Tunnel of Guangdong Higher Education Institutes, Shantou 515063, China
[3] Department of Engineering, University of Messina, 98166 Messina, Italy
[4] State Key Laboratory of Subtropical Building Science, South China University of Technology, Guangzhou 510640, China; znxie@scut.edu.cn
* Correspondence: dario.dedomenico@unime.it; Tel.: +39-090-6765921

Received: 25 October 2019; Accepted: 18 November 2019; Published: 22 November 2019

Abstract: The beneficial mass-amplification effect induced by the inerter can be conveniently used in enhanced variants of the traditional Tuned Mass Damper (TMD), namely the Tuned Mass-Damper-Inerter (TMDI) and its special case of Tuned Inerter Damper (TID). In this paper, these inerter-based vibration absorbers are studied for mitigating the wind-induced response of high-rise buildings, with particular emphasis on a 340 m tall building analyzed as case study. To adopt a realistic wind-excitation model, the analysis is based on aerodynamic forces computed through experimental wind tunnel tests for a scaled prototype of the benchmark building, which accounts for the actual cross-section of the structure and the existing surrounding conditions. Mass and stiffness parameters are extracted from the finite element model of the primary structure. Performance-based optimization of the TMDI and the TID is carried out to find a good trade-off between displacement- and acceleration-response mitigation, with the installation floor being an explicit design variable in addition to frequency and damping ratio. The results corresponding to 24 different wind directions indicate that the best vibration mitigation is achieved with a lower installation floor of the TMDI/TID scheme than the topmost floor. The effects of different parameters of TMD, TMDI and TID on wind-induced displacement and acceleration responses and on the equivalent static wind loads (ESWLs) are comparatively evaluated. It is shown that the optimally designed TMDI/TID can achieve better wind-induced vibration mitigation than the TMD while allocating lower or null attached mass, especially in terms of acceleration response.

Keywords: tuned mass damper; inerter; high-rise buildings; wind tunnel test; wind-induced response; structural control; synchronous multi-point pressure measurement

1. Introduction

In 2018 alone, 143 tall buildings having height of more than 200 m have been constructed, which set up a new record for the annual completion of high-rise buildings around the world. The total number of such buildings reached 1497 up to now based on the statistics from the Council on Tall Buildings and Urban Habitat [1]. These high-rise buildings are very sensitive to wind loads especially in hurricane prone regions. Wind loads may induce large displacement and acceleration responses, which may cause higher stresses in the structural members and discomfort to building occupants. Shape optimization was put forward to improve aerodynamic performance of tall buildings and suppress wind-induced responses [2–6]. However, methods of structural modification sometimes limit the usage of building space (setback of cross-section) [7]. As an alternative, installation of passive vibration control devices, e.g., fluid-viscous dampers [8–13], Tuned Mass Damper (TMD) [14,15], Multi Tuned Mass Damper (MTMD) [16] distributed TMD [17,18], and Tuned Liquid Column Damper (TLCD) [19–21], was widely used to suppress wind-induced responses. Other attractive and effective implementations of un-conventional TMD schemes were also recently proposed that take advantage from claddings and facades in buildings [22,23]. Further researches based on the wind tunnel test [14–31] and full-scale monitoring [27,32] were performed to evaluate the mitigation effect of TMD on wind-induced responses.

It is widely recognized that the effectiveness of TMD to mitigate vibrations depends heavily on its mass. In general, the larger the attached TMD mass that can be accommodated, the more effective and robust the TMD becomes for vibration control [33–35]. However, the attached mass of TMD in high-rise buildings rarely exceeds 0.5% of the total primary mass because of structural and architectural constraints in practical projects [36]. For example, TMD systems of Ping-An Finance Center in Shenzhen City, China, have a weight of 1000 t including mass blocks and supporting-frame structures [37]. The mass of TMD installed in Taipei 101 Tower reaches 660 t. It has the diameter of 6 m and occupies the space from the 87th to 91st floor, including the supporting cables [38]. The TMD not only occupies valuable space of top floors of high-rise buildings but also increases construction cost because of its enormous mass. Motivated by these practical aspects, inerter-based vibration absorbers, e.g., TID and TMDI, were recently proposed to mitigate the vibrations of structure. Lazar et al. [39,40] presented a novel inerter-based vibration absorber system termed as tuned inerter damper (TID). The TID takes advantage of the "mass amplification effect" of the inerter, a two-terminal device of negligible mass/weight whose internal force can reasonably be assumed proportional to the relative acceleration of its two terminals [41]. Acting as an additional, apparent mass, the inerter can modify the inertial properties of the system. Therefore, the TID represents a lower-mass and more effective alternative to the TMD, as it can achieve comparable or even higher vibration suppression level by significantly reducing the attached mass. Marian and Giaralis [42] unified both TMD and TID scheme by proposing an effective passive control system termed Tuned Mass-Damper-Inerter (TMDI). The TMDI scheme will degenerate into TMD and TID by decreasing the inertance ratio and mass ratio to zero, respectively. Most of the recent research has been directed towards optimal design and performance evaluation of inerter-based systems for seismic protection of building structures [39,43–51], wind turbine towers [52] and storage tanks [53], for vibration suppression of cables [40], and for mitigation of vortex-induced vibration in long-span bridges [54]. A few earlier studies also suggested the use of TID [55] and TMDI [36,56] to suppress wind-induced vibration in high-rise buildings. Giaralis and Petrini [36] investigated wind-induced vibration mitigation of a 74-story benchmark building equipped with TMDI using a frequency-domain stochastic approach, based on empirical power spectral density (PSD) matrix of across-wind aerodynamic force [57]. Their results indicated that the TMDI reduced the peak top-floor acceleration more effectively than the TMD but employing smaller attached-mass values, especially for some selected topologies of installation. Additionally, they showed that the inclusion of the inerter dramatically reduced the TMD stroke. In this regard, it is worth noting that across-wind aerodynamic forces in high-rise buildings are induced by vortex shedding which highly rely on the actual cross section of the building [57]. Moreover, the empirical PSD of across-wind aerodynamic force adopted in [36] is not applicable when surrounding buildings exist, such that wind directions are not consistent with coordinates of structures.

This research work falls into the same research line as the previous papers, but it uses a different wind-excitation model underlying an alternative time-domain analysis perspective. More specifically, in this paper synchronous multi-point pressure measurements from wind tunnel tests of a scaled high-rise building of height 340 m are carried out. This allows the definition of a more appropriate set of aerodynamic forces that are consistent with the actual cross-section of the benchmark building and with the existing surrounding conditions. The time histories of aerodynamic forces at each story are determined along 24 different wind directions from 0° to 345° at an interval of 15°. Performance-oriented optimization of parameters of both the TMDI and the TID are carried out to find out the optimal parameters of corresponding vibration mitigation device to suppress wind-induced responses with an eye for practical aspects like frequency ratio, damping ratio and placement of TID/TMDI. Combining aerodynamic forces from wind tunnel tests, structural dynamic characteristics (mass and stiffness of the building) extracted from the finite element model of the primary structure and parameters of optimally-designed TMDI and TID, a time-domain mathematical model of the benchmark building is used in this study for analyzing the wind-induced responses under the assumption of linear elastic behaviors. The effects of TMDI and TID on wind-induced displacement and acceleration responses and on ESWLs are studied and compared with results of the benchmark building equipped with classical TMD that shares the same physical mass as the former two inerter-based vibration absorbers (the physical mass denotes the attached mass in the TMD case, and the sum of mass of inerter devices and attached mass in the inerter-based vibration absorbers).

2. Equations of Motion and Preliminary Concepts

According to the sketch in Figure 1, let us consider a high-rise building modeled as a lumped-mass system, equipped with a linear TMDI comprising an attached mass m_t that is connected to the primary structure via linear spring and dashpot elements, k_t and c_t, respectively. The TMDI mass is placed in series with one or several linear inerters (of inertance b in sum) whose two terminals are denoted as 1 and 2, respectively. The first terminal 1 is connected to the attached mass m_t, while the second terminal 2 can be attached to another floor along the building height. The building is subject to a wind-induced excitation field represented by a set of forces acting at each floor of the discretized system. Adopting a standard matrix-vector notation, the equations of motion of this n-story lumped-mass system can be written in the following form

$$\mathbf{M}\ddot{\mathbf{u}}(t) + \mathbf{C}\dot{\mathbf{u}}(t) + \mathbf{K}\mathbf{u}(t) = \mathbf{p}(t) \tag{1}$$

where $\ddot{\mathbf{u}}(t), \dot{\mathbf{u}}(t), \mathbf{u}(t)$ represent the acceleration, velocity, and displacement vectors, respectively, and $\mathbf{p}(t)$ is the vector of the corresponding aerodynamic forces applied at the center of mass of each floor slab. All these vectors are $(n + 1)$-dimensional, as this is the total number of degrees of freedom (DOFs) of the n-story building equipped with TMDI. We assume all these vectors are augmented by one last (bottom) row containing the kinematic terms related to the TMDI DOF for $\ddot{\mathbf{u}}(t), \dot{\mathbf{u}}(t), \mathbf{u}(t)$ and containing a zero entry for the aerodynamic force vector $\mathbf{p}(t)$. Aerodynamic loading can be obtained from wind tunnel test measurements by synchronous multi-point scanning of pressures on the high-rise building model. In Equation (1), $\mathbf{M}, \mathbf{C}, \mathbf{K}$ are the mass, damping and stiffness matrices of the TMDI-equipped structure, respectively. When the TMDI is installed at the tth floor and has a so-called "$-p$" topology (meaning that the second terminal of the inerter is attached to a floor $t-p$), these matrices can be expressed as:

$$\begin{aligned}
\mathbf{M} &= \mathbf{M}_s^{n+1} + (m_t + b)\mathbf{1}_{\mathbf{n+1}}\mathbf{1}_{\mathbf{n+1}}^T + b\,\mathbf{1}_{\mathbf{t-p}}\mathbf{1}_{\mathbf{t-p}}^T - b\left(\mathbf{1}_{\mathbf{n+1}}\mathbf{1}_{\mathbf{t-p}}^T + \mathbf{1}_{\mathbf{t-p}}\mathbf{1}_{\mathbf{n+1}}^T\right) \\
\mathbf{C} &= \mathbf{C}_s^{n+1} + c_t\left(\mathbf{1}_{\mathbf{n+1}}\mathbf{1}_{\mathbf{n+1}}^T + \mathbf{1}_{\mathbf{t}}\mathbf{1}_{\mathbf{t}}^T - \mathbf{1}_{\mathbf{n+1}}\mathbf{1}_{\mathbf{t}}^T - \mathbf{1}_{\mathbf{t}}\mathbf{1}_{\mathbf{n+1}}^T\right) \\
\mathbf{K} &= \mathbf{K}_s^{n+1} + k_t\left(\mathbf{1}_{\mathbf{n+1}}\mathbf{1}_{\mathbf{n+1}}^T + \mathbf{1}_{\mathbf{t}}\mathbf{1}_{\mathbf{t}}^T - \mathbf{1}_{\mathbf{n+1}}\mathbf{1}_{\mathbf{t}}^T - \mathbf{1}_{\mathbf{t}}\mathbf{1}_{\mathbf{n+1}}^T\right)
\end{aligned} \tag{2}$$

where $\mathbf{M}_s^{n+1}, \mathbf{C}_s^{n+1}, \mathbf{K}_s^{n+1} \in \mathbb{R}^{(n+1)\times(n+1)}$ represent the augmented mass, damping and stiffness matrices of the primary structure, respectively, constructed by adding one last (bottom) row with zero entries and one last (rightmost) column of zero entries in the original matrices $\mathbf{M}_s, \mathbf{C}_s, \mathbf{K}_s \in \mathbb{R}^{n\times n}$. All the vectors $\mathbf{1}_\mathbf{j} \in \mathbb{R}^{(n+1)\times 1}$ are constructed such that only the jth entry is equal to one while all the remaining entries are equal to zero (the superscript T indicates transpose operator). The above equations also apply to a TMD-equipped structure (without inerter), which is retrieved by setting $b = 0$ in the mass matrix. In the same way, a TID scheme as series-Parallel Layout 1 Inerter system proposed in [50] can be obtained by setting $m_t = 0$ (without attached mass). Besides the $(t-p)^{\text{th}}$ entry in the diagonal of the augmented mass matrix, the presence of the inerter modifies the mass matrix by introducing certain non-diagonal inertial coupling terms between the $(n+1)^{\text{th}}$ DOF of the attached mass and the DOF of the $(t-p)^{\text{th}}$ floor. Inerter topologies in which the inerter spans more than one story ($p > 1$)

may be realized with pendulum-like implementations like in the Taipei 101 skyscraper ($p = 4$ for 87th floor to 91st floor). Such inerter topologies or installation configurations, earlier studied in [36], were found to achieve higher vibration control effectiveness. This was intuitively motivated by the fact that the motion of two non-consecutive floors is seemingly less correlated: therefore, the inerter is likely to undergo higher relative accelerations at its two terminals than if it were installed between two consecutive floors, thus experiencing higher engagement (larger forces for equal inertance value b).

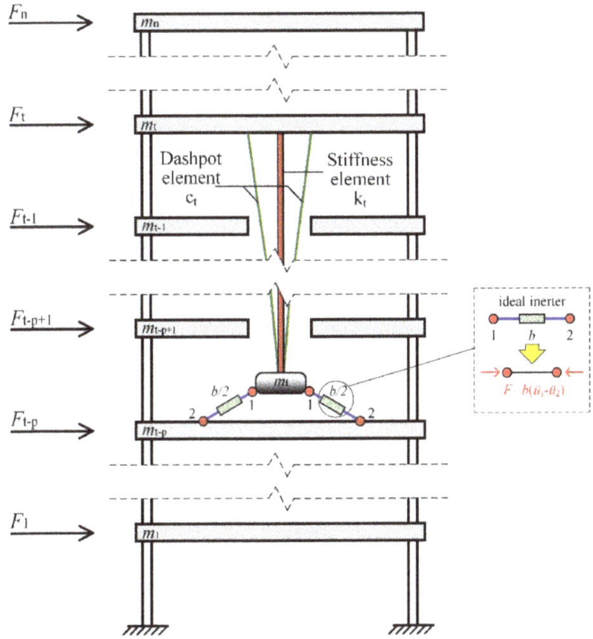

Figure 1. Sketch of Tuned Mass Damper-Inerter (TMDI)-equipped high-rise building modeled as a linear lumped mass system.

By inspection of Equations (1) and (2) it is worth pointing out the following distinctive aspects in comparison with the earlier study by Giaralis and Petrini [36]: (i) The analysis is here conducted in the time domain since the aerodynamic forces are identified by wind tunnel tests of the benchmark building, whereas Giaralis and Petrini operated on a frequency-domain stochastic approach based on an empirical PSD for the across-wind force field; (ii) The formulation in Equation (2) slightly extends that developed by Giaralis and Petrini since the TMDI should not necessarily be placed at the topmost floor, but it can installed at a generic t^{th} floor. Although it is customary to attach the TMD at the top floor because of its widely recognized effectiveness to control the fundamental mode in multistory buildings [58], this installation configuration might be not feasible in some practical projects because of potential structural or architectural constraints. The installation of traditional TMD at different floors, not just at the topmost floor, was recently investigated by Elias and Matsagar [29]. For TMDI, Ruiz et al. [59] and Giaralis and Taflandis [48] also assessed the influence of installation floor on its performance, although they did not explicitly consider the installation floor as one of the variables to be optimized. Additionally, the inerter might be unable to exert its due performance because the relative peak acceleration does not always occur between the top floor and the $(n-p)^{\text{th}}$ floor. This is why in this study we have directly set the installation floor as one of the design variables of the inerter-based vibration absorber within parametric optimizations, which represents another novel aspect in comparison with definitions of optimization problem in previous literature studies.

The TMDI-equipped structure may have non-proportional damping. Therefore, it is convenient to transform Equation (1) into state space variable form that is more suitable for complex modal analysis

$$\mathbf{A}\dot{\mathbf{z}}(t) + \mathbf{B}\mathbf{z}(t) = \mathbf{f}(t) \tag{3}$$

where $\mathbf{z}(t) = [\mathbf{u}(t), \dot{\mathbf{u}}(t)]^T$ is the state space variable vector and the matrices $\mathbf{A}, \mathbf{B}$ and the vector $\mathbf{f}(t)$ are expressed as

$$\mathbf{A} = \begin{bmatrix} \mathbf{C} & \mathbf{M} \\ \mathbf{M} & 0 \end{bmatrix}, \mathbf{B} = \begin{bmatrix} \mathbf{K} & 0 \\ 0 & -\mathbf{M} \end{bmatrix}, \mathbf{f}(t) = \begin{bmatrix} \mathbf{p}(t) \\ 0 \end{bmatrix}. \tag{4}$$

From Equations (3) and (4), after performing complex modal analysis it is possible to determine the transfer function of the system response to assess wind-induced response mitigation induced by the TMD\TMDI\TID. In particular, the transfer function of the displacement and acceleration response at the pth DOF induced by forces at the qth DOF can be expressed in the following forms, respectively

$$\begin{aligned} H_D^{pq}(j\omega) &= \sum_{k=1}^{n+1}\left(\frac{\varphi_{pk}\varphi_{qk}}{a_k(j\omega - s_k)} + \frac{\varphi_{pk}^*\varphi_{qk}^*}{a_k^*(j\omega - s_k^*)}\right) \\ H_A^{pq}(j\omega) &= \sum_{k=1}^{n+1}\left(\frac{-\omega^2\varphi_{pk}\varphi_{qk}}{a_k(j\omega - s_k)} + \frac{-\omega^2\varphi_{pk}^*\varphi_{qk}^*}{a_k^*(j\omega - s_k^*)}\right) \end{aligned} \tag{5}$$

where φ_{pk}, φ_{qk} are the values of the pth and q^{th} degree of freedom, respectively, in the kth complex mode shape, while φ_{pk}^* and φ_{qk}^* represent the conjugate values of φ_{pk} and φ_{qk}, respectively. In Equation (5) s_k and s_k^* denote the complex eigenvalue and its conjugate value, respectively, while a_k and a_k^* are coefficients determined as follows

$$\mathbf{\Phi}^T\mathbf{A}\mathbf{\Phi} = \text{diag}\{\, a_1 \ \cdots \ a_k \ \cdots \ a_{n+1} \ \ a_1^* \ \cdots \ a_k^* \ \cdots \ a_{n+1}^* \,\} \tag{6}$$

where $\mathbf{\Phi}$ is the modal shape matrix collecting the eigenvectors $\boldsymbol{\phi}_i$. According to the complex mode superposition approach, the state space vector response $\mathbf{z}(t)$ (under the assumption of zero initial conditions) can be expressed as

$$\begin{aligned} \mathbf{z}(t) &= \sum_{i=1}^{2n+2} \boldsymbol{\phi}_i\, \mathbf{q}_i(t) \\ \mathbf{z}(t) &= \sum_{i=1}^{n+1}\left(\frac{\boldsymbol{\phi}_i}{a_i}\int_0^t F_i(\tau)e^{s_i(t-\tau)}\mathrm{d}\tau + \frac{\boldsymbol{\phi}_i^*}{a_i^*}\int_0^t F_i^*(\tau)e^{s_i^*(t-\tau)}\mathrm{d}\tau\right) \end{aligned} \tag{7}$$

where $\boldsymbol{\phi}_i$ and $\boldsymbol{\phi}_i^*$ represent the ith mode shape and its conjugate mode, respectively. The ith generalized force and its conjugate are expressed as

$$\begin{aligned} F_i(t) &= \boldsymbol{\phi}_i{}^T\mathbf{f}(t) \\ F_i^*(t) &= \boldsymbol{\phi}_i^{*T}\mathbf{f}(t) \end{aligned} \tag{8}$$

Once the wind-induced displacement responses are calculated, they are utilized to predict equivalent static wind loads by the method of Displacement Gust Loads Factor (DGLF) [60]:

$$F_{eswl}(z) = G(z)\overline{P}(z) \tag{9}$$

where $\overline{P}(z)$ is the mean wind load, which can be obtained by pressure measurements from wind tunnel test, and $G(z)$ is the DGLF, which considers effects of structural dynamic characteristics on response. In the DGLF method, $G(z)$ is evaluated in terms of the expected extreme and mean displacement:

$$G(z) = \frac{\hat{D}(z)}{\overline{D}(z)} \tag{10}$$

where $\overline{D}(z)$ and $\hat{D}(z)$ are the mean and expected extreme displacement, respectively, at the structural height z. The expected values of extreme displacement and acceleration used are $\hat{D}(z) = \mu_{dis} + g\sigma_{dis}$ and $\hat{D}_{acc}(z) = g\sigma_{acc}$, respectively, where $g = 3.5$ is the peak factor estimated from the widely used empirical formula given by Davenport [61].

3. Description of the 340 m Tall Building and Wind Tunnel Testing

The primary structure has 69 stories and total height of 340 m. This benchmark building represents the *Qiaokou* tower, located in Wuhan City, China, built in 2012. The photograph of the building and the main peculiarities of its structural configuration are presented in Figure 2.

Figure 2. Photograph of the 340 m tall building (**left**) and plan view (**right**).

The mass distribution of the primary structure, including dead load and live load, has been extracted from the finite element model of the building and is reported in Figure 3a. Similarly, the distribution of the lateral stiffness along the x-axis, which is much smaller than that along the y-axis, has been extracted from the finite element model as well, and is presented in Figure 3b. Mass and stiffness matrices can be established in terms of mass and lateral stiffness distributions. It is worth noting that the developed model of the case study building accounts for the primary structural elements only. The presence of secondary structural components (here not considered for simplicity) could slightly modify the results in terms of interstory displacements and could add some stiffening contributions in the overall building model.

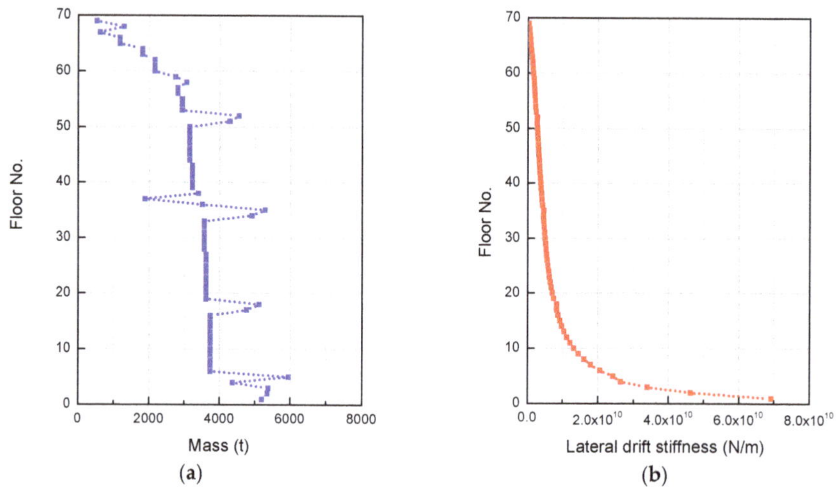

Figure 3. Dynamic properties of the benchmark building extracted from the finite element model: (**a**) mass distribution; (**b**) lateral stiffness distribution.

The full damping matrix $\mathbf{C}_s$ of the original structure (without vibration absorber) has been calculated from the modal damping matrix $\mathbf{C}_{\text{mod}}$ [62]:

$$\mathbf{C}_s = \left(\mathbf{\Phi}^T\right)^{-1}\mathbf{C}_{\text{mod}}(\mathbf{\Phi})^{-1} \tag{11}$$

where $\mathbf{\Phi}$ is the modal shape matrix of the original structure. The modal damping matrix $\mathbf{C}_{\text{mod}} \in \mathbb{R}^{n\times n}$ is a diagonal matrix collecting the modal damping ratios and can be calculated as follows

$$\mathbf{C}_{\text{mod}}(k,k) = 2\xi_k\omega_k\left(\boldsymbol{\varphi}_k^T\mathbf{M}_s\boldsymbol{\varphi}_k\right);\ k = 1,2,\ldots,69 \tag{12}$$

where ω_k,$\boldsymbol{\varphi}_k$ are the kth natural frequency and vibration mode, respectively. The kth modal damping ratio of the system ξ_k is taken equal to 1% for k = 1, 2, 3; 4% for k = 4, 5, 6; 6% for k = 7, 8, 9, 10; 9% for k = 11, 12, ... , 20; 12% for k = 21, 22, ... , 40; 15% for k = 41, 42, ... , 60; 18% for k = 61, 62, ... , 69. These values were selected based on available field-recorded of high-rise steel framed buildings in the [0–7] Hz frequency range [63]. The main dynamic parameters of the primary structure are listed in Table 1.

Table 1. Main dynamic parameters of the primary structure.

Total Mass M	First-Order Natural Frequency along x-axis ω_1	First-Order Generalized Mass	First-Order Damping Ratio (Assumed)
231,659 t	0.176 Hz	61,287 t	1%

The synchronous multi-point pressure tests of the building with existing surrounding conditions were performed in boundary layer wind tunnel tests (shown in Figure 4) under a simulated C type wind field corresponding to China load code for the design of building structures [64], which reflects the characteristics of the wind field in urban areas. The profiles of mean wind speed and turbulence intensity are shown in Figure 5a. The reference coordinates of the wind tunnel test and wind-induced response analysis are shown in Figure 5b. 24 wind directions are considered in this study, which are identified by a β_w angle (between wind axis and x-axis) ranging from 0° to 345° at an interval of 15°.

Figure 4. The rigid building model with existing surrounding conditions mounted in the wind tunnel lab of Shantou University from two different perspectives.

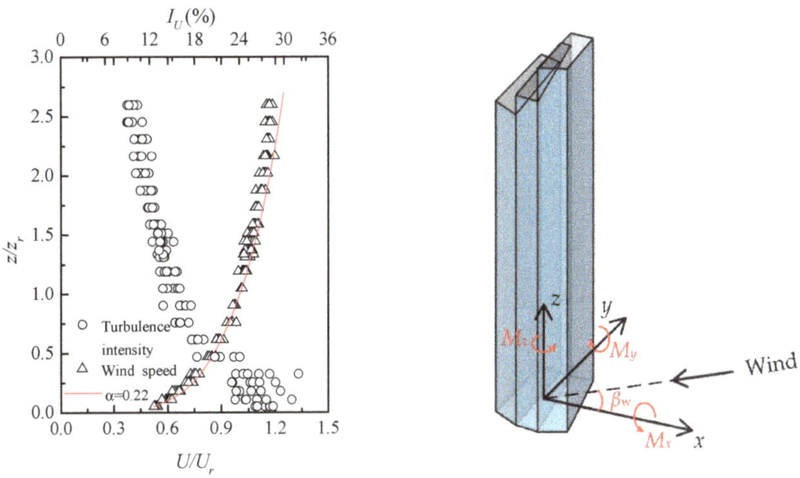

(a) (b)

Figure 5. Wind model characteristics: (**a**) wind profile; (**b**) definition of coordinates.

When the wind direction β_w is 0° and 90°, the wind is blowing from the positive direction of the x and y axes, respectively. The parameters of wind tunnel tests are listed in Table 2. By properly scaling wind tunnel test results, the wind pressure coefficients firstly were transferred into wind aerodynamic pressure on the full-scale building, and then aerodynamic pressure in the prototype building was integrated at the base of the tributary area of each pressure tap to obtain the aerodynamic force component of each floor along the x-axis.

Table 2. Wind tunnel test parameters.

Geometric Scale	Wind Speed	Sampling Frequency	Sampling Length	Incremental Step	Measuring Taps
1:350	12 m/s	312.5 Hz	20,480	15°	471

In Figure 5a, α (the exponent of power law formulation for vertical mean wind profile) corresponding to C type wind field is 0.22 in China load code for the design of building structures [64], I_U represents the turbulence intensity, U is the wind speed, and U_r is the wind speed at the reference height.

Samples of aerodynamic force time histories acting on two stories (30th and 50th) along the x-axis are depicted in Figure 6 corresponding to 90° wind direction and wind velocity equal to 42.02 m/s. Figure 6 indicates that mean wind loads approach to zero, which is expected for across-wind aerodynamic forces mainly induced by vortex shedding.

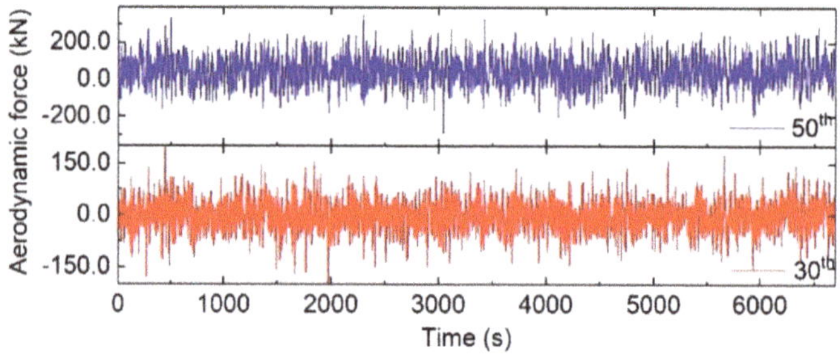

Figure 6. Time histories of aerodynamic forces on typical stories corresponding to 90° wind direction.

4. Wind-Induced Response Mitigation Using Inerter-Based Vibration Absorbers

4.1. Optimization of the Parameters of the TMDI/TID Scheme Applied to the Benchmark Building

To compare the vibration mitigation effect of optimal TMD, TMDI and TID, performance-based optimizations for parameters of TMDI and TID were conducted to obtain the best TMDI/TID scheme. At the same time, the influence of variations of parameters in preset intervals on the vibration mitigation effect is also investigated.

In the mathematical model of TMD/TMDI/TID-equipped structure as described in Equations (1) and (2), vibration absorbers are introduced to control the wind-induced response along the x-axis, which is the most critical direction due to the lower lateral stiffness of the building (higher oscillations are expected). As the largest displacement and acceleration responses at the top floor of primary structure occurs at wind direction of 90°, the peak displacement and acceleration at the top floor induced by aerodynamic forces at this wind direction are selected to be the two individual objective functions.

From Equation (2), there are totally six parameters of TMDI, i.e., μ, β, υ, ζ, p and t, that need to be fixed to calculate the wind-induced responses of TMDI-equipped structure (5 parameters for TID). The two mass related notation, i.e., mass ratio and inertance ratio, are defined as $\mu = m_t/M$ and $\beta = b/M$, respectively. The frequency ratio is defined as:

$$\upsilon = \frac{\omega_t}{\omega_1} = \sqrt{\frac{k_t}{(m_t + b)}}/\omega_1 \tag{13}$$

where ω_t is the circular frequency of TMD, TMDI or TID, and ω_1 is the first order circular frequency of the primary structure. The damping ratio is defined as:

$$\zeta = \frac{c_t}{2\sqrt{(m_t + b)k_t}} \tag{14}$$

The other two discrete parameters are the topologies of inerter and the installation floor of vibration absorber as denoted in Equation (2).

To shed light on the better vibration mitigation effect of inerter-based vibration absorbers, the physical mass ratio μ_{phy} of TMD, TMDI and TID are defined below and fixed to the same value equal to 0.5%:

$$\mu_{phy} = \mu + \beta/200 = m_{\text{TMD, TMDI or TID}}/M = 0.5\% \tag{15}$$

In the above-introduced physical mass ratio, not only the physical mass of the TMD (named as the attached mass), but also the physical mass of the inerter device is taken into consideration in the TMDI and TID schemes. Indeed, when dealing with large values of inertance (apparent mass of the inerter) in the order of tons, which might be the case for high-rise buildings, the physical mass of the inerter turns out to be non-negligible, whereas the majority of the literature studies ignored this term. In Equation (15), the physical mass ratio is constrained to be 0.5% based on the same threshold of TMD mass ratio proposed in [36], and the ratio of the inertance coefficient and physical mass of inertance devices is assumed to be 200 following the previous research about the "mass-amplification" effect of inerter [65]. Therefore, the mass-related parameter of the three vibration absorbers can be determined, i.e., TMD (μ = 0.5%), TMDI (μ = 0.25%, β = 50%) and TID (β = 100%). This makes it possible to compare three different configuration schemes sharing a common physical mass ratio for wind-induced response mitigation of the benchmark building. According to the conclusion from previous research that inerter devices spanning more stories lead to a better mitigation effect of the TMDI, the value of the topologies is determined to be p = 4 in the optimization procedure, based on practical considerations like in the pendulum-like TMD scheme implemented in the Taipei 101 (spanning from 87th floor to 91st floor).

4.1.1. Optimization of Parameters of TMDI

As stated above, the displacement- and acceleration-based optimization of the three parameters, i.e., frequency ratio, damping ratio and floor of installation, can be expressed as Equations (16) and (17), respectively. The preset intervals of three parameters are determined based on practical considerations and results of previous researches [35,50]

$$\begin{cases} \text{minimize } f_1(v,\zeta,t) = \hat{D}_{dis} \\ s.t. \quad \mu = 0.25\%, \beta = 50\%, -p = -4, \\ \quad v \in [0.7, 1.2], \zeta \in [0, 20\%], \\ \quad t \in [30, 58] \end{cases} \tag{16}$$

$$\begin{cases} \text{minimize } f_2(v,\zeta,t) = \hat{D}_{acc} \\ s.t. \quad \mu = 0.25\%, \beta = 50\%, -p = -4, \\ \quad v \in [0.7, 1.2], \zeta \in [0, 20\%], \\ \quad t \in [30, 58] \end{cases} \tag{17}$$

where $\hat{D}_{dis}$ and $\hat{D}_{acc}$ are the peak displacement and acceleration at the top floor at wind direction of 90°. It is worth noting that the installation floor t represents an explicit design variable of the constrained optimization problem stated in Equations (16) and (17).

For such an optimization of three variables, i.e., v, ζ and t, a three-dimensional space representation is proposed where the three variables are set to be the orthogonal axes and the value of corresponding object (peak responses) is expressed by different colors. To present a clear vision of the distributions of the colors in a 3D space, sliced contours from three aspects were plotted as shown in Figure 7.

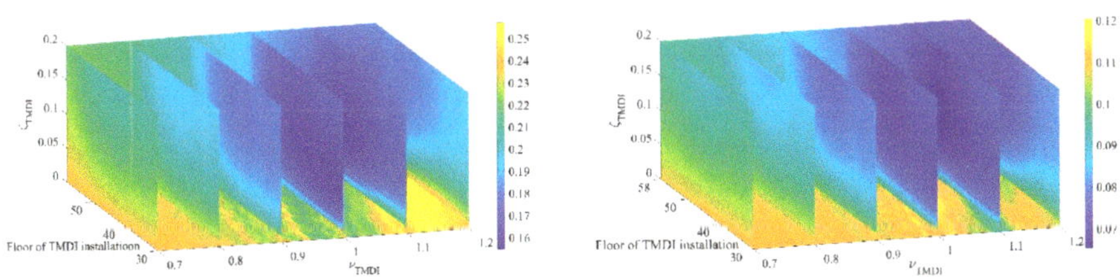

(a)

Figure 7. *Cont.*

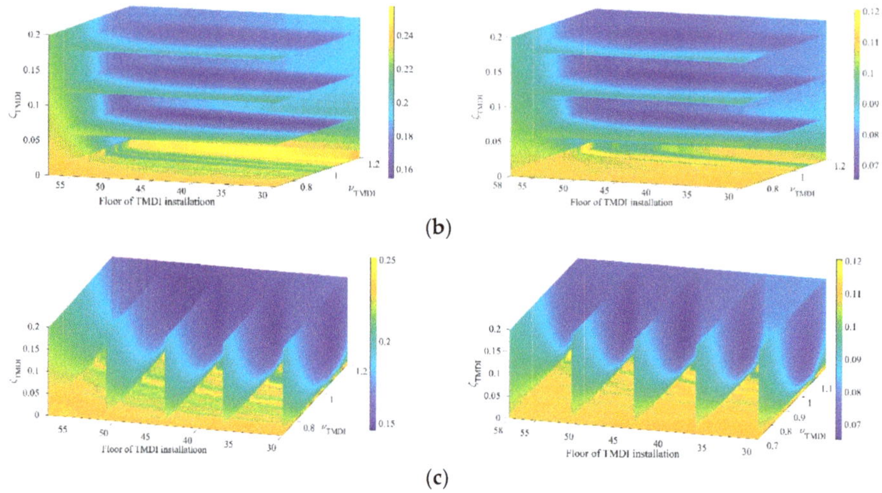

Figure 7. Peak displacement (**left**) and acceleration (**right**) at top floor of TMDI-equipped structure at wind direction of 90° in the $\upsilon - \zeta - t$ space. (**a**) Distributions of peak displacement (**left**) and acceleration (**right**) at constant frequency ratios (υ = 0.7, 0.8, 0.9, 1.0, 1.1, 1.2). (**b**) Distributions of peak displacement (**left**) and acceleration (**right**) at constant damping ratios (ζ = 6%, 12%, 18%). (**c**) Distributions of peak displacement (**left**) and acceleration (**right**) at constant floor of TMDI installation (t = 30, 37, 44, 51, 58).

From the sliced contour shown in Figure 7, the cool-color area represents the most efficient combination of parameters, leading to the minimum value of the two specific response indicators. The following conclusions can be drawn:

- For TMDIs which efficiently mitigate the wind-induced displacement responses, the optimal frequency ratio lies around 1.1, which indicates that the frequency of optimal TMDI is close to the first order frequency of the primary structure;
- The minimal peak displacement and acceleration are achieved when the damping ratios are 7% and 10%, respectively;
- As for the optimal floor of installation of TMDI, it can be seen that the best vibration mitigation effect is achieved when the TMDI is installed at the middle-upper portion of the benchmark building (around 44th floor), and not in the conventional configuration of TMD, i.e., at the topmost floor;
- For acceleration mitigation purpose, the optimal frequency ratio and installation floor of TMDI is slightly larger than that of displacement-oriented optimization. Such differences may be justified in view of the fact that the transfer function of acceleration is $-\omega^2$ times that of displacement, which means that a better mitigation effect of acceleration can be realized by decreasing the value of transfer function at higher frequency around the first peak under the same fluctuating wind excitations.

Through the optimization results of TMDI at wind direction of 90° (the most adverse conditions), a set of optimal tuning parameters of TMDI with good trade-off between displacement mitigation and acceleration mitigation was selected by approximately averaging the two corresponding parameters due to the smooth gradient between two optimal schemes of TMDI (as shown in Figure 8). The notations Dis_{min} and Acc_{min} represent two configurations of TMDI that achieve the best displacement and acceleration mitigation effect, respectively.

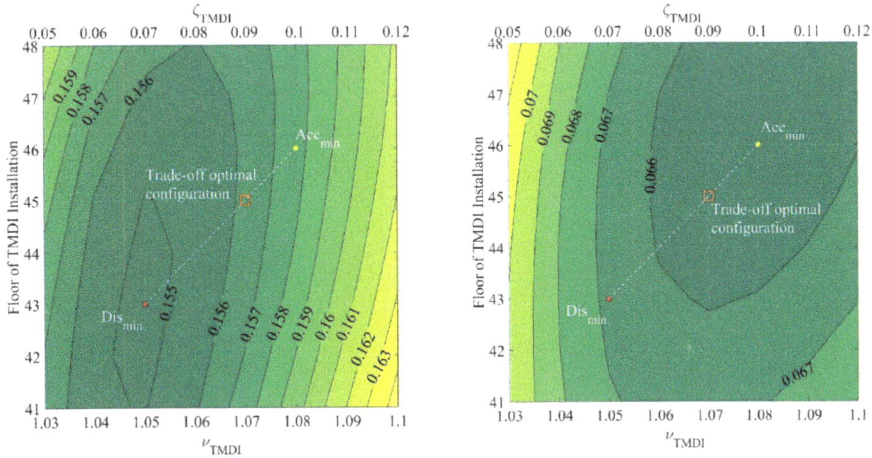

(a) (b)

Figure 8. Contour of (**a**) peak displacement (**b**) peak acceleration with variable frequency ratio, damping ratio and floor of TMDI installation.

4.1.2. Optimization of Parameters of TID

Similar to the configurations of the optimization in Section 4.1.1, the optimization of parameters of TID for mitigating peak displacement and acceleration at top floor can be expressed as Equations (18) and (19), respectively.

$$\begin{cases} \text{minimize } f_1(v,\zeta,t) = \hat{D}_{dis} \\ s.t. \quad \beta = 100\%, -p = -4, \\ \quad\quad v \in [0.7, 1.2], \zeta \in [0, 20\%], \\ \quad\quad t \in [30, 58] \end{cases} \tag{18}$$

$$\begin{cases} \text{minimize } f_2(v,\zeta,t) = \hat{D}_{acc} \\ s.t. \quad \beta = 100\%, -p = -4, \\ \quad\quad v \in [0.7, 1.2], \zeta \in [0, 20\%], \\ \quad\quad t \in [30, 58] \end{cases} \tag{19}$$

The results of optimization of TID in $v - \zeta - t$ space are displayed in Figure 9:

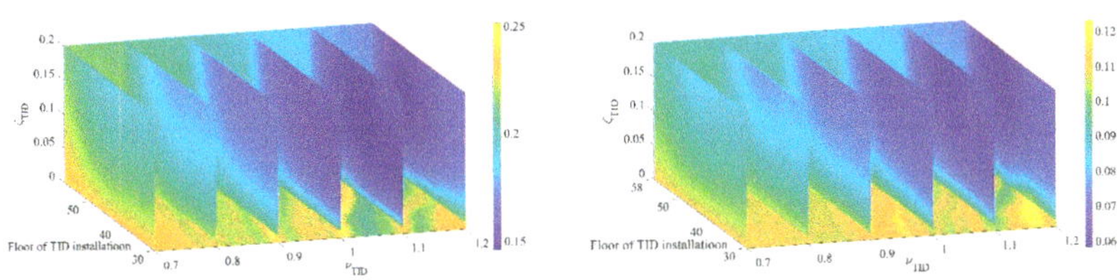

(a)

Figure 9. *Cont.*

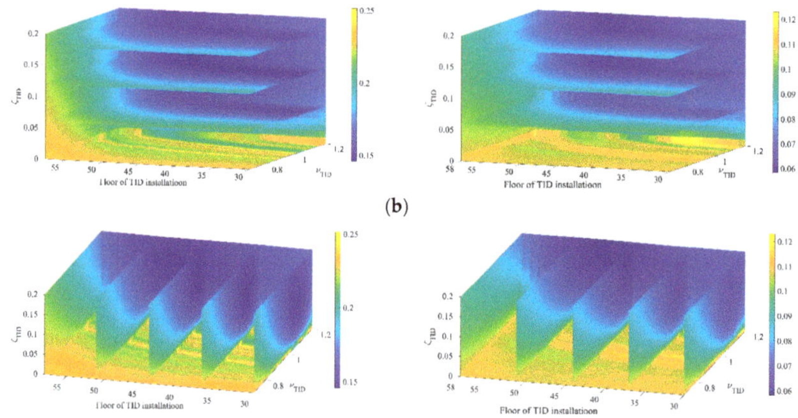

(b)

(c)

Figure 9. Peak acceleration at top floor of Tuned Inerter Damper (TID)-equipped structure at wind direction of 90° in the $\upsilon - \zeta - t$ space. (**a**) Distributions of peak displacement (**left**) and acceleration (**right**) at constant frequency ratios (υ = 0.7, 0.8, 0.9, 1.0, 1.1, 1.2). (**b**) Distributions of peak displacement (**left**) and acceleration (**right**) at constant damping ratios (ζ = 6%, 12%, 18%). (**c**) Distributions of peak displacement (**left**) and acceleration (**right**) at constant floor of TMDI installation (t = 30, 37, 44, 51, 58).

The similar trends of the distribution of the parameters of TID as that of TMDI can be observed by comparing the Figures 7 and 9. For TID having inertance ratio of 100%, which is twice that of TMDI, the highest reduction of displacement and acceleration responses is achieved when the damping ratios are 12% and 19%, respectively, which are almost twice those of TMDI. The inerter devices in TID scheme produce larger inertia, which corresponds to a better ability to store energy, thus the corresponding requirement for dissipating rate of energy stored in both inerter and attached mass increases at the same time. Based on the same considerations of determining the optimal parameters of TMDI as shown in Figure 10, the configuration of the optimally-designed TID is determined and listed in Table 3. Slightly different from the result of that of TMDI, the point inside the orange square, which represents the selected configuration of TID, lies close to the dashed line due to the limitation of discrete parameter, i.e., floor of TMDI installation.

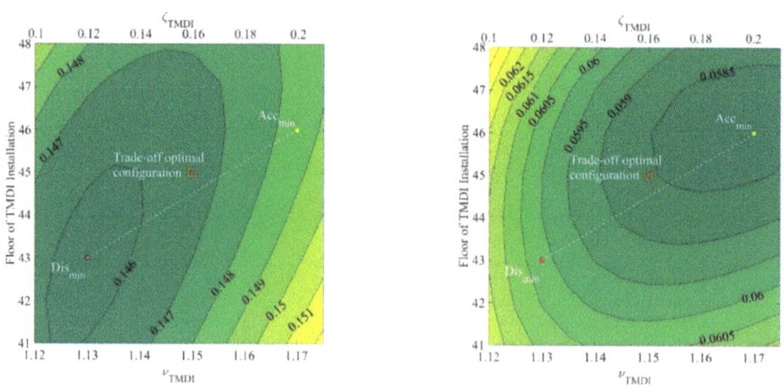

(a) (b)

Figure 10. Contour of (**a**) peak displacement (**b**) peak acceleration with variable frequency ratio, damping ratio and floor of TID installation.

Table 3. Design parameters of Tuned Mass Damper (TMD), Tuned Mass-Damper-Inerter (TMDI) and Tuned Inerter Damper (TID) used for the comparative study.

Parameters	TMD	TMDI	TID
TMD installation floor	58th floor	45th floor	45th floor
Effective mass ratio $\mu_{eff} = \mu + \beta$	0.5%	50.5%	100%
Physical mass ratio $\mu_{phy} = \mu + \beta/200$	0.5%	0.5%	0.5%
Mass ratio μ	0.5%	0.25%	\
Inertance ratio β	\	50%	100%
Frequency ratio υ	0.99	1.07	1.15
Damping ratio ζ	7%	9%	16%
TMDI topology $-p$	\	−4	−4

Apart from the two inerter-based vibration absorbers, the optimal parameters of TMD with fixed mass ratio equal to 0.5% (and, thus, equal physical mass ratio to that of the TMDI and TID) are determined by performing the same optimization procedure.

As stated above, the optimal parameters of TMD, TMDI and TID in a comparison group are listed in Table 3. It can be seen that the TMDI and TID scheme benefit a lot from the inerter system on the effective mass.

Once the optimal parameters of the TMD/TMDI/TMD are selected, the corresponding mass, damping and stiffness matrices can be determined according to Equation (2). Therefore, the transfer function of displacement and acceleration response as per Equation (5) can be computed to assess the effects of TMD/TMDI/TID in mitigating the wind-induced response. Figures 11 and 12 present the modulus of the transfer function of displacement and acceleration responses, respectively, at the 69th story (top floor) of the benchmark 340 m tall building subject to the aerodynamic forces consistent with the wind tunnel test measurements.

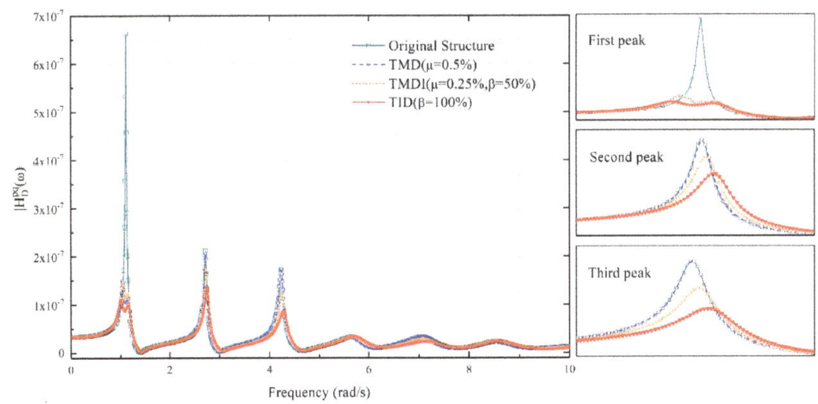

Figure 11. Displacement transfer function at top floor for different structural control schemes.

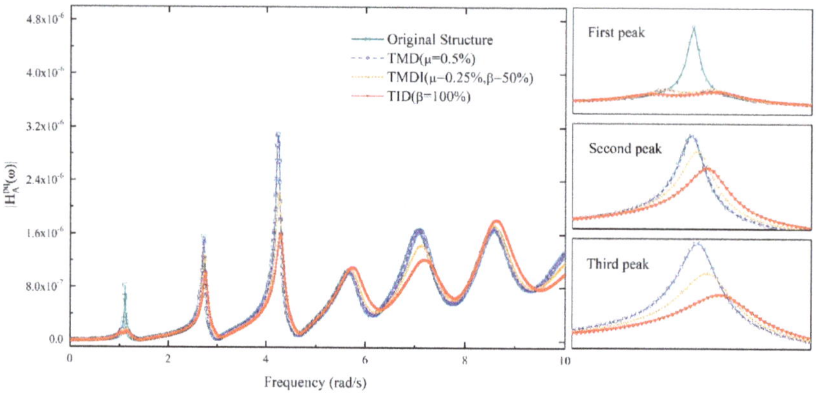

Figure 12. Acceleration transfer function at the top floor for different structural control schemes.

In Figure 11, the first peaks of displacement transfer function of the original structure, structure with TMD, TMDI and TID appear around the first natural frequency (1.10 rad/s, consistent with 0.176 Hz reported in Table 1). The peak of the original structure (in dark cyan) is much higher than that of three other vibration absorber equipped structures, i.e., TMD-equipped structure (in blue), TMDI-equipped structure (in orange) and the TID-equipped structure (in red). Overall, Figure 11 indicates that TMD, TMDI and TID mitigate the displacement response corresponding to the first vibration mode, and the mitigation effects of the TID are better than that of TMD and TMDI whose physical mass ratios are the same as that of TID. Around the second and third natural frequencies, the transfer function of the original structure overlaps with that of structure with TMD, while the transfer function of the structure with TMDI is slightly lower, and the TID achieves the best mitigation effect. For higher natural frequencies, transfer functions of the four cases are almost identical.

In Figure 12, the highest peak of the acceleration transfer function is observed around the third natural frequency. All three vibration absorbers efficiently suppress (in a comparable manner) wind-induced acceleration response corresponding to the first natural frequency with the optimal frequency ratio between 0.99 and 1.15. These graphs demonstrate the advantages of the inerter-based vibration absorbers in achieving a considerable wind-induced vibration mitigation in comparison to the TMD by employing the same physical mass ratio.

4.2. Effects of the Inerter-Based Vibration Absorbers on Wind-Induced Displacements

After setting the parameters of the TMDI and TID systems, the $\mathbf{M}, \mathbf{C}, \mathbf{K}$ matrices can be calculated by using Equation (2), respectively, hence the time histories of wind-induced displacements of the TMDI- and TID-equipped benchmark buildings can be analyzed corresponding to wind speed of 40.07 m/s (50-years return period stipulated by survivability limit state design) for each of the 24 wind directions from 0° to 345° according to Equation (8). Figure 13 represents a segment of the displacement time-history response (for an overall duration of 20 min) at the 69th story corresponding to 90° wind direction.

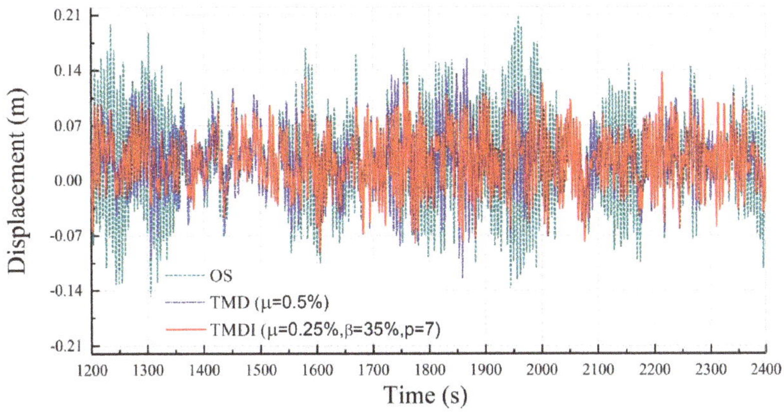

Figure 13. Time histories of displacement at the 69th story along the x-axis corresponding to 90° wind direction.

Extreme wind-induced displacement response can be evaluated in terms of mean and Root Mean Square (RMS) value for each of the 24 wind directions considered in this study. Figure 14 presents variation of extreme displacement responses (as per Equation (10)) along the building height corresponding to three typical wind directions (0°, 45° and 90°). By inspection of Figure 14, it appears clear that TMD, TMDI and TID significantly decrease the wind-induced extreme displacement responses. The mitigation effects of TID are slightly better than that of TMD and TMDI. This demonstrates that the inerter plays a significant role in the vibration mitigation of the structure.

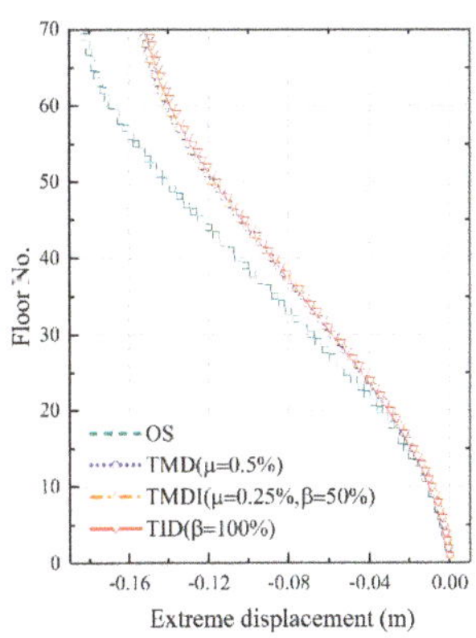

(**a**)

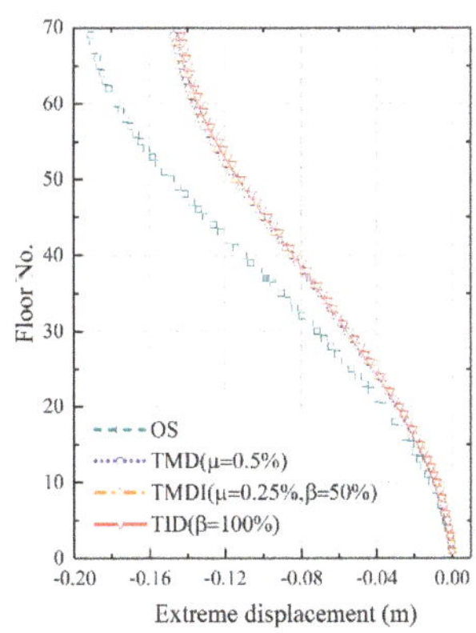

(**b**)

Figure 14. *Cont.*

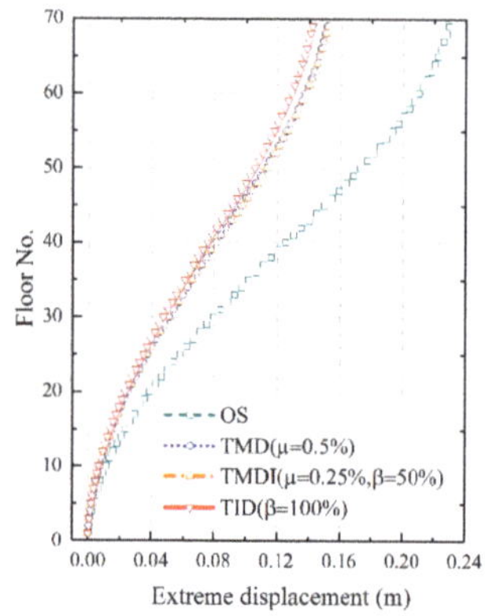

(c)

Figure 14. Profile of extreme wind-induced displacement response at the x-axis along the building height corresponding to three different wind directions: (**a**) 0° wind direction. (**b**) 45° wind direction. (**c**) 90° wind direction.

Figure 15 shows the variation of mean and extreme top-floor displacements corresponding to a variety of wind directions ranging from 0° to 345°. The maximum absolute displacement is smaller than 1/1500 of the height of the benchmark building, which justifies the linear elastic behavior assumption made in this paper for the building dynamic model. Figure 15 indicates that three vibration absorbers have no effects on mean displacement responses, which in fact coincide with those of the OS. The mean displacement responses approach zero for wind direction corresponding to 90° and 270° wind-direction, because this corresponds to the across-wind response induced by the vortices shedding at both edges of the windward side. For any other wind direction, TMD, TMDI and TID significantly suppress wind-induced top-floor displacement responses along the x-axis. Based on Figures 13–15, it can be concluded that TID with same physical mass ratio achieved a slightly better vibration mitigation effect to the TMD and TMDI in terms of displacement response.

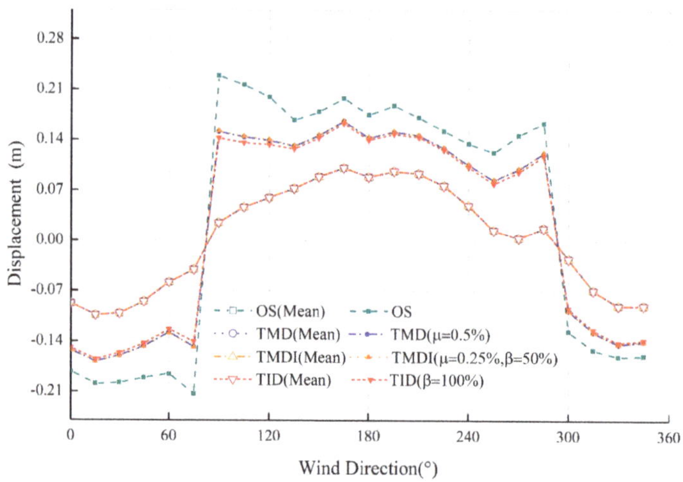

Figure 15. Mean and extreme top-floor displacement response at the x-axis corresponding to wind directions ranging from 0° to 345°.

To quantify the mitigation effect of TMD, TMDI and TID, a factor of vibration-absorbing F_{va} is defined as follows

$$F_{va}(\%) = \left|\frac{R_{OS} - R_T}{R_{OS}}\right| \times 100 \tag{20}$$

where R_{OS} represents the response of the original structure, while R_T denotes the corresponding response indicator for the structure with TMD\TMDI\TID. Some relevant results for a few emblematic wind directions and loading conditions are listed in Table 4. In particular, Table 4 presents the vibration-absorbing factor for some typical conditions. For example, for wind direction of 90° the TID has shown the best vibration mitigation effect among the three vibration absorbers: the F_{va} is equal to 38.02% for the TID, 34.01% for the TMD and 33.74% for the TMDI. The worst vibration-absorbing effect of TID takes place for 330° wind direction, where the F_{va} with TID is only 11.94% and F_{va} with TMD and TMDI are only 10.56% and 10.52%, respectively. The two inerter-based vibration absorbers share a similar variation (vary synchronously) against wind excitations at different directions as it can be observed from the curvilinear shapes of both TMDI and TID in Figure 15. This leads to the result that the 1st and the 3rd, the 2nd and the 4th rows are identical. Generally, structural engineers are mainly concerned about the maximum absolute value of extreme displacement. For example, in this study, the maximum positive extreme top-floor displacement of the original structure is 0.23 m corresponding to 90° wind direction. The positive extreme displacement decreases from 0.23 m to around 0.14 m when the TID is installed on the original structure. The negative extreme top-floor displacement occurs for 75° wind direction. The displacement drops from −0.21 m to −0.14 m when the TID is used to mitigate the wind-induced vibration of the primary structure. The corresponding factor of vibration-absorbing is 34.07%.

Table 4. Vibration-mitigation effects in terms of top-floor displacement response for some loading configurations.

Selected Condition	Wind Direction (°)	u_{OS}^{top}(m)	u_{TMD}^{top}(m)	u_{TMDI}^{top}(m)	u_{TID}^{top}(m)	$F_{va}^{(TMD)}$	$F_{va}^{(TMDI)}$	$F_{va}^{(TID)}$
Minimum F_{va} (TMDI)	330	−0.1608	−0.1438	−0.1438	−0.1416	10.56	10.52	11.94
Maximum F_{va} (TMDI)	90	0.2293	0.1513	0.1519	0.1421	34.01	33.74	38.02
Minimum F_{va} (TID)	330	−0.1608	−0.1438	−0.1438	−0.1416	10.56	10.52	11.94
Maximum F_{va} (TID)	90	0.2293	0.1513	0.1519	0.1421	34.01	33.74	38.02
Max positive displacement OS	90	0.2293	0.1560	0.1519	0.1421	34.01	33.74	38.02
Max negative displacement OS	75	−0.2130	−0.1476	−0.1482	−0.1404	30.71	30.44	34.07

4.3. *Effects of the Inerter-Based Vibration Absorbers on Wind-Induced Accelerations*

Excessive wind-induced acceleration response may cause discomfort to building occupants and poses serious serviceability issues [66]. Wind-induced accelerations of the benchmark 340 m tall building, together with the TMD-, TMDI- and the TID-equipped building, are analyzed for a 33.86 m/s wind speed (10-years return period related to serviceability limit state) for each of the 24 wind directions. Figure 16 illustrates a segment of the top-floor time-history acceleration response (for an overall duration of 20 min) corresponding to 90° wind direction.

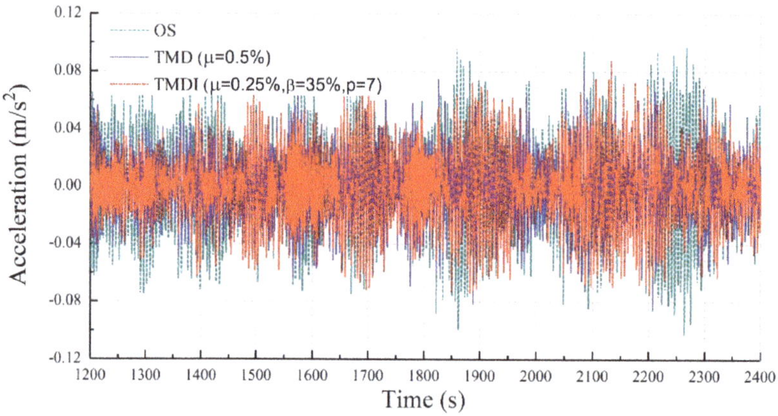

Figure 16. Time histories of acceleration at the 69th story along the x-axis corresponding to 90° wind direction.

Following time histories of acceleration response, extreme acceleration responses can be obtained in terms of RMS value. Figure 17 presents variation of extreme acceleration responses along the building height corresponding to three typical wind directions, namely 0°, 45° and 90° wind directions. Figure 17 demonstrates that TMD, TMDI and TID have significant vibration absorbing effects on wind-induced acceleration response. In general, the vibration-mitigation effects of the TID are better than those of the TMD and TMDI, especially for wind direction of 45°. As stated previously, the three vibration absorbers share the same physical mass ratio. The vibration-mitigation effects are not only related to the parameters of TMD, TMDI or TID, but also depend on the predominant frequency components of aerodynamic forces. At 0° and 90° wind direction (cf. Figure 17a,c) the acceleration response at the x-axis are mainly induced by incoming turbulence flow and vortex shedding effects, respectively. At 45° wind direction, the acceleration response at the x-axis is affected by a combination of incoming turbulence flow and vortex shedding. Hence, it can be concluded that the vibration mitigation effects of TID is the best among three vibration absorbers, and the performance of TMDI is better than that of TMD for specific wind direction of 45°.

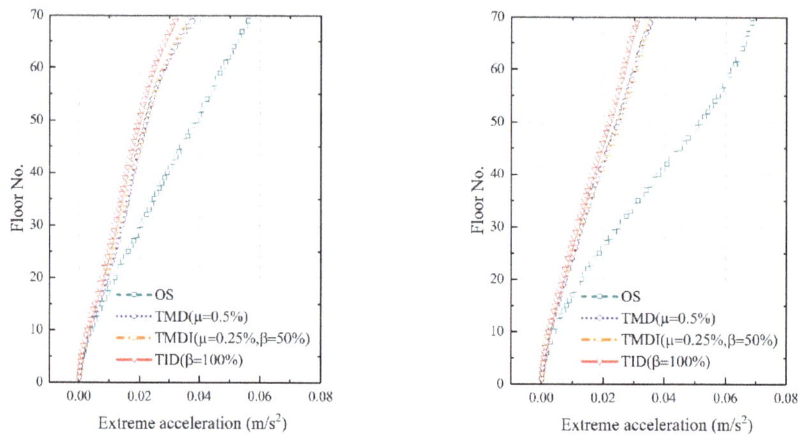

(a) (b)

Figure 17. *Cont.*

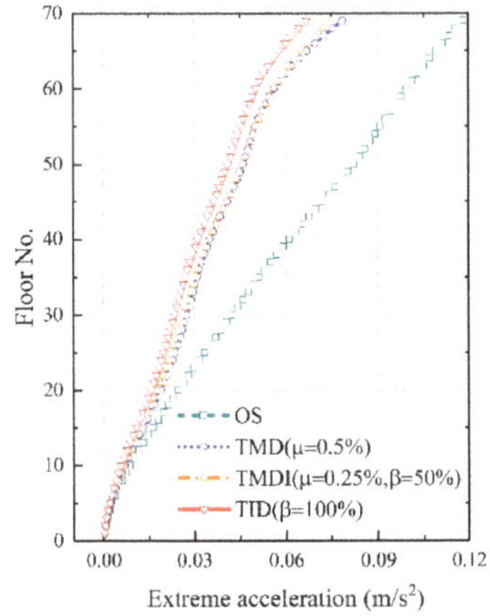

(c)

Figure 17. Profile of extreme wind-induced acceleration response at the x-axis along the building height corresponding to three different wind directions: (**a**) 0° wind direction; (**b**) 45° wind direction; (**c**) 90° wind direction.

Figure 18 shows the variation of extreme top-floor accelerations corresponding to a variety of wind directions ranging from 0° to 345°. From Figure 18 we can see that TMD, TMDI and TID significantly control extreme acceleration responses, especially at 45° wind direction. For this wind direction, the response decreased from 0.0687 m/s^2 (original structure) to 0.0307 m/s^2 (TID-equipped structure).

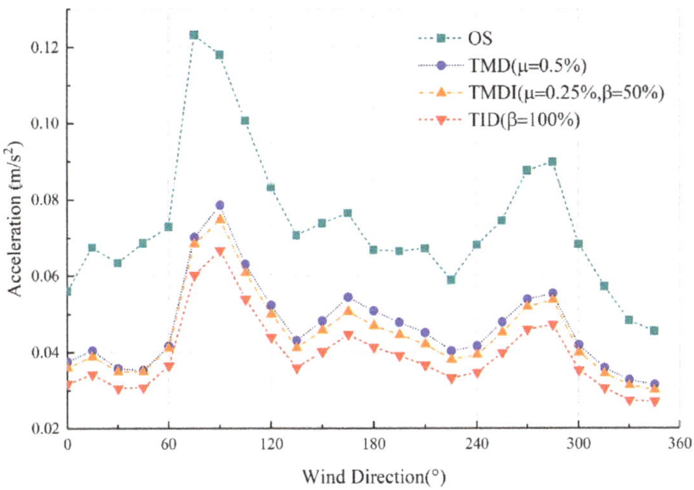

Figure 18. Extreme top-floor acceleration response at the x-axis corresponding to wind directions ranging from 0° to 345°.

Based on Equation (20), the factor of vibration absorbing F_{va} related to the top floor acceleration response is evaluated and listed in Table 5 for a few emblematic wind directions and loading conditions.

Table 5. Vibration-mitigation effects in terms of top-floor acceleration response for some loading configurations.

Selected Condition	Wind Direction (°)	$\ddot{u}_{OS}^{top}$ (m/s²)	$\ddot{u}_{TMD}^{top}$ (m/s²)	$\ddot{u}_{TMDI}^{top}$ (m/s²)	$\ddot{u}_{TID}^{top}$ (m/s²)	$F_{va}^{(TMD)}$	$F_{va}^{(TMDI)}$	$F_{va}^{(TID)}$
Minimum F_{va} (TMDI)	180	0.0667	0.0508	0.0469	0.0413	23.88	29.78	38.09
Maximum F_{va} (TMDI)	45	0.0687	0.0353	0.0350	0.0307	48.63	49.06	55.26
Minimum F_{va} (TID)	180	0.0667	0.0508	0.0469	0.0413	23.88	29.78	38.09
Maximum F_{va} (TID)	45	0.0687	0.0353	0.0350	0.0307	48.63	49.06	55.26
Maximum acceleration OS	75	0.1232	0.0701	0.0685	0.0602	43.04	44.40	51.12

In all wind directions, the TID dramatically reduces wind-induced top-floor accelerations. The worst vibration-absorbing effect of TMDI takes place for 180° wind direction, where the corresponding F_{va} is 38.09% and is 1.6 times than that of the TMD having same physical mass. Generally, structural engineers are mainly concerned about the maximum extreme acceleration response of top floor, which may cause discomfort to residents. The maximum value of the acceleration of the original structure is 0.1232 m/s^2 corresponding to a wind direction of 75°. The TMD, TMDI and TID can reduce such extreme acceleration value of more than 40%, namely from 0.1232 m/s^2 to 0.0701 m/s^2, 0.0685 m/s^2 and 0.0602 m/s^2, respectively. These results show that the TID has a significant acceleration-reduction effect due to the enormous inertia benefitting from the inerter device, despite employing the same physical mass of the TMD. An important aspect for practical implementation of TMDI/TID systems is the force generated by the inerter, as bigger inertances bring also higher forces that are difficult to handle in a conventional structure [59]. In the present example, the maximum resistance force produced by inerter is 5319 kN at wind direction of 330°. Such requirement for inerter force can be practically implemented by installing several parallel inerter devices as shown in Figure 1.

4.4. Effects of the Inerter-Based Vibration Absorbers on ESWLs

ESWLs are important parameters used by structural engineers for limit-state design as well as for assessing the bearing capacity of structures. Displacement Gust Loads Factor (DGLF) method [59], Moment-Based Gust Loads Factor (MGLF) method [66,67], Load-Response Correlation (LRC) method [68], and Weighted Combination of Modal Inertial Load Component (WCMILC) method [69] have been proposed to calculate ESWLs of high-rise buildings. Among them, the DGLF method is widely used in practical projects owing to its simplicity and for this reason it is adopted in this paper.

After the calculation of mean and extreme values of wind-induced profiles of displacements along the building height corresponding to 24 wind directions, ESWLs can be obtained from Equations (9) and (10). We here describe variation of ESWLs along the building height at a wind speed of 42.02 m/s (100-years return period).

Figure 19a,b show the profiles of ESWLs of original structure, structure with TMD, TMDI and TID along the building height, corresponding to wind directions of 0° and 45°, respectively. It is noted that the ESWLs of the original structure are larger than those of structure with vibration absorbers for every story. In this case, the performance of the three vibration absorbers is more or less comparable. As the ESWL is calculated by Equations (9) and (10) based on the extreme displacement, the mitigation effects of three vibration absorbers on ESWL are similar to that on extreme displacement. Above the 60th floor, the mean wind loads become smaller because floors above 60th floor gradually draw back in plane as already illustrated in Figure 2.

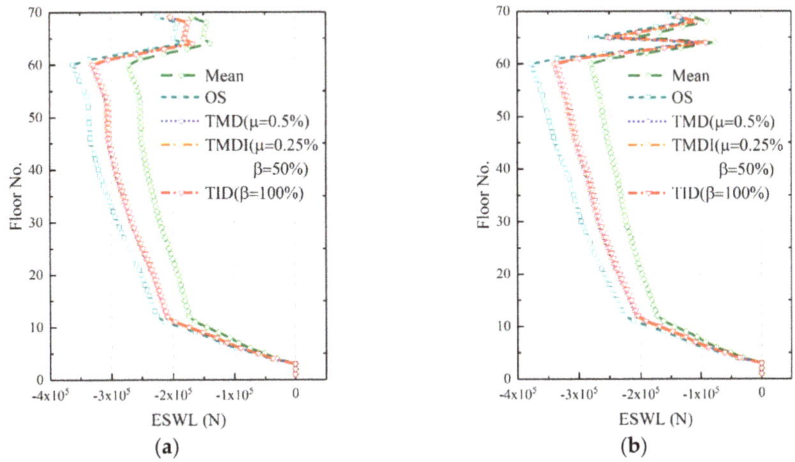

Figure 19. Variation of ESWLs along the building height corresponding to two wind directions: (**a**) 0° wind direction; (**b**) 45° wind direction.

5. Conclusions

The wind-induced response of a benchmark 340 m tall building equipped with inerter-based vibration absorber, i.e., TMDI and TID, has been investigated. The analysis has been carried out in the time-domain, by considering the time histories of aerodynamic forces computed from synchronous multi-point pressure measurements in wind tunnel tests, which accounts for the actual cross section of the building and the existing surrounding conditions. The results have been analyzed in terms of wind-induced displacement and acceleration response as well as ESWLs on the original structure, and comparatively on the building equipped with the TMD, TMDI and TID corresponding to 24 different wind directions (from 0° to 345° at an interval of 15°).

The main contents and findings of the present work are summarized as follows:

1. Displacement- and acceleration-based optimizations have been performed to obtain the best parameters of the TMD, TMDI and TID in a 3D design space, including the installation floor, the frequency ratio and the damping ratio as explicit design variables. The proposed procedure attempts to find a good trade-off between displacement mitigation and acceleration mitigation, considering results from a constrained optimization problem in which the installation floor represents a design variable being incorporated in the optimization procedure;
2. Both wind-induced extreme top-floor displacement and acceleration responses of the benchmark building can be effectively mitigated by the TMDI and TID. Among the three vibration absorbers, the TID outperforms the TMDI and the TMD, and the acceleration mitigation effect of the TMDI is better than that of the TMD. The extreme displacement and acceleration response of the original structure are 0.2293 m and 0.1232 m/s^2, respectively. The installation of the TMDI has reduced these response values to 0.1519 m and 0.0685 m/s^2, respectively, with a resulting factor of vibration absorbing F_{va} equal to 33.74% and 44.40%, respectively. The best vibration mitigation effect is achieved by the TID, which reduces the extreme displacement and acceleration to 0.1421 m and 0.0602 m/s^2, respectively, corresponding to F_{va} of 38.02% and 51.12%, respectively;
3. Comparison among the three different vibration absorbers has shown that the TID with same physical mass ratio as the TMD and TMDI can achieve better vibration mitigation effects in terms of displacement and acceleration responses. In particular, the factors of vibration absorbing F_{va} of TMD, TMDI and TID for extreme displacement have been 34.01%, 33.74% and 38.02%, respectively, and the analogous factors for extreme acceleration have been 43.04%, 44.40% and 51.12%, respectively. The performance of TID slightly outperforms the other two vibration

absorbers in terms of mitigating ESWLs, which is consistent with the results of displacement mitigation effect;

4. Optimizations of parameters have demonstrated that the TMDI and TID achieve the best vibration mitigation effects when the first terminal is not installed at the top floor, but at the mid-upper place of the primary structure with TMDI/TID topologies such that the inerter spans four stories. In this configuration, the TID can achieve better wind-induced vibration mitigation than the TMD employing the same physical mass ratio as that of the corresponding TMD (thus implying a significant reduction in terms of physical mass actually allocated due to the mass-amplification effect of the inerter when the TID scheme is designed to achieve the same vibration mitigation effect as that of TMD);
5. The TID having the same physical mass as the TMD (meaning that the inertia is entirely provided by the inerter, with ideally null attached mass) can achieve much better vibration mitigation effects than the TMD in terms of acceleration response when the frequency ratio υ and damping ratio ζ of the TID are tuned to be around 1.15 and 16%, respectively, and the TID is installed at the 45th floor. A slightly better displacement mitigation effect can be achieved by adopting a relative smaller frequency ratio, damping ratio and lower installation floor, e.g., 1.13, 12% and 43rd floor, respectively.

The present study has focused on the design and optimization of the vibration absorbers based on the expected wind pressure, thus emphasizing the effects of wind loading on the high-rise building. Future investigations concerning the analysis of the proposed structural control systems against other types of dynamic loads, such as earthquake excitations, are currently underway.

Author Contributions: Conceptualization, Q.W. and H.Q.; methodology, Q.W., H.Q. and D.D.D.; software, H.Q.; validation, H.Q., Z.Z. and Z.X.; formal analysis, Q.W. and D.D.D.; investigation, H.Q., D.D.D. and Q.W.; writing—original draft preparation, H.Q. and D.D.D.; writing—review and editing, Q.W. and D.D.D.; supervision, D.D.D.; funding acquisition, Q.W.

Acknowledgments: This research was funded by Key Laboratory of Structure and Wind Tunnel of Guangdong Higher Education Institutes. This support is gratefully acknowledged.

Conflicts of Interest: The authors declare no conflict of interest.

References

1. CTBUH. The Global Tall Buildings Database of the CTBUH. 2019. Available online: http://www.skyscrapercenter.com/ (accessed on 22 June 2019).
2. Kareem, A.; Kijewski, T.; Tamura, Y. Mitigation of motions of tall buildings with specific examples of recent applications. *Wind Struct. Int. J.* **1999**, *2*, 201–251. [CrossRef]
3. Kim, Y.; Kanda, J. Characteristics of aerodynamic forces and pressures on square plan buildings with height variations. *J. Wind Eng. Ind. Aerodyn.* **2010**, *98*, 449–465. [CrossRef]
4. Kim, Y.C.; Kanda, J. Wind pressures on tapered and set-back tall buildings. *J. Fluids Struct.* **2013**, *39*, 306–321. [CrossRef]
5. Tamura, T.; Miyagi, T.; Kitagishi, T. Numerical prediction of unsteady pressures on a square cylinder with various corner shapes. *J. Wind Eng. Ind. Aerodyn.* **1998**, *74*, 531–542. [CrossRef]
6. Zheng, C.; Xie, Y.; Khan, M.; Wu, Y.; Liu, J. Wind-induced responses of tall buildings under combined aerodynamic control. *Eng. Struct.* **2018**, *175*, 86–100. [CrossRef]
7. Sharma, A.; Mittal, H.; Gairola, A. Mitigation of wind load on tall buildings through aerodynamic modifications: Review. *J. Build. Eng.* **2018**, *18*, 180–194. [CrossRef]
8. Zeng, X.; Peng, Y.; Chen, J. Serviceability-based damping optimization of randomly wind-excited high-rise buildings. *Struct. Des. Tall Special Build.* **2017**, *26*, e1371. [CrossRef]
9. Chen, J.; Zeng, X.; Peng, Y. Probabilistic analysis of wind-induced vibration mitigation of structures by fluid viscous dampers. *J. Sound Vib.* **2017**, *409*, 287–305. [CrossRef]
10. De Domenico, D.; Ricciardi, G. Earthquake protection of structures with nonlinear viscous dampers optimized through an energy-based stochastic approach. *Eng. Struct.* **2019**, *179*, 523–539. [CrossRef]

11. Aydin, E. Optimal damper placement based on base moment in steel building frames. *J. Constr. Steel Res.* **2012**, *79*, 216–225. [CrossRef]
12. Aydin, E. A simple damper optimization algorithm for both target added damping ratio and interstorey drift ratio. *Earth Struct.* **2013**, *88*, 83–109. [CrossRef]
13. De Domenico, D.; Ricciardi, G.; Takewaki, I. Design strategies of viscous dampers for seismic protection of building structures: A review. *Soil Dyn. Earthq. Eng.* **2019**, *118*, 144–165. [CrossRef]
14. Warburton, G.B. Optimum Absorber Parameters for Various Combinations of Response and Excitation Parameters. *Earthq. Eng. Struct. Dyn.* **1982**, *10*, 381–401. [CrossRef]
15. Takewaki, I. Soil—Structure random response reduction via TMD-VD simultaneous use. *Comput. Methods Appl. Mech. Eng.* **2000**, *190*, 677–690. [CrossRef]
16. Stanikzai, M.H.; Elias, S.; Matsagar, V.A.; Jain, A.K. Seismic response control of base-isolated buildings using multiple tuned mass dampers. *Struct. Des. Tall Special Build.* **2019**, *28*, e1576. [CrossRef]
17. Elias, S.; Matsagar, V. Distributed Multiple Tuned Mass Dampers for Wind Vibration Response Control of High-Rise Building. *J. Eng.* **2014**, *2014*, 1–11. [CrossRef]
18. Elias, S.; Matsagar, V.; Datta, T.K. Along-wind response control of chimneys with distributed multiple tuned mass dampers. *Struct. Control Health Monit.* **2019**, *26*, e2275. [CrossRef]
19. Min, K.W.; Kim, H.S.; Lee, S.H.; Kim, H.; Kyung Ahn, S. Performance evaluation of tuned liquid column dampers for response control of a 76-story benchmark building. *Eng. Struct.* **2005**, *27*, 1101–1112. [CrossRef]
20. Di Matteo, A.; Pirrotta, A.; Tumminelli, S. Combining TMD and TLCD: Analytical and experimental studies. *J. Wind Eng. Ind. Aerodyn.* **2017**, *167*, 101–113. [CrossRef]
21. Di Matteo, A.; Furtmüller, T.; Adam, C.; Pirrotta, A. Optimal design of tuned liquid column dampers for seismic response control of base-isolated structures. *Acta Mech.* **2018**, *229*, 437–454. [CrossRef]
22. Bedon, C.; Amadio, C. Numerical assessment of vibration control systems for multi-hazard design and mitigation of glass curtain walls. *J. Build. Eng.* **2018**, *15*, 1–13. [CrossRef]
23. Pipitone, G.; Barone, G.; Palmeri, A. Optimal design of double-skin façades as vibration absorbers. *Struct. Control Health Monit.* **2018**, *25*, e2086. [CrossRef]
24. Tanaka, H.; Mak, C.Y. Effect of tuned mass dampers on wind induced response of tall buildings. *J. Wind Eng. Ind. Aerodyn.* **1983**, *14*, 357–368. [CrossRef]
25. Xu, Y.L.; Kwok, K.C.S.; Samali, B. The effect of tuned mass dampers and liquid dampers on cross-wind response of tall/slender structures. *J. Wind Eng. Ind. Aerodyn.* **1992**, *40*, 33–54. [CrossRef]
26. Xu, Y.L.; Kwok, K.C.S.; Samali, B. Torsion response and vibration suppression of wind-excited buildings. *J. Wind Eng. Ind. Aerodyn.* **1992**, *43*, 1997–2008. [CrossRef]
27. Kwok, K.C.S.; Samali, B. Performance of tuned mass dampers under wind loads. *Eng. Struct.* **1995**, *17*, 655–667. [CrossRef]
28. Kim, Y.M.; You, K.P.; Kim, H.Y. Wind-induced excitation control of a tall building with tuned mass dampers. *Struct. Des. Tall Special Build.* **2008**, *17*, 669–682. [CrossRef]
29. Elias, S.; Matsagar, V. Windresponse control of tall buildings with a tuned mass damper. *J. Build. Eng.* **2018**, *15*, 51–60. [CrossRef]
30. Wang, Q.; Li, Z.; Garg, A.; Hazra, B.; Xie, Z. Effects of TMD on correlation of wind-induced responses and combination coefficients of ESWLs of high-rise buildings. *Struct. Des. Tall Special Build.* **2019**, *28*, e1597. [CrossRef]
31. Wang, Q.; Fu, W.; Yu, S.; Allan, L.; Garg, A.; Gu, M. Mathematical model and case study of wind-induced responses for a vertical forest. *J. Wind Eng. Ind. Aerodyn.* **2018**, *179*, 260–272. [CrossRef]
32. Love, J.S.; Haskett, T.C.; Morava, B. Effectiveness of dynamic vibration absorbers implemented in tall buildings. *Eng. Struct.* **2018**, *176*, 776–784. [CrossRef]
33. De Domenico, D.; Ricciardi, G. Optimal design and seismic performance of tuned mass damper inerter (TMDI) for structures with nonlinear base isolation systems. *Earthq. Eng. Struct. Dyn.* **2018**, *47*, 2539–2560.
34. De Domenico, D.; Ricciardi, G. Earthquake-resilient design of base isolated buildings with TMD at basement: Application to a case study. *Soil Dyn. Earthq. Eng.* **2018**, *113*, 503–521. [CrossRef]
35. De Domenico, D.; Ricciardi, G. An enhanced base isolation system equipped with optimal tuned mass damper inerter (TMDI). *Earthq. Eng. Struct. Dyn.* **2018**, *47*, 1169–1192. [CrossRef]
36. Giaralis, A.; Petrini, F. Wind-Induced Vibration Mitigation in Tall Buildings Using the Tuned Mass-Damper-Inerter. *J. Struct. Eng.* **2017**, *143*, 11. [CrossRef]

37. Li, Q.; He, Y.; Zhou, K.; Han, X.; He, Y.; Shu, Z. Structural health monitoring for a 600 m high skyscraper. *Struct. Des. Tall Special Build.* **2018**, *27*, e1490. [CrossRef]
38. Li, Q.S.; Zhi, L.H.; Tuan, A.Y.; Kao, C.S.; Su, S.C.; Wu, C.F. Dynamic behavior of Taipei 101 tower: Field measurement and numerical analysis. *J. Struct. Eng.* **2011**, *137*, 143–155. [CrossRef]
39. Lazar, I.F.; Neild, S.A.; Wagg, D.J. Using an inerter-based device for structural vibration suppression. *Earthq. Eng. Struct. Dyn.* **2014**, *43*, 1129–1147. [CrossRef]
40. Lazar, I.F.; Neild, S.A.; Wagg, D.J. Vibrationsuppression of cables using tuned inerter dampers. *Eng. Struct.* **2016**, *122*, 62–71. [CrossRef]
41. Smith, M.C. Synthesis of mechanical networks: The inerter. *IEEE Trans. Autom. Control* **2002**, *47*, 1648–1662. [CrossRef]
42. Marian, L.; Giaralis, A. Optimal design of a novel tuned mass-damper-inerter (TMDI) passive vibration control configuration for stochastically support-excited structural systems. *Probab. Eng. Mech.* **2015**, *38*, 156–164. [CrossRef]
43. Mirza Hessabi, R.; Mercan, O. Investigations of the application of gyro-mass dampers with various types of supplemental dampers for vibration control of building structures. *Eng. Struct.* **2016**, *126*, 174–186. [CrossRef]
44. Pietrosanti, D.; De Angelis, M.; Basili, M. Optimal design and performance evaluation of systems with Tuned Mass Damper Inerter (TMDI). *Earthq. Eng. Struct. Dyn.* **2017**, *46*, 1367–1388. [CrossRef]
45. De Domenico, D.; Impollonia, N.; Ricciardi, G. Soil-dependent optimum design of a new passive vibration control system combining seismic base isolation with tuned inerter damper. *Soil Dyn. Earthq. Eng.* **2018**, *105*, 37–53. [CrossRef]
46. De Domenico, D.; Ricciardi, G. Improving the dynamic performance of base-isolated structures via tuned mass damper and inerter devices: A comparative study. *Struct. Control Health Monit.* **2018**, *25*, e2234. [CrossRef]
47. De Domenico, D.; Deastra, P.; Ricciardi, G.; Sims, N.D.; Wagg, D.J. Novel fluid inerter based tuned mass dampers for optimised structural control of base-isolated buildings. *J. Frankl. Inst.* **2019**, *356*, 7626–7649. [CrossRef]
48. Giaralis, A.; Taflanidis, A.A. Optimal tuned mass-damper-inerter (TMDI) design for seismically excited MDOF structures with model uncertainties based on reliability criteria. *Struct. Control Health Monit.* **2018**, *25*, e2082. [CrossRef]
49. Zhao, Z.; Zhang, R.; Jiang, Y.; Pan, C. Seismic response mitigation of structures with a friction pendulum inerter system. *Eng. Struct.* **2019**, *193*, 110–120. [CrossRef]
50. Pan, C.; Zhang, R. Design of structure with inerter system based on stochastic response mitigation ratio. *Struct. Control Health Monit.* **2018**, *25*, e2169. [CrossRef]
51. Pan, C.; Zhang, R.; Luo, H.; Li, C.; Shen, H. Demand-based optimal design of oscillator with parallel-layout viscous inerter damper. *Struct. Control Health Monit.* **2018**, *25*, e2051. [CrossRef]
52. Zhang, R.; Zhao, Z.; Dai, K. Seismic response mitigation of a wind turbine tower using a tuned parallel inerter mass system. *Eng. Struct.* **2019**, *180*, 29–39. [CrossRef]
53. Zhang, R.; Zhao, Z.; Pan, C. Influence of mechanical layout of inerter systems on seismic mitigation of storage tanks. *Soil Dyn. Earthq. Eng.* **2018**, *114*, 639–649. [CrossRef]
54. Xu, K.; Bi, K.; Han, Q.; Li, X.; Du, X. Using tuned mass damper inerter to mitigate vortex-induced vibration of long-span bridges: Analytical study. *Eng. Struct.* **2019**, *182*, 101–111. [CrossRef]
55. Xia, Y.; Zhang, R.F.; Friswell, M.I.; Cao, Y. Suppression of the wind-induced vibration of high-rise buildings with inerter systems. In Proceedings of the International Conference on Noise and Vibration Engineering, Leuven, Belgium, 17–19 September 2018.
56. Luo, Y.; Sun, H.; Wang, X.; Zuo, L.; Chen, N. Wind Induced Vibration Control and Energy Harvesting of Electromagnetic Resonant Shunt Tuned Mass-Damper-Inerter for Building Structures. *Shock Vib.* **2017**, *2017*, 4180134.
57. Liang, S.; Liu, S.; Li, Q.S.; Zhang, L.; Gu, M. Mathematical model of acrosswind dynamic loads on rectangular tall buildings. *J. Wind Eng. Ind. Aerodyn.* **2002**, *90*, 1757–1770. [CrossRef]
58. Rana, R.; Soong, T.T. Parametric study and simplified design of TMDs. *Eng. Struct.* **1998**, *20*, 193–204. [CrossRef]

59. Ruiz, R.; Taflanidis, A.; Giaralis, A.; Lopez-Garcia, D. Risk-informed optimization of the tuned mass-damper-inerter (TMDI) for the seismic protection of multi-storey building structures. *Eng. Struct.* **2018**, *177*, 836–850. [CrossRef]
60. Davenport, A.G. Gust loading factors. *J. Struct. Division* **1967**, *93*, 11–34.
61. Davenport, A.G. Note on the Random Distribution of the Largest Value of a Random Function with Application to Gust Loading. *Proc. Inst. Civ. Eng.* **1964**, *24*, 187–196. [CrossRef]
62. Chopra, A.K. *Dynamics of Structures: Theory and Applications to Earthquake Engineering*, 2nd ed.; Prentice-Hall: Englewood Cliffs, NJ, USA, 2000.
63. Satake, N.; Suda, K.I.; Arakawa, T.; Sasaki, A.; Tamura, Y. Damping evaluation using full-scale data of buildings in Japan. *J. Struct. Eng.* **2003**, *129*, 470–477. [CrossRef]
64. China, M.o.H.a.U.R.D.o.t.P.s.R.o. *Chinese Code GB50009-2012, Load Code for the Design of Building Structures*; China Architecture & Building Press: Beijing, China, 2012.
65. Papageorgiou, C.; Smith, M.C. Laboratory experimental testing of inerters. In Proceedings of the 44th IEEE Conference on Decision and Control, Cdc-Ecc '05, Seville, Spain, 12–15 December 2005.
66. Kareem, A.; Zhou, Y. Gust loading factor—Past, present and future. *J. Wind Eng. Ind. Aerodyn.* **2003**, *91*, 1301–1328. [CrossRef]
67. Zhou, Y.; Kareem, A. Gust Loading Factor: New Model. *J. Struct. Eng.* **2001**, *127*, 168–175. [CrossRef]
68. Kasperski, M. Extreme wind load distributions for linear and nonlinear design. *Eng. Struct.* **1992**, *14*, 27–34. [CrossRef]
69. Chen, X.; Kareem, A. Equivalent Static Wind Loads for Buffeting Response of Bridges. *J. Struct. Eng.* **2001**, *127*, 1467–1475. [CrossRef]

applied sciences

MDPI

Article

Displacement-Dependent Damping Inerter System for Seismic Response Control

Zhipeng Zhao [1], Ruifu Zhang [1,*], Yiyao Jiang [1], Dario De Domenico [2,*] and Chao Pan [3]

[1] Department of Disaster Mitigation for Structures, Tongji University, Shanghai 200092, China; zhaozhipeng@tongji.edu.cn (Z.Z.); yyjiang17@fudan.edu.cn (Y.J.)
[2] Department of Engineering, University of Messina, Messina 98166, Italy
[3] College of Civil Engineering, Yantai University, Yantai 264005, China; panchao@ytu.edy.cn
* Correspondence: zhangruifu@tongji.edu.cn (R.Z.); dario.dedomenico@unime.it (D.D.D.)

Received: 23 November 2019; Accepted: 26 December 2019; Published: 29 December 2019

Abstract: Various inerter systems utilizing velocity-dependent damping for vibration control have been developed. However, a velocity-dependent damping element may exhibit relatively poor performance compared to a displacement-dependent damping element (DDE) of equivalent damping ratio, when the structural deformation is small in the early stage of the seismic response. To address this issue, the advantage of DDE in generating a larger control force in the early stage of excitation is promoted here and enhanced by a supplemental inerter-spring-system, thus realizing a proposed novel displacement-dependent damping inerter system (DDIS). First, the influence of various DDIS-parameters is carried out by resorting to the stochastic linearization method to handle non-linear terms. Then, seismic responses of the DDIS-controlled system are evaluated in the time domain taking the non-linearity into account, thus validating the accuracy of the stochastic dynamic analysis. Several design cases are considered, all of which demonstrated damping enhancement and timely control achieved by the DDIS. The results show that the energy dissipation as well as reduction of structural displacement and acceleration accomplished by the proposed system are significant. DDIS suppresses structural responses in a timely manner, as soon as the peak excitation occurs. In addition, it is demonstrated that interactions among the inerter, spring, and DDE, which constitute the damping-enhancement mechanism, lead to a higher energy-dissipation efficiency compared to the DDE alone.

Keywords: inerter; seismic protection; passive vibration control; displacement-dependent damping; stochastic dynamic analysis

---

1. Introduction

Structural control technology is proven to be effective in suppressing structural responses with the aid of various control devices and methods. Among these devices, inerter systems have been found to be very effective owing to their tuning frequency, mass enhancement, and damping enhancement mechanisms [1–3]. The performance evaluation and benefits of inerter-based systems for the protection of building structures [4–8], storage tanks [9–11], wind turbine towers [12–14], semi-submersible platform [15], and for vibration suppression of cables [16,17], machine [18] and suspension systems [19,20] have been studied in recent literature. An inerter is a mechanical element with two terminals [21–25] and ideally produces a resistive force proportional to its inner relative acceleration and large apparent mass designated as inertance. An essential property of the inerter is that a large inertance can be produced by devices with negligible physical masses. In the past, Kawamata [26] developed a vibration control device that used fluid inertance, which is designated as a mass pump, to suppress the seismic responses of a structure; this device has an inertial mass enhancement mechanism. Arakaki et al. [27,28] utilized the rotation mechanism to amplify the effective damping force of a viscous damper, which is a type of velocity-dependent damping element (VDE). However, these devices did not explicitly use the mass enhancement effect until Ikago et al. [1] proposed the tuned viscous mass damper, which belongs to a type of velocity-dependent damping inerter system (VDIS). The performance of the tuned viscous mass damper control system was subsequently investigated via shaking table tests conducted on single-story systems equipped with scaled-down versions of the damper [29]. Garrido et al. [30] proposed a rotational inertia double-tuned mass damper by replacing the viscous damping of the tuned mass damper with a tuned viscous mass damper, which achieved significantly greater control than the tuned mass damper with similar additional mass ratio. Through the incorporation of an electromagnetic damper, which is a type of VDE, Nakamura et al. [31] developed an electromagnetic inerter mass damper with variable damping force. Asai et al. [32,33] achieved enhanced energy-harvesting performance using a tuned inerter. Zhang et al. [10,34] proposed an isolation inerter system that used an inerter and a VDE to mitigate the vibration of a storage tank. The effect of the mechanical layout of the system was also investigated and considered in the development of a demand-based optimal design method for the system. Ikago et al. [1] presented the closed-form optimal design formulae for a VDIS based on the fixed-point method (an optimal design method to minimize the H_∞ norm of the transfer function). On the basis of consideration of both the response mitigation effect and control cost, Pan et al. [35] proposed a demand-performance-based optimal design methodology for a structure equipped with a VDIS to achieve the desired seismic performance level from the primary structure. Pan and Zhang [36] subsequently derived a closed-form expression of the root-mean-square (RMS) response of a single-degree-of-freedom (SDOF) structure with three representative VDISs, namely, the series layout inerter system, series–parallel layout I inerter system (SPIS-I), and series–parallel layout II inerter system. Chen et al. [37] analyzed the effect of the soil-structure interaction on the vibration mitigation effect of a VDIS. Based on the determined effect, they proposed an optimal design method that utilized a simulated annealing algorithm.

As mentioned above, recent studies in the field have mainly focused on the development and optimal design of inerter systems that utilize VDEs for energy dissipation. As a classic example, the ball-screw inerter system utilizes fluid viscous damping, with the fluid fully filled into a small gap between the fixed inner cylinder and rotating outer cylinder connected to a ball screw [38]. However, the inertance of such a device cannot be adjusted or replaced once it is manufactured even if the mass enhancement ratio is found to be insufficient for vibration control afterwards. In addition, there is also a concern about the leakage risk of the damping fluid, with high pressure within the rotating cylindrical tube requiring thorough sealing of the device [39]. From the perspective of the constitutive relationship of the mechanical behavior, viscous damping and electromagnetic damping (mentioned earlier) are both VDE mechanisms, which means that the damping force is in-phase with the relative velocity of the VDE. Because the maximum displacement and maximum velocity never occur at the same time, there is an inevitable time difference between the maximum damping force and displacement [40,41]. Owing to this difference, the VDE would be incapable of mitigating the structural response greatly in the time domain if a peak response occurs in the early stage of excitation, which is generally true in most earthquake events. This sometimes unfavorable velocity dependence of a VDE can be replaced by a displacement-dependent damper that uses a displacement-dependent damping element (DDE) for energy dissipation, with the damping force directly depending on the relative deformation of the device [42]. Such systems are widely used because of their stable hysteretic behavior and relatively low production and maintenance costs. Kelly et al. [43] were the first to propose a displacement-dependent damper for vibration energy dissipation. Since then, considerable efforts [44–47] have been devoted to analyzing the dynamic behavior of structures controlled by displacement-dependent dampers and the design of these systems. There is no doubt that a displacement-dependent damper is effective for a vibration mitigation, theoretically as well as experimentally. In particular, this damper is preferentially chosen at the beginning of the excitation, based on the fact that displacement-dependent dampers reduce substantial displacements more effectively compared to viscous dampers with equivalent damping ratios (meaning that the energy dissipated per cycle for DDE and VDE [42] is the same). However, the energy dissipation effect of the damper is limited when the deformation in a DDE is small, especially in the early stage of shaking, in a seismic event. The main contribution of the tuned inerter in a displacement-dependent damping inerter system (DDIS) is to amplify the deformation in a DDE to ensure efficient energy dissipation during the early stage of excitation.

Based on these motivations, a novel DDIS is developed that overcomes the aforementioned drawbacks of the VDIS. The proposed DDIS consists of an inerter, a DDE, and a spring. Stochastic dynamic analysis of a structure equipped with the DDIS system is performed, considering a base acceleration modeled with a Kanai–Tajimi power spectral density function. The stochastic linearization technique is used to handle the non-linear terms. It is demonstrated that DDIS combines the advantageous properties of DDE and inerter in a single layout. The structural displacement, acceleration, and energy dissipation mitigation indices are determined from the stochastic analysis results and used to establish an evaluation method for the dynamic performance of the DDIS-equipped structure. A broad parametric analysis is performed within the evaluation framework, taking into consideration the effects of non-linearity of the DDIS and variability of its parameters. Finally, several design cases in the time domain are presented to illustrate the damping enhancement and quick-control provided by the proposed DDIS.

2. Theoretical Analysis of Displacement-Dependent Damping Inerter System (DDIS)

2.1. Mechanical Model

The inerter (in Figure 1) is a mechanical element that ideally provides a force proportional to its inertance m_{in} (having dimensions of mass) and relative acceleration between its two terminals. Experimental research works show that the inertance can be amplified thousands of times that of the physical mass of the inerter [48], proving that the inerter is a very effective vibration mitigation device owing to this mass enhancement effect. A series of physical realizations have been proposed for the construction of the inerter, such as the rack-pinion [21], the ball screw [1], the fluid [49], and the electromagnetic mechanisms [50]. In the ideal model of the inerter, the output force of the inerter F_{in} is given by [51–56]:

$$F_{in} = m_{in}\left(\ddot{u}_1 - \ddot{u}_2\right) \tag{1}$$

where $\ddot{u}_1$ and $\ddot{u}_2$ are the respective accelerations of the two terminals.

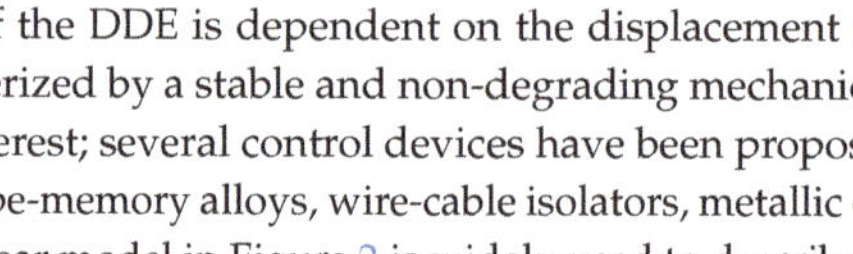

Figure 1. Schematic of the inerter.

2.1.1. Model of Displacement-Dependent Damping Element (DDE)

The damping force of the DDE is dependent on the displacement rather than on the velocity response, and it is characterized by a stable and non-degrading mechanical behavior. The DDE is of significant engineering interest; several control devices have been proposed, which exploit the DDE characteristics (such as shape-memory alloys, wire-cable isolators, metallic dampers, and friction-based dampers) [45,46]. The bilinear model in Figure 2 is widely used to describe the DDE and give it a clear and simple physical meaning. When this model is used to represent the force-deformation relationship of the DDE, the restoring force F_d is given by:

$$F_d(u_d, z_d) = \alpha k_d u_d + (1 - \alpha) k_d z_h \tag{2}$$

where k_d is the initial stiffness of the DDE, α is the ratio of the post-yielding stiffness to pre-yielding stiffness, u_{dy} is the elastic limit displacement or yielding displacement of the corresponding bilinear model, and u_d and z_h are the deformation of the DDE and the hysteretic variable, respectively, which are related as follows:

$$\dot{z}_h = \dot{u}_d\left[\left(1 - \mathrm{sgn}\left(\dot{u}_d\right)\mathrm{sgn}\left(z_h - u_{dy}\right) - \mathrm{sgn}\left(-\dot{u}_d\right)\mathrm{sgn}\left(-z_h - u_{dy}\right)\right)\right] \tag{3}$$

where $\mathrm{sgn}(\cdot)$ is the signum function.

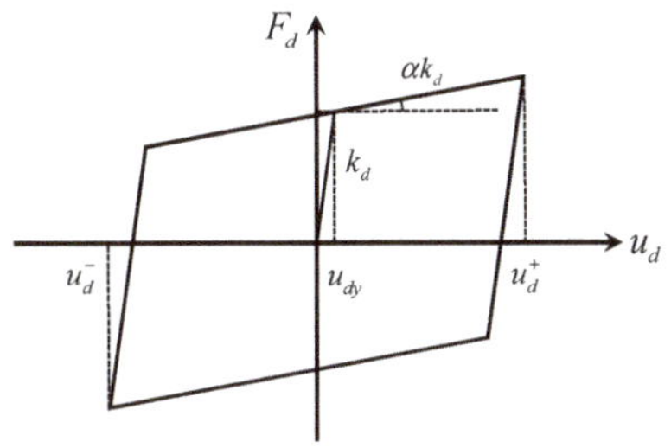

Figure 2. Hysteresis curve of the displacement-dependent damping element (DDE).

2.1.2. Model of DDIS

As depicted in Figure 3, the proposed DDIS consists of an inerter, a DDE, and a spring, where k_t denotes the stiffness coefficient of the spring. In the DDIS, the inerter and DDE are mounted in parallel and are deployed in series with the spring. The supplemental vibratory system consisting of the spring and inerter is tuned to the primary structural frequency, resulting in enhanced energy absorption. Unlike the conventional VDIS in which the viscous damping element is used only for energy dissipation, the DDE of the DDIS provides both energy dissipation and additional stiffness, which further reduces the structural response. The mass enhancement effect [1] of the inerter ensures efficient energy absorption in the DDE without implying an additional weight. And the implements of the spring and inerter constitute the damping enhancement mechanism to dissipate more energy since the DDE deformation of DDIS is amplified and larger than the displacement of controlled structure (Figure 3). The detailed explanation of the intrinsic benefit of the DDIS is given in Sections 3.2 and 4.1.

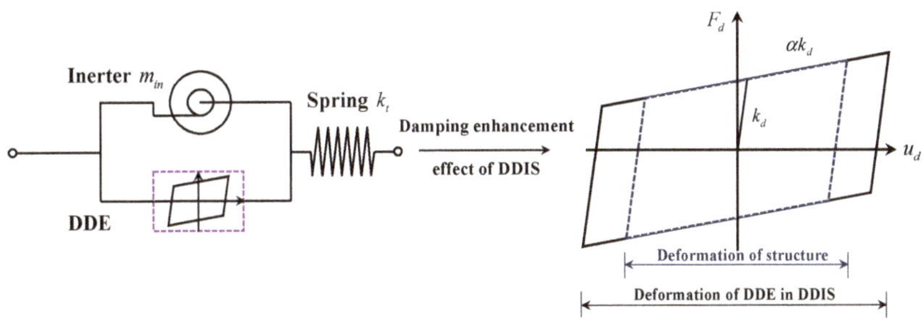

Figure 3. Schematic of the proposed displacement-dependent damping inerter system (DDIS) and its damping-enhancement effect.

2.2. Equations of Motion

As shown in Figure 4, a DDIS is incorporated into an SDOF frame structure having mass m, stiffness k, and damping coefficient c. The governing equation for the present system can be expressed as follows:

$$\begin{cases} m\ddot{u}(t) + c\dot{u}(t) + ku(t) + k_t\left[u(t) - u_d(t)\right] = -m\ddot{u}_g(t) \\ m_{in}\ddot{u}_d(t) + F_d(t) = k_t\left[u(t) - u_d(t)\right] \end{cases} \tag{4}$$

where $\ddot{u}_g(t)$ represents the acceleration of the ground motion, and $u(t)$ and $u_d(t)$ are respectively the displacement of the primary structure relative to the ground and deformation of the DDE. The force F_d is given by Equation (2).

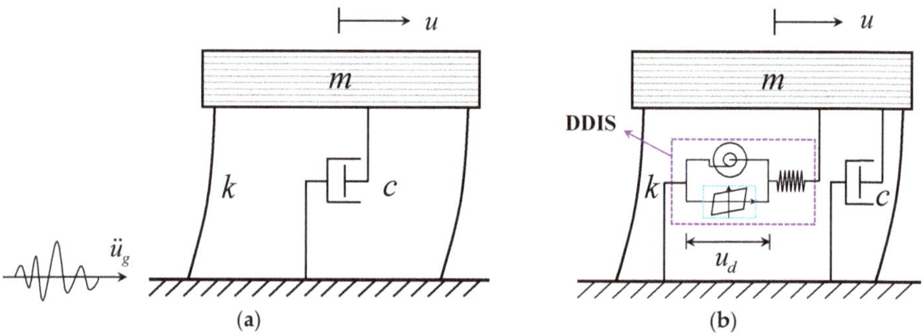

Figure 4. Mechanical model of the single-degree-of-freedom (SDOF) frame structure (**a**) without and (**b**) with the DDIS.

2.3. Stochastic Dynamic Analysis and Stochastic Linearization Method

Considering the uncertain nature of the seismic excitation, the acceleration of the applied ground motion $\ddot{u}_g$ in Equation (4) is modeled as a random process. The Kanai–Tajimi filtered white-noise process is adopted as excitation, as it has been widely employed in previous studies [35,57]. Denoting the bedrock white noise process as $W(t)$, the differential equation of the Kanai–Tajimi model can be established in time domain as follows:

$$\ddot{u}_g(t) = -\omega_g^2 u_g(t) - 2\xi_g\omega_g\dot{u}_g(t) + W(t) \tag{5}$$

where $\omega_g(t)$ and $\xi_g(t)$ are, respectively, the fundamental circular frequency and damping ratio related to the soil characteristics. The power spectral density function corresponding to the Kanai–Tajimi model is as follows:

$$S_{\ddot{u}_g}(\omega) = \frac{\omega_g^4 + 4\xi_g^2\omega_g^2\omega^2}{\left(\omega_g^2 - \omega^2\right)^2 + 4\xi_g^2\omega_g^2\omega^2} S_w \tag{6}$$

where S_w is the power spectral density of the white noise excitation [58]. A relationship between S_w and the peak ground acceleration was introduced in [59].

The solution of the governing equations of motion in Equation (4) is a non-linear problem owing to the non-linear restoring force of the bilinear model. Linear random vibration theory is thus not applicable. In this section, the stochastic linearization method (SLM) [60] is applied to solve Equation (4) using an equivalent linear–viscous damping c_{eq} and a linear stiffness coefficient k_{eq} to represent the bilinear model. Equation (4) can thus be rewritten in terms of SLM as follows:

$$\begin{cases} \mathbf{M}\ddot{\mathbf{u}} + \mathbf{C}\dot{\mathbf{u}} + \mathbf{K}\mathbf{u} + \mathbf{r} = \mathbf{E}\mathbf{M}\ddot{u}_g(t) \\ \dot{z}_h(t) + c_{eq}\dot{u}_d(t) + k_{eq}z_h(t) = 0 \end{cases} \tag{7}$$

where,

$$\mathbf{M} = \begin{bmatrix} m & 0 \\ 0 & m_{in} \end{bmatrix}, \mathbf{C} = \begin{bmatrix} c & 0 \\ 0 & 0 \end{bmatrix}, \mathbf{K} = \begin{bmatrix} k + k_t & -k_t \\ -k_t & \alpha k_d + k_t \end{bmatrix}, \\ \mathbf{r} = \begin{bmatrix} 0 \\ (1-\alpha)k_d z_h \end{bmatrix}, \mathbf{E} = \begin{bmatrix} 1 \\ 0 \end{bmatrix} \tag{8}$$

and $\mathbf{u} = \{u, u_d\}^T$ is the displacement vector. Here, c_{eq} and k_{eq} are determined by an equivalence principle [61,62], minimizing the squared error between the linearized Equation (7) and non-linear Equation (4), which leads to:

$$c_{eq} = \mathrm{E}\left[\frac{\partial F_d(u_d, z_h)}{\partial \dot{u}_d}\right], \; k_{eq} = \mathrm{E}\left[\frac{\partial F_d(u_d, z_h)}{\partial z_h}\right] \tag{9}$$

where E[ ] denotes the expectation operator to obtain the mean value. The coefficients c_{eq} and k_{eq} of the bilinear model resulting from Equation (9) are given by [61,62]

$$\begin{aligned} c_{eq} &= \left\{ \begin{array}{l} -\frac{1}{2}\left[1 + erf\left(\frac{u_{dy}}{\sqrt{2}\sigma_{z_h}}\right)\right] \\ + \frac{1}{\sigma_{z_h}\sqrt{2\pi}} \int_{u_{dy}/(\sqrt{2}\sigma_{z_h})}^{\infty} erf\left(\frac{\rho_{\dot{u}_d z_h} z_h/(\sqrt{2}\sigma_{z_h})}{\sqrt{1-\rho_{\dot{u}_d z_h}}}\right) \cdot \exp\left[-\frac{z_h^2}{2\sigma_{z_h}^2}\right] dz_h \end{array} \right\} \\ k_{eq} &= \left\{ \begin{array}{l} \frac{\rho_{\dot{u}_d z_h}\sigma_{\dot{u}_d} u_{dy}}{\sqrt{2\pi}\sigma_{z_h}^2} \sigma_{z_h} \exp\left[-\frac{u_{dy}^2}{2\sigma_{z_h}^2}\right] \left\{1 + erf\left(\frac{\rho_{\dot{u}_d z_h} u_{dy}}{\sigma_{z_h}\sqrt{2\left(1-\rho_{\dot{u}_d z_h}^2\right)}}\right)\right\} \\ + \frac{\sqrt{2\left(1-\rho_{\dot{u}_d z_h}^2\right)}}{\pi} \frac{\sigma_{\dot{u}_d}}{\sigma_{z_h}} \exp\left[-\frac{u_{dy}^2}{2\sigma_{z_h}^2}\right] \exp\left[\frac{\rho_{\dot{u}_d z_h}^2 u_{dy}^2}{\sigma_{z_h}^2 2\left(1-\rho_{\dot{u}_d z_h}^2\right)}\right] \end{array} \right\} \end{aligned} \tag{10}$$

where σ_{z_h} and $\sigma_{\dot{u}_d}$ are respectively the RMS of hysteretic deformation z_h and the RMS of the DDE velocity $\dot{u}_d$; $\rho_{\dot{u}_d z_h}$ is the correlation ratio of $\dot{u}_d$ and z_h, and erf is the error function given by

$$erf(y) = \frac{2}{\sqrt{\pi}} \int_0^y e^{-s^2} ds \tag{11}$$

Using the SLM, the expected stochastic response can be approximated as a zero-mean Gaussian process under Gaussian excitation. Based on the Laplace transformation, the governing equation in Equation (7) can be rewritten as:

$$\left\{ \begin{array}{l} \left(-\omega^2\mathbf{M} + \mathrm{i}\omega\mathbf{C} + \mathbf{K}\right)\mathbf{U} + \mathbf{R} = \mathbf{EM}A_g(\mathrm{i}\omega) \\ Z_h(\mathrm{i}\omega)\mathrm{i}\omega + c_{eq}U_d(\mathrm{i}\omega)\mathrm{i}\omega + k_{eq}Z_h(\mathrm{i}\omega) = 0 \end{array} \right. \tag{12}$$

where $\mathbf{U} = \{U, U_d\}^T$ and $\mathbf{R} = \{0, (1-\alpha)k_d Z_h\}^T$ are the Laplace transformations of $\mathbf{u}$ and $\mathbf{r}$, respectively; U_d, A_g, and Z_h are the Laplace transformations of u_d, $\ddot{u}_g$, and z_h, respectively. An inspection of Equations (10) and (12) reveals that the determination of c_{eq} and k_{eq} is not straightforward, since they implicitly depend upon the unknowns σ_{z_h}, $\sigma_{\dot{u}_d}$, and $\rho_{\dot{u}_d z_h}$. Therefore, the values of c_{eq} and k_{eq} are determined by an interactive process, wherein Equation (12) is solved in an iterative manner given an initial guess.

2.4. Energy Balance Analysis

From the perspective of energy balance evaluation, a part of the excitation-induced input vibrational energy is dissipated by the primary structure, and the rest is dissipated by the DDIS via the DDE. Equation (7) can be used to calculate the energy balance in the SDOF structure equipped with the DDIS by premultiplying with $\mathbf{u}^T$ and integrating over the time domain. The energy balance equation in the time domain is given by:

$$e_{total}(t) = e_k(t) + e_{e,s}(t) + e_d(t) + e_{DDIS}(t) \tag{13}$$

where the total input energy $e_{total}(t)$ is composed of structural kinetic energy $e_k(t)$, structural elastic strain energy $e_{e,s}(t)$, structural inherent damping dissipated energy $e_d(t)$, and DDIS-dissipated energy $e_{HDIS}(t)$. With particular regard to the three elements of the DDIS, $e_{DDIS}(t)$ is the result of kinetic energy $e_{k,DDIS}(t)$, elastic strain energy $e_{e,DDIS}(t)$, and DDE-dissipated energy $e_{d,DDIS}(t)$. Because the input excitation is assumed to be a stochastic process, the energy response should be evaluated stochastically by applying the expectation operator to Equation (13). On the basis of the stationarity hypothesis, the expected values of $e_{e,DDIS}(t)$, $e_{k,DDIS}(t)$, $e_k(t)$, and $e_{e,s}(t)$ are all zero. Hence, the stochastically expected value of Equation (13) is given by:

$$\mathrm{E}[e_{total}(t)] = \mathrm{E}[e_d(t)] + \mathrm{E}\Big[e_{d,DDIS}(t)\Big] = c\sigma_{\dot{u}}^2 + (1-\alpha)k_d\sigma_{z_h\dot{u}_d} \tag{14}$$

where $\sigma_{\dot{u}}^2$ and $\sigma_{z_h\dot{u}_d}$ are respectively the variance of the structural velocity and the cross-variance of z_h and $\dot{u}_d$.

3. Characteristics of DDIS

3.1. Stochastic Performance Indices

To characterize the dynamic performance of the SDOF structure equipped with DDIS and further evaluate the vibration mitigation effect of the DDIS, different stochastic performance indices are introduced. First, the dimensionless parameters of the DDIS are defined, including the inertance–mass ratio μ, DDE stiffness ratio κ, and stiffness ratio λ:

$$\mu = \frac{m_{in}}{m}, \kappa = \frac{k_t}{k}, \lambda = \frac{k_d}{k} \tag{15}$$

The commonly used index is the displacement mitigation ratio γ_{Dis}, which is the ratio of the RMS displacement of the structure–DDIS system $\sigma_{Dis,SDOF-DDIS}$ to that of the original structure $\sigma_{Dis,SDOF}$. Another index is the acceleration mitigation ratio γ_{Acc}, which is the ratio of the RMS acceleration of the structure–DDIS system $\sigma_{Acc,SDOF-DDIS}$ to that of the original structure $\sigma_{Acc,SDOF}$. The parameters γ_{Dis} and γ_{Acc} are, therefore, defined as follows:

$$\begin{aligned} \gamma_{Dis}\Big(\mu,\kappa,\lambda,\alpha,u_{dy}\Big) &= \frac{\sigma_{Dis,SDOF-DDIS}}{\sigma_{Dis,SDOF}}, \\ \gamma_{Acc}\Big(\mu,\kappa,\lambda,\alpha,u_{dy}\Big) &= \frac{\sigma_{Acc,SDOF-DDIS}}{\sigma_{Acc,SDOF}} \end{aligned} \tag{16}$$

However, an energy-based performance index is more robust against various types of input ground motion than displacement- and acceleration-based performance indices. Referring to the energy balance in Equation (14), the excitation-induced input energy is eventually dissipated by the structural inherent damping and DDE of the DDIS. The portion of the energy dissipated by the DDE is maximized, whereas the portion filtered into the primary structure and dissipated by the structural inherent damping should be minimized. A filtered energy ratio γ_E is thus defined here to quantify the portion of energy dissipated by DDIS out of the total input energy:

$$\begin{aligned} \gamma_E\Big(\mu,\kappa,\lambda,\alpha,u_{dy}\Big) &= \frac{\mathrm{E}[e_{total}(t)] - \mathrm{E}\Big[e_{d,DDIS}(t)\Big]}{\mathrm{E}[e_{total}(t)]} \\ &= 1 - \frac{(1-\alpha)k_d\sigma_{z_h\dot{u}_d}}{c\sigma_{\dot{u}}^2 + (1-\alpha)k_d\sigma_{z_h\dot{u}_d}} \end{aligned} \tag{17}$$

A low value of γ_E indicates that only a small portion of the vibrational energy is filtered into the primary structure, indicating the energy dissipation efficiency of the DDIS.

3.2. Parametric Analysis of DDIS

A thorough parametric analysis is conducted using γ_{Dis}, γ_{Acc}, and γ_E as performance evaluation indices. The three indices are directly related to the design parameters of the DDIS, which contain μ, κ, λ, α, and u_{dy}.

3.2.1. Influence of Excitation Severity

The performance evaluation of the structure equipped with the DDIS is a non-linear problem because of the signum function, and it is definitely influenced by the severity of the external excitation. In this section, the power spectrum density is considered as the Kanai–Tajimi model; whereas, the magnitude of S_w varies in a wide range. A single-span, single-floor, SDOF frame structure is used for the parametric analysis, which is characterized by mass $m = 20$ ton, fundamental natural period $T = 0.54$ s, and inherent damping ratio $\zeta = 0.02$. Regarding the bilinear model of the DDIS, α and u_{dy} were assumed to be 0.02 and 0.001 m [44], respectively.

Figure 5 illustrates the variation pattern of the effect of γ_{Dis} and γ_{Acc} against the change of S_w, where μ of DDIS increases from 0 to 0.50. The increase of S_w affects the structural response and the extent to which the non-linear effects of DDIS are triggered, which leads to the increase of γ_{Dis} and γ_{Acc}. In terms of the special case, i.e., a stiff-supported DDE ($\mu = 0$), the correspondingly minimum γ_{Dis} and γ_{Acc} are quantified as 0.68 and 0.85, respectively. For any specific value of S_w, the proposed DDIS exhibits an improved vibration mitigation effect compared with the displacement-dependent damper (which is represented by the limit case of DDIS having $\mu = 0$). Although γ_{Dis} and γ_{Acc} of DDIS vary with the excitation severity, the advantage of DDIS over the displacement-dependent damper is definitely true for different levels of excitation severity. Furthermore, once the DDIS meets the target demand of vibration control under the excitation for the upper limit of concerned severity, it will be more effective for vibration mitigation when subjected to the excitation with lower severity (S_w > 0.02).

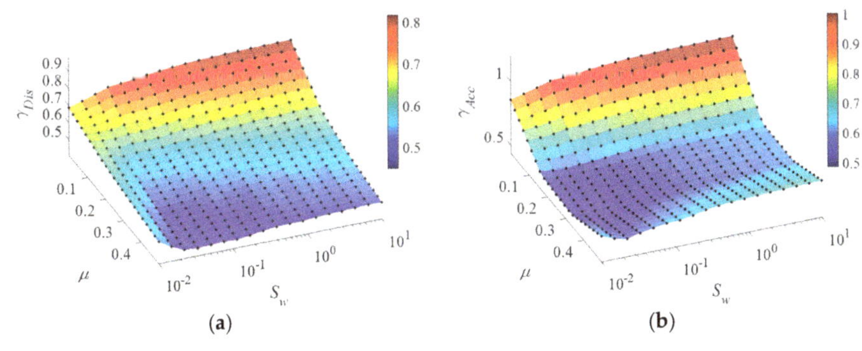

Figure 5. Influence analysis of γ_{Dis} and γ_{Acc} against the variation of S_w: (**a**) γ_{Dis} and (**b**) γ_{Acc}.

3.2.2. Influence of DDIS Parameters

In this section, several cases are considered to demonstrate the vibration mitigation effect of the DDIS, with κ varying in the range of 10^{-2}–1.00, and μ varying in the range of 10^{-3}–3.00. λ is set to 0.20, 1.00, and 1000.00 to simulate low, medium, and high stiffness of the support (i.e., the spring), respectively. A broad-band stochastic excitation is adopted, with ω_g = 15, and ξ_g = 0.60 [57,63] to simulate the commonly used firm soil condition. The analysis results of γ_{Dis},γ_{Acc}, and γ_E are summarized in Figures 6–8, respectively.

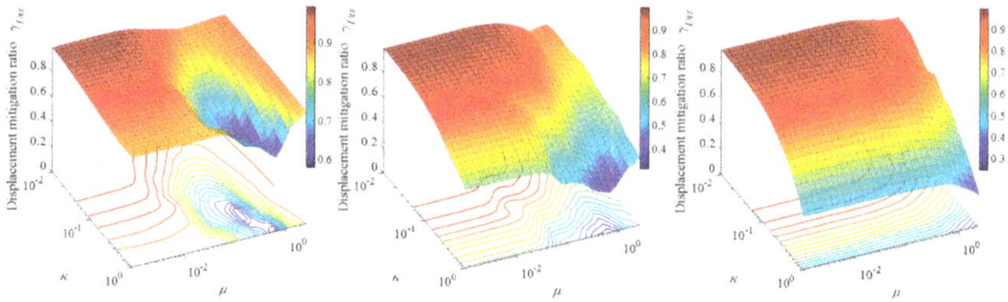

(**a**) (**b**) (**c**)

Figure 6. Variation of γ_{Dis} for: (**a**) $\lambda = 0.2$, (**b**) $\lambda = 1.00$, and (**c**) $\lambda = 1000$.

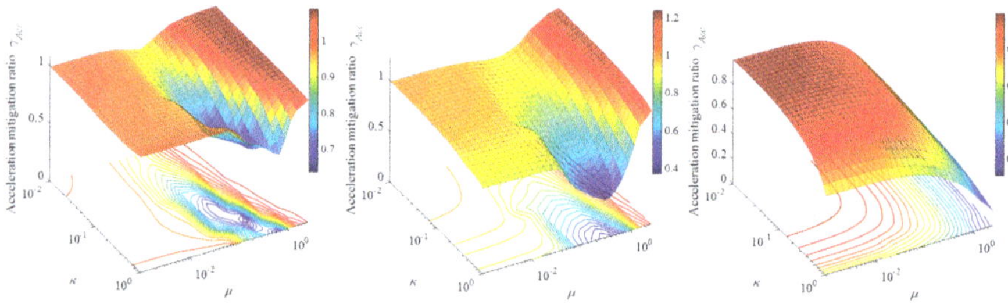

(**a**) (**b**) (**c**)

Figure 7. Variation of γ_{Acc} for: (**a**) $\lambda = 0.2$, (**b**) $\lambda = 1.00$, and (**c**) $\lambda = 1000$.

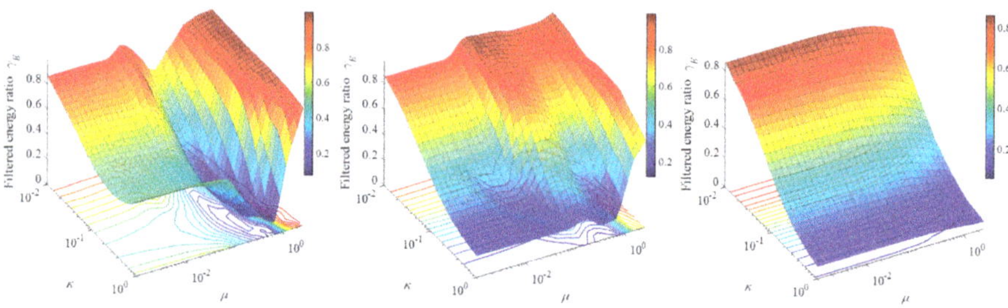

(**a**) (**b**) (**c**)

Figure 8. Variation of γ_E for: (**a**) $\lambda = 0.2$, (**b**) $\lambda = 1.00$, and (**c**) $\lambda = 1000$.

As expected, the displacement performance of the structure-DDIS system, γ_{Dis}, gradually decreases with the increase in hysteretic stiffness ratio κ, especially for medium and rigid stiffness of the support. Similar to a conventional displacement-dependent damper, an increase in κ implies greater stiffness of the primary structure to resist vibration, and hence, decreased structural displacement through increased non-linear damping. As can be observed from the surf plots in Figure 6, a decrease in μ is accompanied by an increase in γ_{Dis}. Considering the specific case in which μ decreases to zero, the DDIS degenerates into a conventional displacement-dependent damper with a stiff support. The proposed DDIS thus provides much greater displacement mitigation through the tuning effect of the spring and the energy storage effect of the inerter.

Regarding the acceleration mitigation ratio γ_{Acc}, Figure 7 shows that, for the DDIS with a tunable spring (Figure 7a,b), a medium inertance ($\mu \approx 0.50$) is beneficial to the mitigation of the acceleration response, while a small κ (< 0.05) and large inertance ($\mu > 1.0$) have a detrimental effect. This can be attributed to a weak DDE being incapable of timely dissipation of the large amount of energy stored by a large inerter, with the energy filtering into the primary structure. Figure 7a shows that when the DDE and inerter are set in parallel with a low-stiffness spring, the supplemental inerter substantially decreases the structural acceleration effectively. For the DDIS without a tunable spring (Figure 7c), the significant reduction of structural acceleration (small γ_{Acc}) is reached at a high price of large inertance ($\mu > 1.0$).

With regard to γ_E, Figure 8 shows that an increase in κ ($\kappa \in [0.01,\ 1.0]$) causes a decrease in γ_E, implying greater dissipation of vibrational energy by a large DDE ($\kappa > 0.50$), with less energy filtering into the primary structure. Regarding the effect of the added inerter, an increase in inertance decreases γ_E, especially when the DDIS springs are tunable stiffness ($\lambda < 1.0$). From the perspective of energy dissipation, the application of the inerter enhances the energy dissipation and vibration mitigation effect of the DDIS compared with only a stiff-supported DDE identical to that used in the system. For the DDIS without a tunable spring (Figure 8c), this benefit of the inerter is not evident. The damping enhancement effect of the proposed DDIS may be attributed to its combination mechanism in which the spring is used to tune the frequency of the inerter to improve energy absorption. This fundamental mechanism of the DDIS provides an alternative explanation of its advantage over a DDE identical to that utilized in the system.

An additional parametric analysis is conducted to investigate the effect of the variation of the mechanical parameters of the DDIS on its performance (evaluated with respect to κ and λ). The corresponding surfs plots are shown in Figures 9 and 10, which consider wide ranges of κ and λ values. The other mechanical parameters are the same as in the earlier parametric analysis. Non-zero values of μ: 0.10, 0.20, and 0.30, are assumed for the DDIS. As the inertance-mass ratio increases from 0.10 to 0.30, the displacement and acceleration mitigation effects of the DDIS substantially increase. The minimum displacement and acceleration responses coincide with the areas that have the upper boundary of κ and $\lambda \approx 1.0$. This is ascribed to the almost negligible deformation of the high-stiffness support (i.e., large λ), resulting in the vibration energy being rapidly and directly transferred to and dissipated by the DDE. It can be deduced from the surf plots that the DDIS is not sensitive to change in κ and λ, because the surf plots in a blue zone that indicates the low value of the displacement and acceleration responses change slowly when the parameters (κ and λ) vary rapidly. This indicates that a decrease in κ and λ does not considerably reduce the structural performance, including the acceleration and displacement responses. Furthermore, increasing the inertance makes the DDIS more effective for vibration control, in positive association with the increased energy storage provided by a larger inertance. Referring to the results of parametric analysis, it is suggested to design the DDIS with a medium inertance-mass ratio ($\mu \approx 0.5$), a large DDE $\kappa \geq 0.5$, and a tunable spring ($\lambda \leq 1.0$).

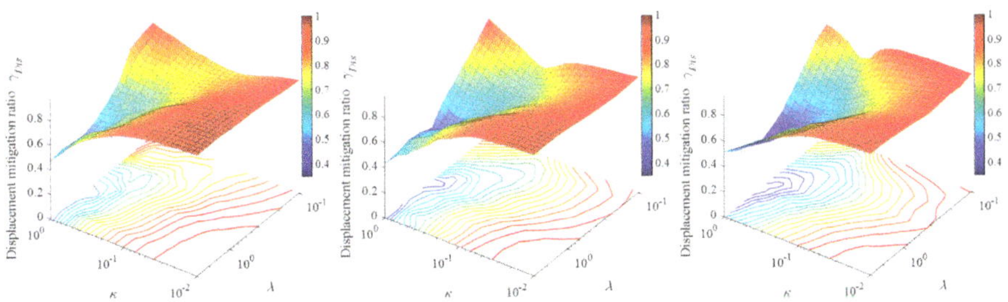

(a) (b) (c)

Figure 9. Variation of γ_{Dis} for (**a**) $\mu = 0.10$, (**b**) $\mu = 0.20$, and (**c**) $\mu = 0.30$.

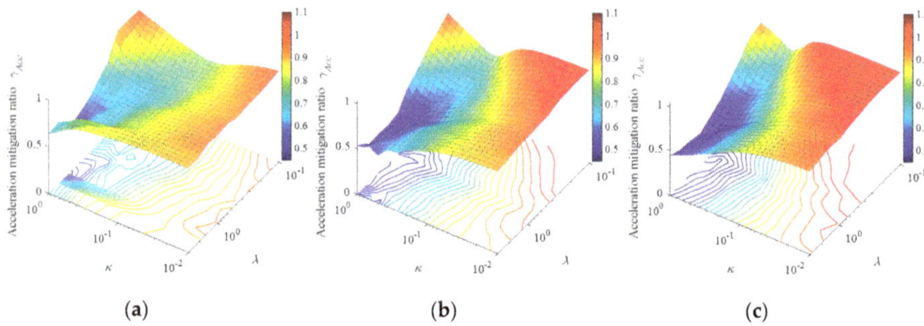

Figure 10. Variation of γ_{Acc} for (a) $\mu = 0.10$, (b) $\mu = 0.20$, and (c) $\mu = 0.30$.

3.2.3. Influence of Structural Parameters

Physical and model uncertainty of the primary structure may cause the practical behavior of the DDIS to deviate from theoretical expectations. The sensitivity of the vibration control effect of the DDIS to uncertainties of the primary structure is also worth analyzing. Therefore, the stiffness of the primary structure was considered to be variable owing to errors yielded by imprecise construction and deterioration during service.

Figure 11 illustrates the effect of the variation of k_{new}/k and ζ on the vibration control effect (i.e., γ_{Dis} and γ_{Acc}) of the DDIS, where k_{new} is the changed stiffness of the SDOF frame structure mentioned in Section 3.2.1. The k_{new} is set in the range of 50% variation of the stiffness k of SDOF frame structure. The increase of μ from 0.00 to 0.40 causes the displacement-dependent damper to evolve into the proposed DDIS with a larger inertance. In accordance with the earlier analysis, the structural responses γ_{Dis} and γ_{Acc} of the SDOF structure equipped with the DDIS are all lower than those equipped with the displacement-dependent damper, implying that the addition of the inerter improves the structural performance. This improvement is observed in structures with different stiffness and is not significantly affected by a rapid change of k_{new}. For instance, in Figures 11 and 12b–d ($\mu = 0.10$, 0.25, 0.40, respectively), the degradation of k_{new} and increase of ζ diminishes the acceleration mitigation effect of the DDIS to some extent; however, γ_{Dis} and γ_{Acc} still remain lower than those for the displacement-dependent damper depicted in Figures 11a and 12a.

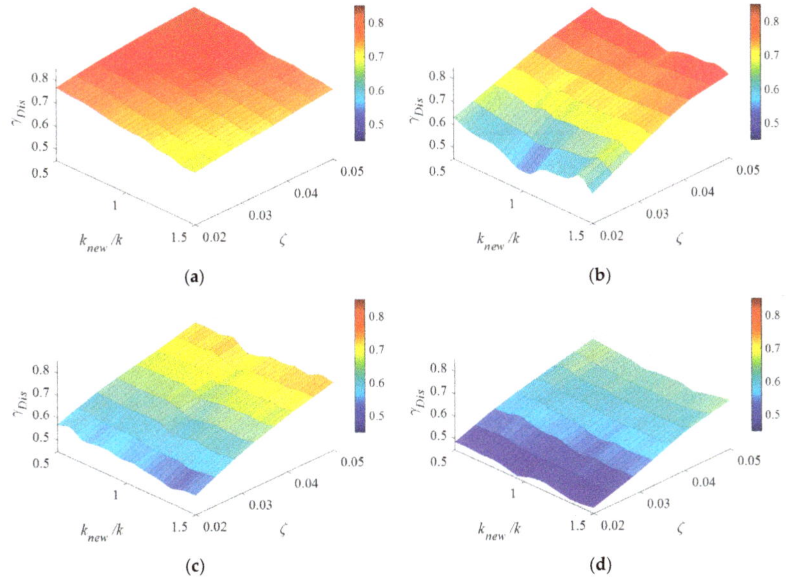

Figure 11. Sensitivity analysis of γ_{Dis} with variation of the structural circular frequency for $\kappa = 0.50$ and $\lambda = 1.00$ with (**a**)$\mu = 0$, (**b**) $\mu = 0.10$, (**c**) $\mu = 0.25$, and (**d**) $\mu = 0.40$.

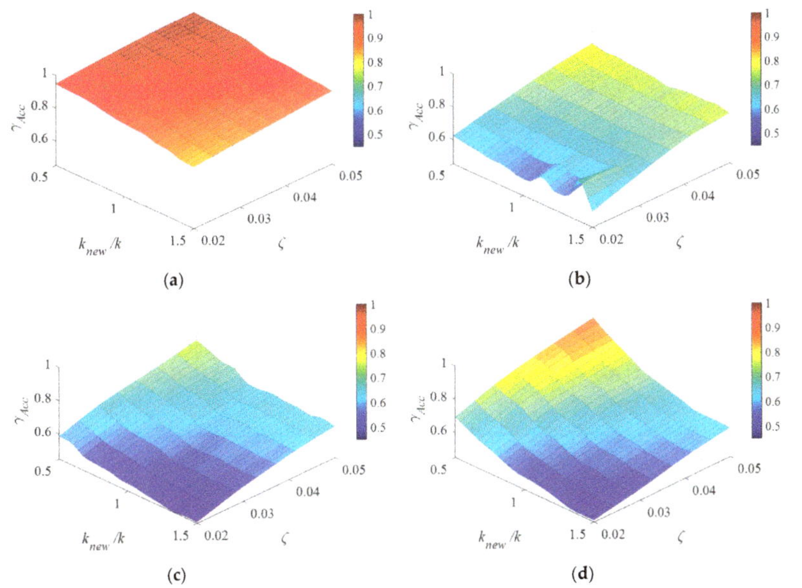

Figure 12. Sensitivity analysis of γ_{Acc} with variation of the structural circular frequency for $\kappa = 0.50$ and $\lambda = 1.00$ with (**a**) $\mu = 0$, (**b**) $\mu = 0.10$, (**c**) $\mu = 0.25$, and (**d**) $\mu = 0.40$.

4. Non-Linear Time-History Analysis and Performance of DDIS

4.1. Displacement Mitigation Effect in Early Stage of Seismic Response

As mentioned above, the classic inerter system utilizes a VDE for energy dissipation. The damping reaction force of the system is proportional to the relative velocity of the damping element; furthermore, there is an inevitable time difference between the times when maximum damping force and maximum displacement are generated. The DDE of the proposed DDIS avoids this problem, as the maximum displacement and maximum damping force occur simultaneously. When an earthquake occurs, the deformation of the DDE induced by the structural vibration directly causes the DDIS to produce a timely reaction force for effective vibration control. This results in significant reduction of the dynamic responses of the structure during the early stage of the seismic response.

To characterize the displacement mitigation effect of the DDIS, its results are compared with those of a conventional VDIS when both were applied to the single-span, single-floor frame structure described in Section 3.2. The viscous damping ratio is considered to be equal to the equivalent damping ratio of the DDE [42,47] to achieve the maximum hysteresis loop area of VDE under the dynamic excitation being equal to that of DDE. The stiffness of the support (λ) is set to infinity to ignore its tuning effect and only compare the VDIS and DDIS in terms of the structural displacement mitigation difference induced by the difference in the damping element. The design parameters of the DDIS μ and κ are both 0.50, a value that is chosen arbitrarily from the parametric analysis results. The characteristics of the DDIS and the benefits produced by the inerter also hold true for the DDIS with other parameter values. Regarding the hysteresis of the DDIS, α and u_{dy} were assumed to be 0.02 and 0.001, respectively. The main conclusions drawn from the test results, as discussed below, are true for other parameter combinations (μ, κ, and λ). The Kanai–Tajimi spectrum is adopted as the input power spectrum, with the predominant frequency of the dynamic excitation, ω_g, set to $0.5\omega_{s0}$, ω_{s0}, and $2.0\omega_{s0}$ (ω_{s0} denoting the circular frequency of the primary structure) to simulate low-frequency seismic (LS) excitation, the severest seismic (SS) excitation (resonance condition), and high-frequency seismic (HS) excitation, respectively, and artificial records are generated accordingly. The well-studied ground motion record of the 1940 Imperial Valley Earthquake recorded at El Centro (N–S component) is also used as a representative natural excitation. Under the four excitations, the equivalent damping ratio $\xi_{VDIS} = c_d/2\sqrt{mk}$ (c_d is the damping coefficient of the VDIS) is designed as 0.025, 0.032, 0.039, and 0.025. The displacement responses of the structures with the DDIS and VDIS under different excitations are shown in Figure 13. The figure shows that for all the excitations, the roof displacement is reduced more effectively by the DDIS compared with the VDIS despite the same equivalent damping ratio. The roof displacements of the structures with the DDIS and VDIS are substantially suppressed after t_{DDIS} and t_{VDIS}, respectively, while negligible reduction is achieved before t_{DDIS} and t_{VDIS}. Furthermore, t_{DDIS} is definitely lower than t_{VDIS}, demonstrating the timely control advantage of the DDIS over the VDIS in the early stage of the seismic response. As shown in the hysteretic curves of the damping element in DDIS and VDIS (Figure 14), at the beginning of the excitation, the energy dissipated by DDIS (dotted line: 1-2-3-4-5-6) is definitely larger than that in the case of VDIS (solid line: 1′-2′-3′-4′). The larger damping force produced by DDIS in the early stage is beneficial for the roof displacement reduction in a timely and more effective manner.

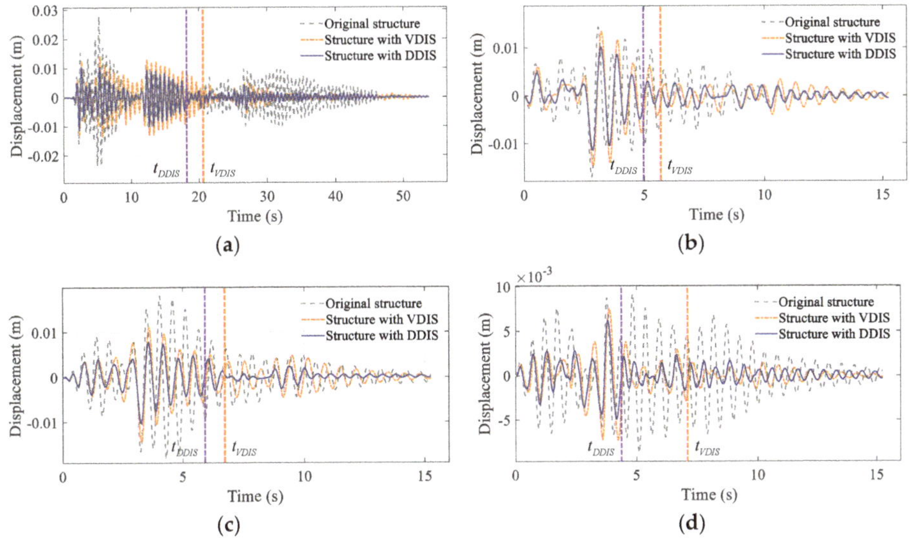

Figure 13. Displacement responses of the structures with the DDIS and velocity-dependent damping inerter system (VDIS) under (**a**) El Centro, (**b**) low-frequency seismic (LS) excitation, (**c**) severest seismic (SS) excitation, and (**d**) high-frequency seismic (HS) excitation.

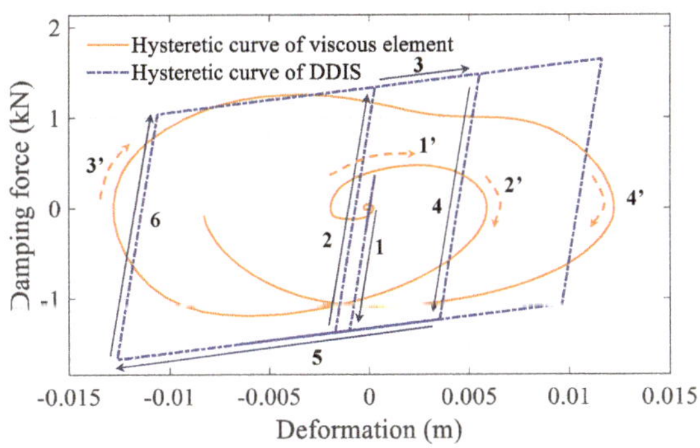

Figure 14. Hysteretic curve of DDIS and VDIS under El Centro in the early stage of the seismic response.

4.2. Seismic Performance of Structure Equipped with DDIS

The time-domain dynamic performance of structures equipped with the DDIS and displacement-dependent damper (as a special case of the DDIS without the inerter) are investigated by applying the DDIS and displacement-dependent damper to the SDOF structure described in Section 3.2.

The values of κ and λ of the damper and DDIS and the value of μ of the DDIS are all set to 0.50. A suite of ground motions, 10 LS waves, SS waves, and HS waves are generated using the Kanai–Tajimi spectrum mentioned in Section 4.1. In addition, El Centro 1940 N–S record is employed as a natural ground excitation. The peak ground acceleration of the input ground motion is scaled to a common value of 0.1 g. The average γ_{Dis}, γ_{Acc}, and γ_E determined by the analyses for the artificial records and El Centro ground motion are presented in Table 1. The results show that for different types of excitations, 10% to 15% of the vibrational energy, as determined by γ_E, is filtered through the DDIS to the primary structure. This indicates that the DDIS considerably reduces the amount of energy that the primary structure needs to dissipate. Furthermore, the displacement and acceleration of the DDIS-fitted structure are suppressed to a low level (reduction ratios ranging from 30% to 60% compared to the uncontrolled case).

Table 1. Analysis results of γ_{Dis}, γ_{Acc}, and γ_E for case in Section 4.2.

Type of Excitation	γ_{Dis} *	γ_{Acc}	γ_E
LS	0.47 (0.45)	0.49 (0.48)	0.11 (0.10)
SS	0.40 (0.37)	0.58 (0.54)	0.15 (0.15)
HS	0.39 (0.40)	0.70 (0.71)	0.12 (0.14)
El Centro	0.41	0.61	0.14

* Values reported in round brackets denote the corresponding standard deviation ratios computed from the stochastic analysis.

Considering the SS and El Centro waves as examples, the displacement response curves of the structures fitted with the displacement-dependent damper and DDIS are shown in Figure 15. The DDIS more effectively reduces the structural displacement compared with the displacement-dependent damper with a DDE identical to that used in the system. To explain the greater displacement reduction ability of the DDIS from the perspective of energy dissipation, Figure 16 shows the hysteresis curves of the DDEs of the displacement-dependent damper and DDIS. The values of the corresponding DDE deformation enhancement ratio, given by

$$\rho = \frac{\max(\text{DDE deformation})}{\max(\text{displacement of strucuture with DDIS})},$$

are indicated in the figure. In agreement with the findings in Section 3.2.2, the DDE deformation of the DDIS is over 60% larger than the structural displacement. The amplified deformation is achieved by the damping enhancement mechanism of the DDIS, in the absence of which the deformation would be equal to the structural displacement, as for the structure with the conventional displacement-dependent damper. The mechanism of the damping enhancement achieved here is similar to that of a VDIS [1].

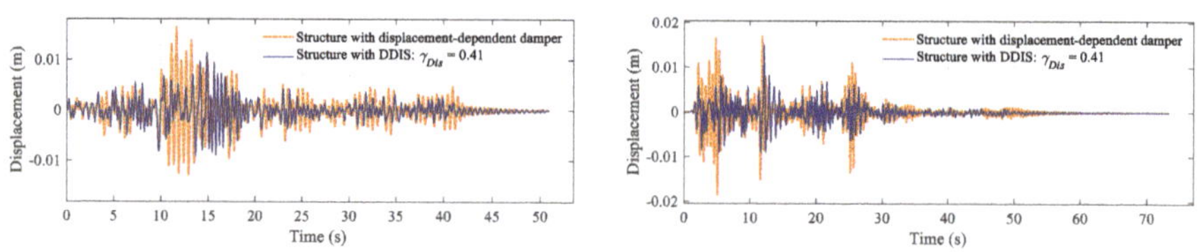

(a) (b)

Figure 15. Displacement responses of structures fitted with the DDIS and displacement-dependent damper under (**a**) El Centro and (**b**) SS excitation.

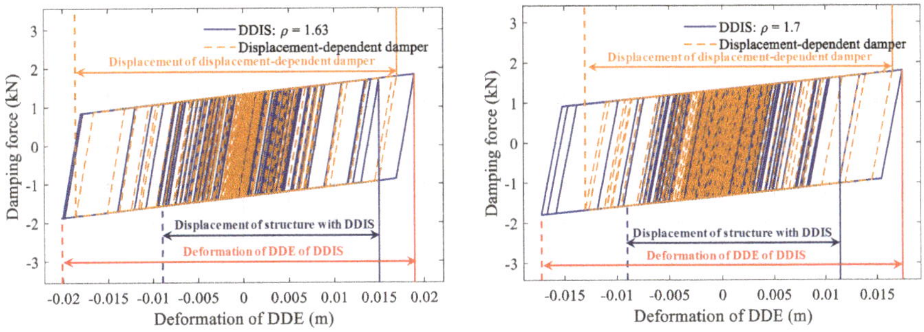

(**a**) (**b**)

Figure 16. Hysteresis curves of the DDEs of the DDIS and displacement-dependent damper under (**a**) El Centro and (**b**) SS excitation.

5. Conclusions

The response deformation of a structure is relatively small during the early stage of the seismic response, which leads to poor performance of velocity-dependent damping compared with the displacement-dependent damping with the same equivalent damper ratio [40,42]. In this study, the ability of the displacement-dependent damper to generate a larger control force during the early stage of the excitation is exploited and further enhanced by a supplemental inerter-spring system tuned to the primary structure, realizing an effective DDIS. A summary of the study and the main conclusions drawn from this investigation are reported below:

1. The DDIS is observed to suppress the structural responses in a timely manner as soon as a peak response occurs during the early stage of the excitation. The proposed equivalent linearization method is effective to conduct the stochastic dynamic analysis of the DDIS-equipped structure. The dynamic response of the DDIS controlled systems are further evaluated in the time domain considering the non-linearity and validating the accuracy of the stochastic analysis.
2. The interaction between the inerter, spring and the DDE constitutes the damping enhancement mechanism of the DDIS. Compared with an identical DDE, the proposed mechanism amplifies the deformation of the DDE in the DDIS by over 60%; thus, the DDIS is characterized by a higher energy dissipation capability.
3. The proposed DDIS considerably reduces the structural displacement and acceleration, which is a result of its damping and mass-enhancement mechanism. For the DDIS with medium value of stiffness ratio and inertance-mass ratio, the displacement and acceleration responses of original uncontrolled structure are reduced by 60% and 40%, respectively.
4. The benefits and characteristics of the proposed DDIS are obtained from an extensive parametric analysis and not limited to any specific optimal design procedures. Further studies should be conducted on the parameter design methodology and practical design procedure of the DDIS. The equivalent linearization method proposed in this paper would help reduce the computational cost required to develop a design method.

Author Contributions: Conceptualization, R.Z.; methodology, R.Z. and Z.Z.; software, Z.Z. and Y.J.; validation, Z.Z., R.Z., D.D.D. and Y.J.; formal analysis, Z.Z. and Y.J.; investigation, Z.Z., Y.J. and C.P.; data curation, Z.Z.; writing—original draft preparation, Z.Z., R.Z., Y.J., D.D.D. and C.P.; writing—review and editing, Z.Z., R.Z., C.P. and D.D.D. All authors have read and agreed to the published version of the manuscript.

Funding: This study was supported by the National Natural Science Foundation of China under Grant (grant no. 51978525 and 51778489), the Natural Science Foundation of Shandong Province (No. ZR2018BEE033); and the Fundamental Research Funds for the Central Universities (grant no. 22120180064).

Acknowledgments: The authors wish to acknowledge Kohju Ikago of Tohoku University for his help in interpreting the significance of the results of this study.

Conflicts of Interest: The authors declare no conflict of interest.

References

1. Ikago, K.; Saito, K.; Inoue, N. Seismic control of single-degree-of-freedom structure using tuned viscous mass damper. *Earthq. Eng. Struct. Dyn.* **2012**, *41*, 453–474. [CrossRef]
2. Zhao, Z.P.; Zhang, R.F.; Lu, Z. A particle inerter system for structural seismic response mitigation. *J. Frankl. Inst.* **2019**, *256*, 7669–7688. [CrossRef]
3. Zhao, Z.P.; Chen, Q.J.; Zhang, R.F.; Pan, C.; Jiang, Y.Y. Optimal design of an inerter isolation system considering the soil condition. *Eng. Struct.* **2019**, *196*, 109324. [CrossRef]
4. Brzeski, P.; Kapitaniak, T.; Perlikowski, P. Novel type of tuned mass damper with inerter which enables changes of inertance. *J. Sound. Vib.* **2015**, *349*, 56–66. [CrossRef]
5. Málaga-Chuquitaype, C.; Menendez-Vicente, C.; Thiers-Moggia, R. Experimental and numerical assessment of the seismic response of steel structures with clutched inerters. *Soil Dyn. Earthq. Eng.* **2019**, *121*, 200–211. [CrossRef]
6. Masri, S.F.; Caffrey, J.P. Transient response of a sdof system with an inerter to nonstationary stochastic excitation. *J. Appl. Mech.-T Asme* **2017**, *84*, 041005. [CrossRef]
7. Taflanidis, A.A.; Giaralis, A.; Patsialis, D. Multi-objective optimal design of inerter-based vibration absorbers for earthquake protection of multi-storey building structures. *J. Frankl. Inst.* **2019**, *356*, 7754–7784. [CrossRef]
8. Petrini, F.; Giaralis, A.; Wang, Z. Optimal tuned mass-damper-inerter (TMDI) design in wind-excited tall buildings for occupants' comfort serviceability performance and energy harvesting. *Eng. Struct.* **2019**, 109904. [CrossRef]
9. Luo, H.; Zhang, R.F.; Weng, D.G. Mitigation of liquid sloshing in storage tanks by using a hybrid control method. *Soil Dyn. Earthq. Eng.* **2016**, *90*, 183–195. [CrossRef]
10. Zhang, R.F.; Zhao, Z.P.; Pan, C. Influence of mechanical layout of inerter systems on seismic mitigation of storage tanks. *Soil Dyn. Earthq. Eng.* **2018**, *114*, 639–649. [CrossRef]
11. Jiang, Y.Y.; Zhao, Z.P.; Zhang, R.F.; De Domenico, D.; Pan, C. Optimal design based on analytical solution for storage tank with inerter isolation system. *Soil Dyn. Earthq. Eng.* **2019**. [CrossRef]
12. Hu, Y.; Wang, J.; Chen, M.Z.Q.; Li, Z.; Sun, Y. Load mitigation for a barge-type floating offshore wind turbine via inerter-based passive structural control. *Eng. Struct.* **2018**, *177*, 198–209. [CrossRef]
13. Zhang, R.F.; Zhao, Z.P.; Dai, K. Seismic response mitigation of a wind turbine tower using a tuned parallel inerter mass system. *Eng. Struct.* **2019**, *180*, 29–39. [CrossRef]
14. Sarkar, S.; Fitzgerald, B. Vibration control of spar-type floating offshore wind turbine towers using a tuned mass-damper-inerter. *Struct. Control Health Monit.* **2020**, *27*, e2471. [CrossRef]
15. Ma, R.; Bi, K.; Hao, H. Mitigation of heave response of semi-submersible platform (SSP) using tuned heave plate inerter (THPI). *Eng. Struct.* **2018**, *177*, 357–373. [CrossRef]
16. Lazar, I.F.; Neild, S.A.; Wagg, D.J. Vibration suppression of cables using tuned inerter dampers. *Eng. Struct.* **2016**, *122*, 62–71. [CrossRef]
17. Luo, J.N.; Jiang, J.Z.; Macdonald, J.H.G. Cable vibration suppression with inerter-based absorbers. *J. Eng. Mech.* **2019**, *145*, 15. [CrossRef]
18. Wang, F.C.; Lee, C.H.; Zheng, R.Q. Benefits of the inerter in vibration suppression of a milling machine. *J. Frankl. Inst.* **2019**, *356*, 766. [CrossRef]
19. Scheibe, F.; Smith, M.C. Analytical solutions for optimal ride comfort and tyre grip for passive vehicle suspensions. *Vehicle. Syst. Dyn.* **2009**, *47*, 1229–1252. [CrossRef]
20. Shen, Y.; Chen, L.; Yang, X.; Shi, D.; Yang, J. Improved design of dynamic vibration absorber by using the inerter and its application in vehicle suspension. *J. Sound. Vib.* **2016**, *361*, 148–158. [CrossRef]
21. Smith, M.C. Synthesis of mechanical networks: The inerter. *IEEE T Automat. Contr.* **2002**, *47*, 1648–1662. [CrossRef]
22. Chen, M.Z.Q.; Hu, Y.; Huang, L.; Chen, G. Influence of inerter on natural frequencies of vibration systems. *J. Sound. Vib.* **2014**, *333*, 1874–1887. [CrossRef]

23. Krenk, S. Resonant inerter based vibration absorbers on flexible structures. *J. Frankl. Inst.* **2019**, *356*, 7704–7730. [CrossRef]
24. Chen, Q.J.; Zhao, Z.P.; Xia, Y.Y.; Pan, C.; Luo, H.; Zhang, R.F. Comfort-based floor design employing tuned inerter mass system. *J. Sound. Vib.* **2019**, *458*, 143–157. [CrossRef]
25. Zhao, Z.P.; Zhang, R.F.; Jiang, Y.Y.; Pan, C. A tuned liquid inerter system for vibration control. *Int. J. Mech. Sci.* **2019**, *164*, 105171. [CrossRef]
26. Kawamata, S. *Development of a Vibration Control System of Structures by Means of Mass Pumps*; Institute of Industrial Science, University of Tokyo: Tokyo, Japan, 1973.
27. Arakaki, T.; Kuroda, H.; Arima, F.; Inoue, Y.; Baba, K. Development of seismic devices applied to ball screw: Part 1 basic performance test of rd-series. *J. Archit. Build. Sci.* **1999**, *5*, 239–244.
28. Arakaki, T.; Kuroda, H.; Arima, F.; Inoue, Y.; Baba, K. Development of seismic devices applied to ball screw: Part 2 performance test and evaluation of rd-series. *J. Archit. Build. Sci.* **1999**, *9*, 265–270.
29. Saito, K.; Inoue, N. A study on optimum response control of passive control systems using viscous damper with inertial mass: Substituting equivalent nonlinear viscous elements for linear viscous elements in optimum control systems. *J. Techn. Des.* **2007**, *13*, 457–462.
30. Garrido, H.; Curadelli, O.; Ambrosini, D. Improvement of tuned mass damper by using rotational inertia through tuned viscous mass damper. *Eng. Struct.* **2013**, *56*, 2149–2153. [CrossRef]
31. Nakamura, Y.; Fukukita, A.; Tamura, K.; Yamazaki, I.; Matsuoka, T.; Hiramoto, K.; Sunakoda, K. Seismic response control using electromagnetic inertial mass dampers. *Earthq. Eng. Struct. Dyn.* **2014**, *43*, 507–527. [CrossRef]
32. Asai, T.; Araki, Y.; Ikago, K. Energy harvesting potential of tuned inertial mass electromagnetic transducers. *Mech. Syst. Signal. Pr.* **2017**, *84*, 659–672. [CrossRef]
33. Asai, T.; Araki, Y.; Ikago, K. Structural control with tuned inertial mass electromagnetic transducers. *Struct Control Health Monit.* **2018**, *25*, e2059. [CrossRef]
34. Zhang, R.F. *Seismic Response Analysis of Base-Isolated Vertical Tank*; Tongji University: Shanghai, China, 2014.
35. Pan, C.; Zhang, R.F.; Luo, H.; Li, C.; Shen, H. Demand-based optimal design of oscillator with parallel-layout viscous inerter damper. *Struct. Control Health Monit.* **2018**, *25*, e2051. [CrossRef]
36. Pan, C.; Zhang, R.F. Design of structure with inerter system based on stochastic response mitigation ratio. *Struct. Control Health Monit.* **2018**, *25*, e2169. [CrossRef]
37. Chen, Q.J.; Zhao, Z.P.; Zhang, R.F.; Pan, C. Impact of soil–structure interaction on structures with inerter system. *J. Sound. Vib.* **2018**, *433*, 1–15. [CrossRef]
38. Kida, H.; Watanabe, Y.; Nakaminami, S.; Tanaka, H.; Sugimura, Y.; Saito, K.; Ikago, K.; Inoue, N. Full-scale dynamic tests of tuned viscous mass damper with force restriction mechanism and its analytical verification. *J. Struct. Constr. Eng.* **2011**, *76*, 1271–1280. [CrossRef]
39. Chen, Y. Leaking analysis of fluid viscous damper for engineering structure. *Steel Constr.* **2008**, *23*, 53–58.
40. Pong, W.S.; Tsai, C.S.; Lee, G.C. *Seismic Study of Building Frames with Added Energy-Absorbing Devices*; State University of New York at Buffalo: Buffalo, NY, USA, 1994; pp. 50–66.
41. Kelly, J.M.; Skinner, M.S. *A Review of Current Uses of Energy-Absorbing Devices*; Earthquake Engineering Research Center, University of California at Berkeley: Berkeley, CA, USA, 1979.
42. Makris, N. Rigidity–plasticity–viscosity: Can electrorheological dampers protect base-isolated structures from near-source ground motions? *Earthq. Eng. Struct. Dyn.* **1997**, *26*, 571–591. [CrossRef]
43. Kelly, J.M.; Skinner, R.I.; Heine, A.J. Mechanics of energy absorption in special devices for use in earthquake resistant structures. *Bull. New Zealand Soc. Earthq. Eng.* **1972**, *5*, 63–88.
44. Moreschi, L.M.; Singh, M.P. Design of yielding metallic and friction dampers for optimal seismic performance. *Earthq. Eng. Struct. Dyn.* **2010**, *32*, 1291–1311. [CrossRef]
45. Sauter, D.; Hagedorn, P. On the hysteresis of wire cables in stockbridge dampers. *Int. J. Non-linear Mech.* **2002**, *37*, 1453–1459. [CrossRef]
46. Carpineto, N.; Lacarbonara, W.; Vestroni, F. Hysteretic tuned mass dampers for structural vibration mitigation. *J. Sound. Vib.* **2014**, *333*, 1302–1318. [CrossRef]
47. Yamamoto, M.; Sone, T. Damping systems that are effective over a wide range of displacement amplitudes using metallic yielding component and viscoelastic damper in series. *Earthq. Eng. Struct. Dyn.* **2015**, *43*, 2097–2114. [CrossRef]

48. Saito, K.; Sugimura, Y.; Inoue, N. A study on response control of a structure using viscous damper with inertial mass. *J. Struct. Eng.* **2008**, *54*, 635–648.
49. De Domenico, D.; Deastra, P.; Ricciardi, G.; Sims, N.D.; Wagg, D.J. Novel fluid inerter based tuned mass dampers for optimised structural control of base-isolated buildings. *J. Frankl. Inst.* **2019**, *356*, 7626–7649. [CrossRef]
50. Gonzalez-Buelga, A.; Clare, L.R.; Neild, S.A.; Jiang, J.Z.; Inman, D.J. An electromagnetic inerter-based vibration suppression device. *Smart Mater. Struct.* **2015**, *24*, 055015. [CrossRef]
51. Takewaki, I.; Murakami, S.; Yoshitomi, S.; Tsuji, M. Fundamental mechanism of earthquake response reduction in building structures with inertial dampers. *Struct. Control Health Monit.* **2012**, *19*, 590–608. [CrossRef]
52. Javidialesaadi, A.; Wierschem, N.E. Optimal design of rotational inertial double tuned mass dampers under random excitation. *Eng. Struct.* **2018**, *165*, 412–421. [CrossRef]
53. Basili, M.; De Angelis, M.; Pietrosanti, D. Defective two adjacent single degree of freedom systems linked by spring-dashpot-inerter for vibration control. *Eng. Struct.* **2019**, *188*, 480–492. [CrossRef]
54. Zhao, Z.P.; Zhang, R.F.; Jiang, Y.Y.; Pan, C. Seismic response mitigation of structures with a friction pendulum inerter system. *Eng. Struct.* **2019**, *193*, 110–120. [CrossRef]
55. Faraj, R.; Jankowski, L.; Graczykowski, C.; Holnicki-Szulc, J. Can the inerter be a successful shock-absorber? The case of a ball-screw inerter with a variable thread lead. *J. Frankl. Inst.* **2019**, *356*, 7855–7872. [CrossRef]
56. Kras, A.; Gardonio, P. Active vibration control unit with a flywheel inertial actuator. *J. Sound Vib.* **2020**, *464*, 114987. [CrossRef]
57. De Domenico, D.; Ricciardi, G. Optimal design and seismic performance of tuned mass damper inerter (tmdi) for structures with nonlinear base isolation systems. *Earthq. Eng. Struct. Dyn.* **2018**, *47*, 2539–2560.
58. De Domenico, D.; Impollonia, N.; Ricciardi, G. Soil-dependent optimum design of a new passive vibration control system combining seismic base isolation with tuned inerter damper. *Soil Dyn. Earthq. Eng.* **2018**, *105*, 37–53. [CrossRef]
59. Buchholdt, H. *Structural Dynamics for Engineers*; Thomas Teldfort: London, UK, 1997.
60. Noori, M.N.; Saffar, A.; Davoodi, H. A comparison between non-gaussian closure and statistical linearization techniques for random vibration of a nonlinear oscillator. *Comput. Struct.* **1987**, *26*, 925–931. [CrossRef]
61. Wen, Y.K. Equivalent linearization for hysteretic systems under random excitation. *J. Appl. Mech.* **1980**, *47*, 150–154. [CrossRef]
62. Roberts, J.B.; Spanos, P.D. *Random Vibration and Statistical Linearization*; Wiley: New York, NY, USA, 1990.
63. Castaldo, P.; Ripani, M.; Priore, R.L. Influence of soil conditions on the optimal sliding friction coefficient for isolated bridges. *Soil Dyn. Earthq. Eng.* **2018**, *111*, 131–148. [CrossRef]

applied sciences

MDPI

Article

Experimental Investigation on the Mechanical Properties of Curved Metallic Plate Dampers

Jie Zheng [1], Chunwei Zhang [2,*] and Aiqun Li [3,4,5]

1 School of Science, Qingdao University of Technology, Qingdao 266033, China; zhengjie@qut.edu.cn
2 School of Civil Engineering, Qingdao University of Technology, Qingdao 266033, China
3 Beijing Advanced Innovation Center for Future Urban Design, Beijing 100044, China; liaiqun@bucea.edu.cn
4 School of Civil and Transportation Engineering, Beijing University of Civil Engineering and Architecture, Beijing 100044, China
5 School of Civil Engineering, Southeast University, Nanjing 210096, China
* Correspondence: zhangchunwei@qut.edu.cn; Tel.: +86-(532)-8507-1693

Received: 11 November 2019; Accepted: 24 December 2019; Published: 30 December 2019

Abstract: This study proposes a novel curved steel plate damper to improve the seismic performance of structures. The theoretical analysis of the curved plate damper was carried out deriving formulas of key parameters of the curved plate damper including elastic lateral stiffness, yield strength, and yield displacement. Moreover, a cyclic loading test of four sets of specimens was conducted, and the hysteretic performance, ductility, energy dissipation performance, and strain of the specimens were studied. The results showed that the initial stiffness of the damper was large, no obvious damage was observed, and the hysteresis loop was full. The tested dampers had good deformation and energy dissipation performance. The stress variable rule of the damper was obtained by stress analysis, where the plastic deformation at the end of the semi-circular arc was large. The formula for various parameters of the damper was compared with experimental and numerical results; thus, the value of the adjustment coefficient was determined reasonable. Meanwhile, the rationality of the finite element model was also verified.

Keywords: metal damper; performance parameter; cyclic loading test; hysteretic behavior; energy dissipation capability

1. Introduction

Earthquakes are sudden natural disasters endangering people's lives and property. The may not only cause housing damage, traffic interruption, water disaster, fire, disease, and other secondary disasters, but can also endanger human life and safety.

To minimize casualties and economic losses, preventing structural collapse and serious damage to the civil infrastructure against seismic hazard has become a challenging task that researchers need to address. Improving the seismic capacity of engineering structures through novel techniques and technical measures is the most effective way to mitigate seismic hazards. The traditional seismic method takes the "anti-seismic" approach as an important way to resist earthquakes by enlarging the section of structures and adding more reinforcements. The result is that the larger the section of structural components is, the greater the stiffness becomes, and the greater the impact of earthquakes. In this vicious circle, it is not only difficult to ensure safety, but construction costs required for earthquake resistance also increase greatly.

An effective way to overcome increased inertia properties is structural vibration control through energy dissipation devices. Energy dissipation devices absorb and dissipate seismic energy, thereby reducing the dynamic response of the main structure under an earthquake. Based on numerous practical applications of energy dissipation devices, the vibration control technique can be used to mitigate seismic hazard and achieve better seismic performance of the structure under vibrations.

The passive energy dissipation method has been widely used in the seismic control of structures since the 1980s [1,2]. Dampers can be classified into metal, viscoelastic [3], viscous [4], tuned absorption dampers [5], smart materials based isolation and suspension system [6,7], recently developed smart materials based damping or negative stiffness structure [8–10], actively tuned damper and driver system [11,12], according to the energy dissipation methods, control force characteristics [13] as well as materials used. The metal damper has a certain stiffness, and its energy dissipation mechanism depends on the elastic-plastic deformation. It is cheap, easy to install, and less affected by temperature. Therefore, it is suitable for all kinds of building structures and has a good economic value and effectiveness for the reinforcement and reconstruction of existing and new buildings. Metallic dampers can further be categorized into axial yield [14,15], shear yield [16], flexural yield [17,18], and combined yield devices [19].

Existing forms of flexural yield dampers include the added damping and stiffness (ADAS) device [20–22], triangular added damping and stiffness (TADAS) device [23,24], knee brace device [25,26], steel-composited wall dampers and B-C-W members [27,28], non-uniform steel strip damper [29], rhombic low yield strength steel plate [30], J-shaped steel hysteresis damper [31,32], U-shaped damper [33–35], pipe damper [36], dual-pipe damper [37], bar-fuse damper [38], accordion metallic damper [39], pipe-fuse damper [40], pure bending yielding dissipater [41], crawler steel damper [42], and hourglass-shaped strip damper [43].

Because the yield load of single-sheet steel is small, some bending plate dampers need to be combined with multiple sheets of the steel plate. In order to facilitate the energy consumption and reduce the stress concentration, it is necessary to optimize the shape of the steel plate, which makes the processing difficult and cannot eliminate the influence of the vertical force on the damper. In view of the above problems, this paper proposes the improvement of a curved plate damper based on the principle of the U-shaped damper. The mechanical characteristics and energy dissipation capacity of the curved plate damper were investigated using theoretical and experimental methods.

2. Performance Parameters of Curved Plate Damper

Three views with dimensioning is shown in Figure 1. The parameters of the curved plate damper mainly include thickness (t), width (b), and radius (R, $R = R' - t/2$). The performance of the damper will vary with the parameters. The AB section and CD section of the damper are straight sections connected with other components, which constitutes the non-energy-consuming part of the damper, as shown in Figure 1a. The half-arc AC and BD segments are the main work parts of the damper.

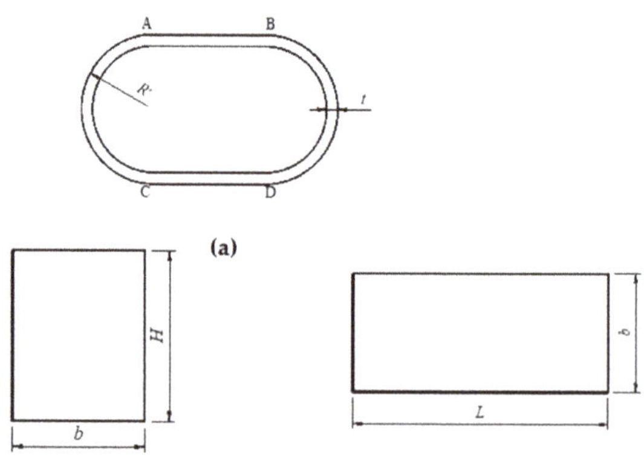

Figure 1. Front (a), lateral (b), and vertical (c) views of the curved plate damper.

2.1. Elastic Stiffness Calculation Formula

A single semi-circular arc steel plate was taken as the research object, which can be regarded as a curved beam with two fixed ends [44], as shown in Figure 2a. Figure 2b shows the deformation sketch and internal force analysis diagram of the central axis of the semi-circular arc steel plate when displacement (Δ = 1) occurs at the support.

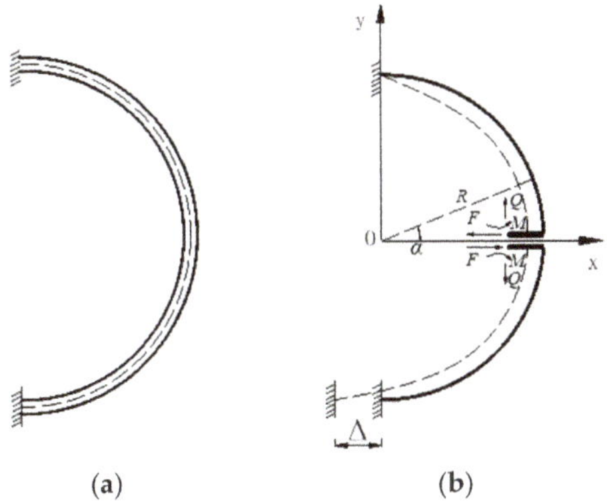

Figure 2. Diagram of a single semi-circular arc steel plate (**a**) and computing model (**b**).

According to the elastic center method, using the symmetry of the structure, M and Q are unknown symmetric forces, F is an unknown anti-symmetric force, and the force method equation can be simplified as follows:

$$\left.\begin{aligned} \delta_{11}M + \delta_{12}Q + \Delta_{1C} = 0 \\ \delta_{21}M + \delta_{22}Q + \Delta_{2C} = 0 \\ \delta_{33}F + \Delta_{3C} = 0 \end{aligned}\right\} \tag{1}$$

Using the rigid arm, the sum of δ_{12} and δ_{21} equals zero. Therefore, the simplified form of the equation is expressed as follows:

$$\left.\begin{aligned} \delta_{11}M + \Delta_{1C} = 0 \\ \delta_{22}Q + \Delta_{2C} = 0 \\ \delta_{33}F + \Delta_{3C} = 0 \end{aligned}\right\} \tag{2}$$

where Δ_{1C} and Δ_{2C} are zero due to the horizontal displacement of the support. In addition, the bending moment (M) and vertical force (Q) at the elastic center are both zero. Thus, the above simplified equation can be written as follows:

$$\delta_{33}F + \Delta_{3C} = 0 \tag{3}$$

where δ_{33} and Δ_{3C} can be written as

$$\begin{aligned} \delta_{33} &= \int \frac{\overline{M}_3^2}{EI} d_s + \int \frac{\overline{F}_3^2}{EA} d_s + \int \frac{\overline{Q}_3^2}{GA} d_s \\ &= \int_{-\frac{\pi}{2}}^{\frac{\pi}{2}} \frac{R^2 \sin^2\alpha}{EI} R d\alpha + \int_{-\frac{\pi}{2}}^{\frac{\pi}{2}} \frac{\sin^2\alpha}{EA} R d\alpha + \int_{-\frac{\pi}{2}}^{\frac{\pi}{2}} \frac{\cos^2\alpha}{GI} R d\alpha \\ &= \frac{R\pi}{2}\left(\frac{R^2}{EI} + \frac{1}{EA} + \frac{1}{GA}\right) \end{aligned} \tag{4}$$

$$\Delta_{3C} - -1 \tag{5}$$

Introducing Equations (4) and (5) into (3), the force can be obtained as

$$F = -\frac{\Delta_{3C}}{\delta_{33}} = \frac{2}{R\pi\left(\frac{R^2}{EI} + \frac{1}{EA} + \frac{1}{GA}\right)} \tag{6}$$

where F is the force at the end of the damper with unit displacement, which is the elastic stiffness of the damper with a semi-curved steel plate, and G is the shear modulus of elasticity ($G = 0.4E$). The elastic stiffness of the damper can be simplified as follows:

$$K = \frac{Ebt^3}{R\pi(6R^2 + 7t^2/4)} \tag{7}$$

where E is the elastic modulus of the material, b is the width of the curved plate damper, and t is the thickness of the curved plate damper.

2.2. Formula for Calculating Yield Strength

From the balance mechanism of the forces, the expression of elastic ultimate strength can be obtained as

$$F_e \cdot 2R = 2M \tag{8}$$

The elastic ultimate strength can also be written as

$$F_e = \frac{M_e}{R} = \frac{f_y W_e}{R} = \frac{f_y bt^2}{6R} \tag{9}$$

where f_y is the yield load, and W_e is the elastic moment of resistance.

The method for determining the yield displacement in this paper is shown in Figure 3. The horizontal and vertical coordinates at the intersection of tangent OA and AB are the yield displacement and yield strength, respectively. Because the yield strength is greater than the elastic ultimate strength, the adjustment coefficient β is introduced.

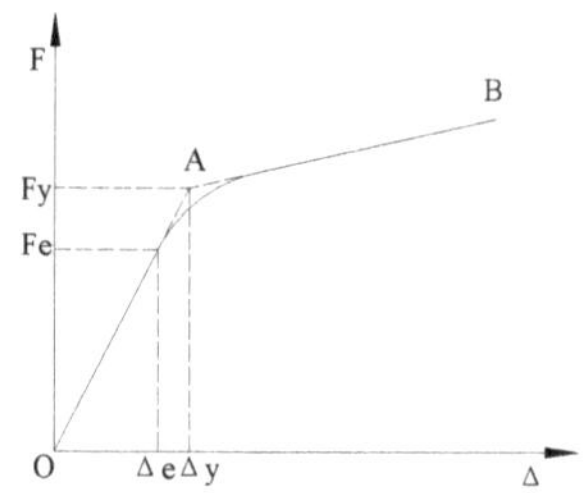

Figure 3. Schematic diagram of yield displacement.

The adjustment coefficient of the yield load can be determined by relevant tests and numerical analysis. The yield load can be written as

$$F_y = \frac{\beta f_y bt^2}{6R} \tag{10}$$

2.3. Formula for Calculating Yield Displacement

The yield displacement can be expressed as

$$\Delta_y = \frac{F_y}{K} = \frac{\beta f_y \pi(6R^2 + 7t^2/4)}{6Et} \tag{11}$$

From Equations (7), (9), and (11), the elastic stiffness, yield strength, and yield displacement of curved plate dampers can be calculated. By adjusting the relevant parameters, a damper was designed to meet the needs of the project.

3. Mechanical Properties of Curved Plate Damper

The curved plate damper was made of Q235B steel. Four groups of specimens were designed for the material properties test. The thickness of steel plates for specimens 1 and 2 was 10 mm, and that for specimens 3 and 4 was 6 mm. The material property test was loaded with a servo actuator (maximum force capacity: 500 kN) in Civil Test Center of Southeast University (Nanjing, China) with a loading rate of 1.2 mm/min. Based on the analysis of stress and strain data, the yield point was taken as the yield strength of the steel. The mechanical properties of steel are shown in Table 1.

Table 1. Mechanical properties of steel.

Specimen	Yield Strength/MPa	Tensile Strength/MPa	Yield Ratio/%	Elongation/%
CSPD (curved steel plate dampers)-1	249.48	375.47	66.44	35.20
CSPD-2	236.64	370.03	63.95	37.71
CSPD-3	245.09	358.03	68.46	37.59
CSPD-4	253.86	372.93	68.07	33.44

The four groups of curved plate dampers were named as CSPD-1, CSPD-2, CSPD-3, and CSPD-4, respectively. The length of the straight section of the four groups was the same. Their geometric dimensions are shown in Table 2.

Table 2. Specimen sizes.

Specimen	b/mm	t/mm	R'/mm	Straight Section/mm
CSPD-1	105	6	65	70
CSPD-2	105	10	65	70
CSPD-3	105	6	43	70
CSPD-4	85	6	65	70

3.1. Loading Device and Test Scheme

The test was carried out in Civil Test Center of Southeast University (Nanjing, China). The loading equipment was a MTS 50-ton fatigue testing machine. The experiment was divided into four groups with eight specimens, with each group having two identical specimens. In order to connect with the servo actuator, a fixture was designed as shown in Figure 4. The inverted T-splint and T-base are equipped with triangular stiffening ribs, and the components of the fixture are connected by the bolts.

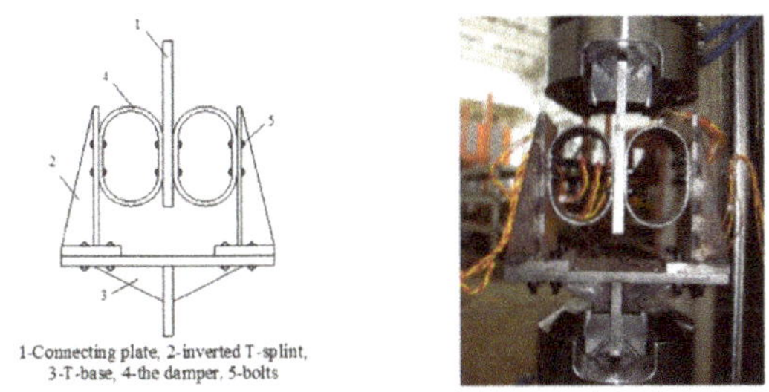

(a) (b)

Figure 4. Schematic diagram of the fixture (**a**) and photograph (**b**) of the testing device.

The test was divided into standard loading and fatigue loading. According to the multiple of the yield displacement of the damper (△), the target displacements of the CSPD-1 specimens were 2 mm

(0.5 △), 4 mm (1 △), 8 mm (2 △), 12 mm (3 △), 16 mm (4 △), 20 mm (5 △), 24 mm (6 △), 28 mm (7 △), 32 mm (8 △), 36 mm (9 △), and 40 mm (10 △), respectively. According to the multiple of yield displacement, all specimens in this test were loaded in 11 stages. The target displacements of CSPD-2 were 1.2, 2.4, 4.8, 7.2, 9.6, 12, 14.4, 16.8, 19.2, 21.6, and 24 mm, respectively. The target displacements of CSPD-3 were 0.9, 1.8, 3.6, 5.4, 7.2, 9, 10.8, 12.6, 14.4, 16.2, and 18 mm, respectively. The target displacements of CSPD-4 were 2.1, 4.2, 8.4, 12.6, 16.8, 21, 25.2, 29.4, 33.6, 37.8, and 42 mm, and the target displacement circulated three times. For fatigue loading, the Code for Seismic Design of Buildings (GB 50011-2010) stipulates that the energy dissipator should circulate 30 times [13] under the designed displacement, and there should be no obvious low-cycle fatigue phenomenon. The fatigue loading displacements of CSPD-1, CSPD-2, CSPD-3, and CSPD-4 were 40, 24, 18, and 42 mm, respectively.

During the test, force and displacement data were automatically recorded by the servo actuator. In order to study the mechanical properties of the damper, strain gauges were attached to the damper. The locations of the measuring points are shown in Figure 5. Two strain gauges were attached to each group of specimens at positions 1, 2, and 3.

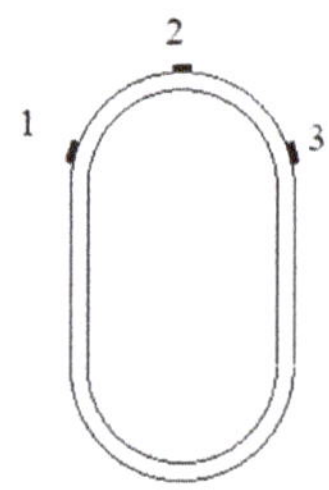

Figure 5. Gauge position.

3.2. Experiment Result Analysis

The described four groups of specimens underwent standard loading. It was observed that when the displacement was small, the damper showed no obvious change. However, with the increase in displacement, the steel oxide layer of the damper's semi-circular arc energy dissipation section appeared to warp and spall with a relatively higher spall at the end of the semi-circular arc. The damper produced plastic deformation during the cyclic loading process. The damper's deformation when it was loaded to the maximum displacement is shown in Figure 6. After 30 cycles of fatigue loading, the damper had obvious deformations but no cracks. No obvious damage was observed and the damper had good integrity.

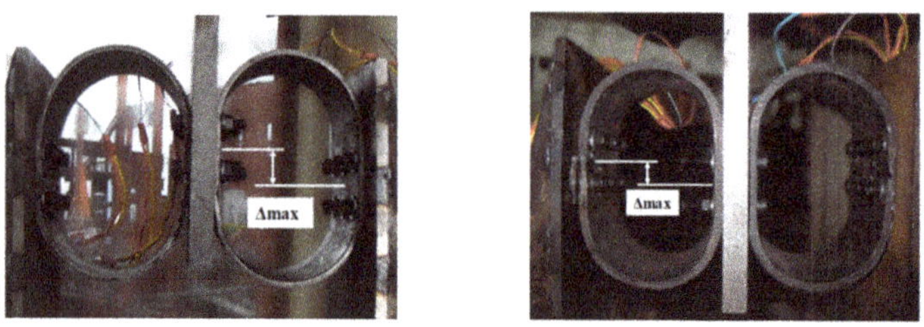

(**a**) (**b**)

Figure 6. *Cont.*

(c) (d)

Figure 6. Deformation of CSPD-1 (**a**), CSPD-2 (**b**), CSPD-3 (**c**), and COPD-4 (**d**) specimens at maximum displacement.

3.2.1. Hysteretic Curve

From the data collected by MTS equipment, the hysteretic curves of curved steel plate dampers are plotted in Figure 7. It can be seen that the hysteretic curves of dampers are full, the elastic stiffness is large, the stiffness of dampers after yielding decreases, and the maximum and minimum loads of dampers are not exactly the same. The reason may be that there were errors in the installation. The four groups of specimens circulated three times at each target displacement, and the three curves basically coincide, which indicates the stability of the damper.

Hysteretic curves under fatigue loading are shown in Figure 8. After 30 cycles of the cyclic loading, the load indices of the specimens CSPD-1 and CSPD-4 decreased, but the attenuation was far less than 15%. However, the load attenuation of the specimens CSPD-2 was higher, about 13.51%. The load attenuation of the specimens CSPD-3 presented positive and negative asymmetric state, which may be due to an installation error. Therefore, the performance of CSPD-2 was slightly worse than that of the other three specimens, but still met the relevant requirements of the code for dampers. It can be concluded that overall, the curved plate damper had good hysteretic performance.

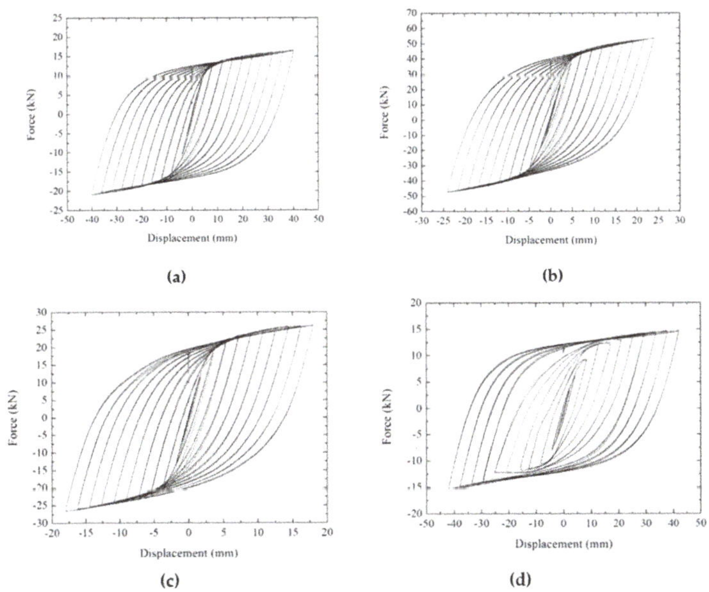

Figure 7. Hysteretic curves of CSPD-1 (**a**), CSPD-2 (**b**), CSPD-3 (**c**), andCSPD-4 (**d**) under standard loading.

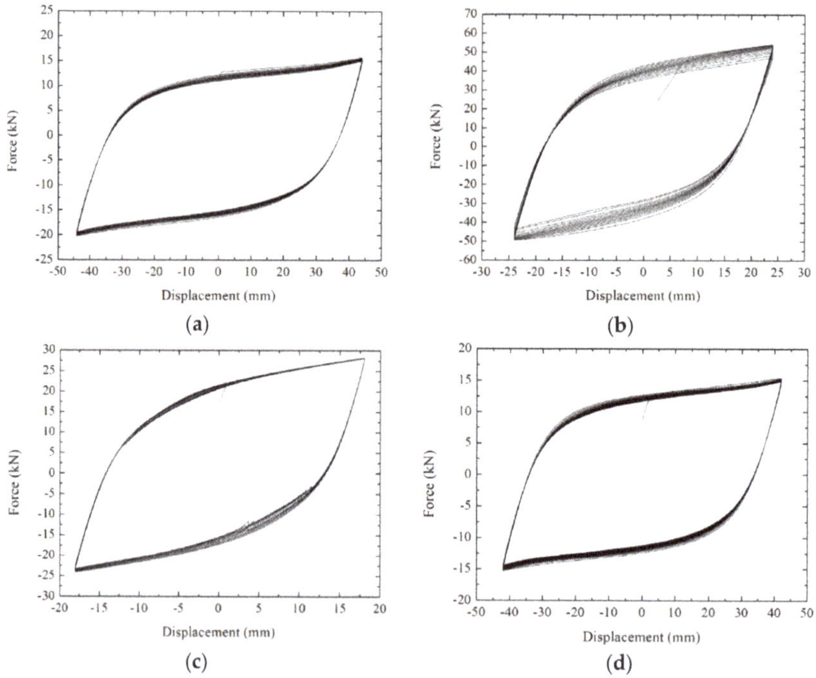

Figure 8. Hysteretic curve of CSPD-1 (**a**), CSPD-2 (**b**), CSPD-3 (**c**), and CSPD-4 (**d**) under fatigue loading.

3.2.2. Ductility Coefficient and Energy Dissipation Coefficient

According to the relevant provisions of test data processing in the code for seismic test methods of buildings (JGJ-96), the ductility coefficient is the ratio of ultimate displacement to yield displacement. The energy dissipation coefficient is measured by the envelope area of the hysteresis curve. The diagram of the hysteresis curve is shown in Figure 9. The ductility coefficient can be written as

$$E = \frac{S_{(ABC+CDA)}}{S_{(OBE+ODF)}} \tag{12}$$

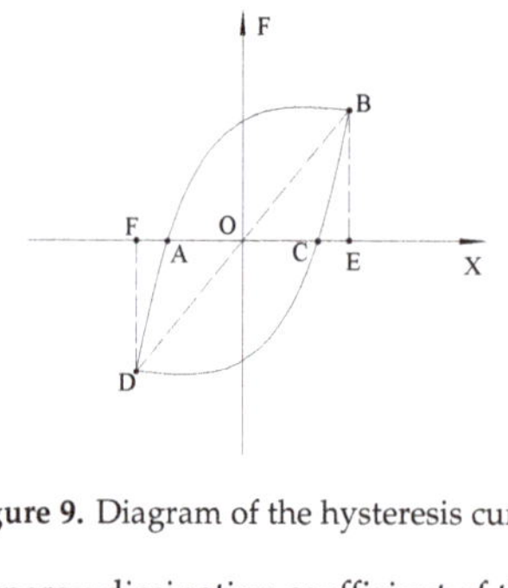

Figure 9. Diagram of the hysteresis curve.

The ductility coefficient and energy dissipation coefficient of the curved plate damper are shown in Table 3. The ductility and energy dissipation capacity of the four groups of specimens were good with few exceptions. Among them, CSPD-2 had the smallest ductility coefficient, and CSPD-3 had the smallest energy dissipation coefficient.

Table 3. Mechanical properties of curved plate damper.

Specimen	Yield Displacement/mm	Limit Displacement/mm	Ductility Coefficient μ	Energy Dissipation Coefficient E
CSPD-1	4.37	40	9.15	2.90
CSPD-2	2.66	24	9.02	2.83
CSPD-3	1.86	18	9.68	2.73
CSPD-4	4.52	42	9.29	2.96

3.2.3. Strain Analysis

The longitudinal coordinate of the strain analysis curve represents the strain value at the measuring point. As shown in Figure 5, strain gauges were attached at three locations on the curved surface steel plate damper. Because the locations of measuring points 1 and 3 were the same, the average values of positive strain extremum at measuring points 1 and 3 were taken. The transverse coordinates were unified as the loading displacement (n × Δ), where Δ is the prediction of the yield displacement.

From Figure 10, it can be seen that the strain at point 1 is much larger than that at point 2, which shows that the plastic deformation at the end of the semi-circular arc was large, and that the stress was also large for curved plate dampers. It is not difficult to see from Figure 10a that when the loading displacement was greater than 3Δ, the strain values of the specimens began to differ significantly. The strains of specimens CSPD-1 and CSPD-4 increased rapidly. At 6Δ, the strain increment of the CSPD-2 specimens slowed down, while the strain of the CSPD-3 specimens increased rapidly from 7Δ.

Figure 10b demonstrates the strain curve at point 2. It can be seen that the strain increment trend of the four groups of specimens was approximately the same. When loaded to 10Δ, the strain values of the other specimens were similar except for the small strain values of CSPD-2 specimens.

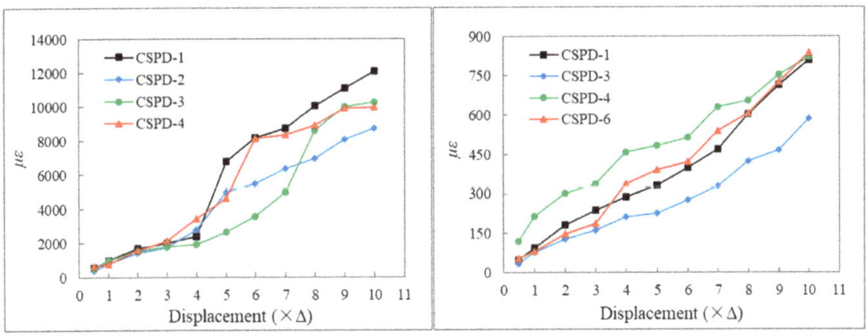

Figure 10. Strain at points 1 (**a**) and 2 (**b**).

3.3. Finite Element Analysis

3.3.1. Hysteresis Curve Analysis

The model shown in Figure 11 was built in general FEM software known as ANSYS. The shell unit shell181 was selected for the model. The constitutive relation of the bilinear follow-up reinforcement model was used for steel material. The elastic modulus of Q235 steel was used as 2.06×10^5 Mpa, Poisson's ratio as 0.3, and yield strength was taken according to Table 1. The model was developed according to the size of CSPD-1, CSPD-2, CSPD-3, and CSPD-4 dampers. The flat section of the bottom of the damper was completely fixed, while the flat section of the top was only horizontally displaced.

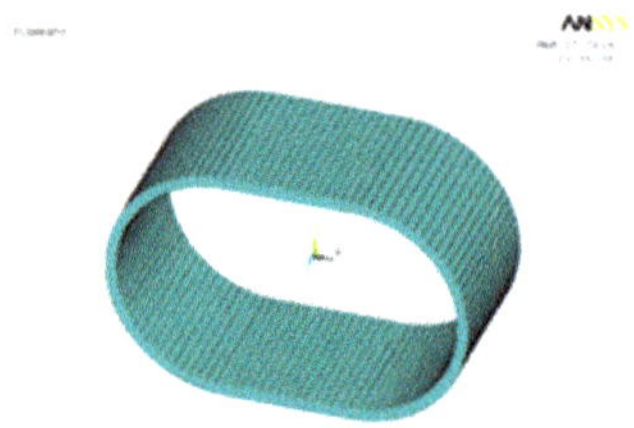

Figure 11. Finite element model.

In the process of standard loading, the hysteretic loops of the double yield displacement and 10 yield displacements were compared with the results of the finite element method. Two specimens were used in each group at the same time. Therefore, the results of the finite element simulation needed to be magnified twice. Figures 12–15 show the hysteretic curves of the four specimens.

(1) In Figure 12, the hysteretic curve coincides well at 4 mm. The positive and negative values are asymmetric in the hysteretic curve at 40 mm. Because the stiffness of CSPD-1 is small, then the horizontal deformation causes an error in the final load value, when the load displacement is large.
(2) In Figure 13, the asymmetry of the hysteretic curve also exists at 2.4 mm. At 24 mm, the maximum load of the test curve and the finite element curve are basically the same, but the envelope area of the finite element curve is slightly larger, and the second stiffness of the damper is smaller.
(3) In Figure 14, the hysteretic curve coincides well at 1.8 mm. The maximum load coincides well at 24 mm, and the envelope area of the finite element curve is slightly larger.
(4) In Figure 15, the hysteretic curve coincides well at 4.2 mm, and the curve of finite element analysis shows an upward trend at 42 mm, which is different from the shape of the test curve. This is because the boundary condition set in the finite element analysis is complete bonding, and a bolt connection is used in the test, which cannot be precisely achieved.

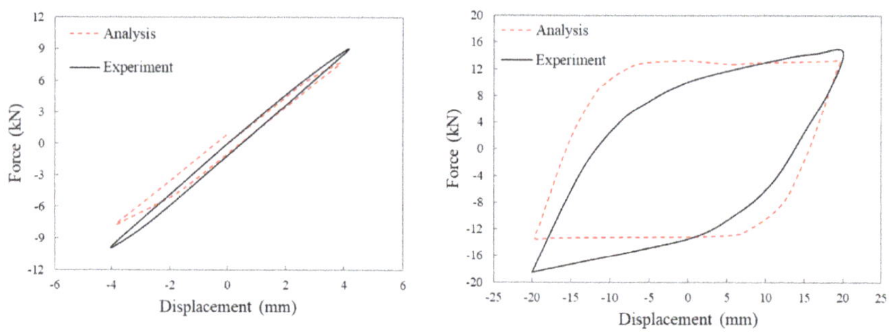

(a) (b)

Figure 12. Hysteresis curve at 4 mm (**a**) and 40 mm (**b**) of CSPD-1.

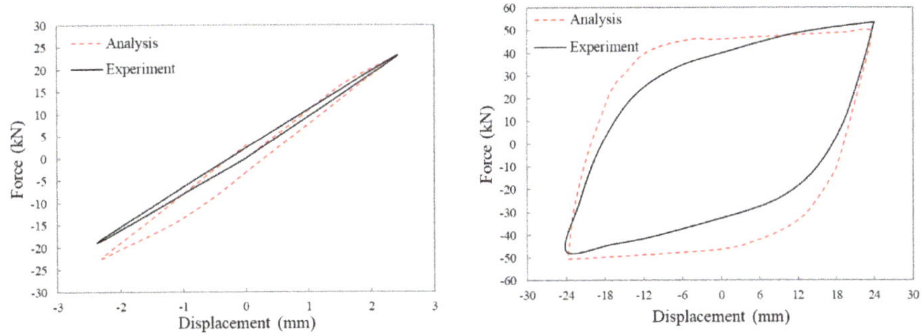

(a) (b)

Figure 13. Hysteresis curve at 2.4 mm (**a**) and 24 mm (**b**) of CSPD-2.

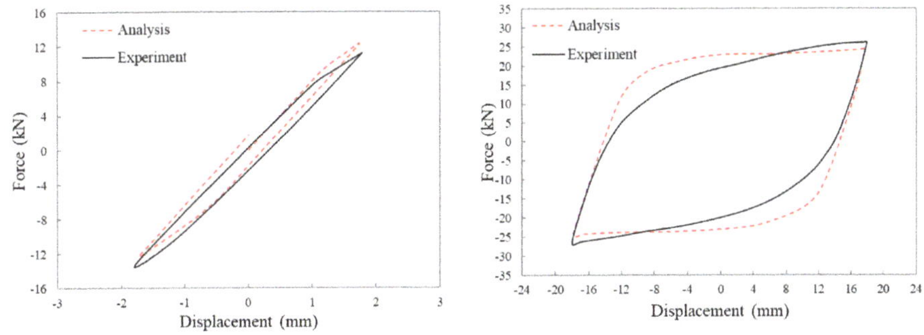

(a) (b)

Figure 14. Hysteresis curve at 1.8 mm (**a**) and 18 mm (**b**) of CSPD-3.

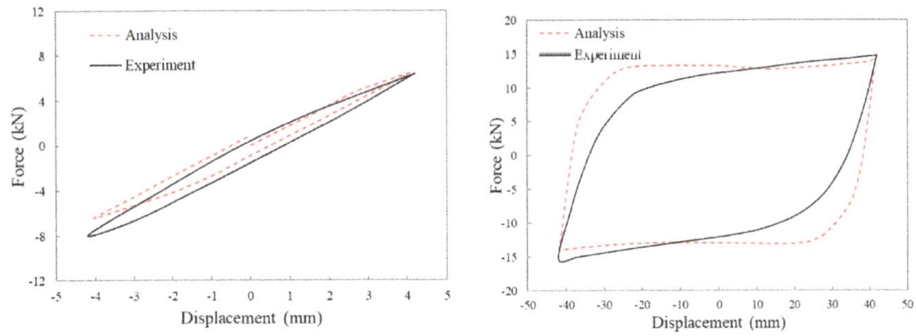

(a) (b)

Figure 15. Hysteresis curve at 4.2 mm (**a**) and 42 mm (**b**) of CSPD-4.

Generally speaking, the curve of the finite element analysis is basically the same as that of the test; therefore, the finite element model is reasonable. However, the ideal elastic-plastic model of material in the finite element is different from the actual material. There are many factors affecting the test process, including processing, installation, and other errors. It is impossible to achieve the ideal state in finite element settings, and some errors are acceptable.

3.3.2. Analysis of Mechanical Property Parameters

According to the theoretical formula of the curved plate damper, combined with the experimental and numerical data presented in this paper, a comparison of mechanical properties of curved plate dampers is provided in Table 4. The coefficient beta (β)in the calculation formula can be determined according to the test value and the finite element value. The yield displacement and yield bearing capacity in the test and the finite element simulation were averaged, and then, correspondingly, the theoretical values were determined to be equal. The error was smaller when the calculation yields a β factor value of 1.78. The maximum error of the mechanical properties is listed in Table 4. Except for some errors of elastic stiffness greater than 10%, the other errors were smaller. This shows that the finite element calculations in conjunction with theoretical formulae can reasonably reflect the performance of the damper.

Table 4. Comparison of mechanical properties of curved plate dampers.

	Items	CSPD-1	CSPD-3	CSPD-4	CSPD-6
Test result	Yield displacement (mm)	4.37	2.66	1.86	4.52
	Yield bearing capacity (kN)	8.78	25.42	12.53	6.97
	Elastic stiffness (kN/mm)	1.92	9.56	6.74	1.54
Finite element simulation	Yield displacement (mm)	4.18	2.44	1.82	4.20
	Yield bearing capacity (kN)	9.34	27.11	14.04	7.56
	Elastic stiffness (kN/mm)	2.13	10.11	7.71	1.80
Theoretical calculation	Yield displacement (mm)	2.48β	1.40β	1.04β	2.47β
	Yield bearing capacity (kN)	5.00β	14.37β	7.76β	4.05β
	Elastic stiffness (kN/mm)	2.02	10.23	7.43	1.64
Maximum error	Yield displacement (mm)	5.50%	9.02%	2.20%	7.62%
	Yield bearing capacity (kN)	6.38%	6.65%	10.75%	7.80%
	Elastic stiffness (kN/mm)	9.86%	7.01%	12.58%	14.44%

4. Conclusions

In this paper, four groups of curved plate dampers were selected for theoretical analysis and mechanical properties testing. The main conclusions are as follows:

(1) No visible cracks were found in the dampers during standard loading and fatigue loading, and no obvious damage was observed. The hysteretic curves of standard loading and fatigue loading were very full. In standard loading, each target displacement cycle had three cycles, and the three curves coincided. After 30 cycles of the fatigue cycle, the attenuation of the load index was less than 15%, which indicates that the damper had stable performance. It can be seen from the ductility and energy dissipation coefficient that the four dampers have good deformation and energy dissipation performance.

(2) Through stress analysis, the strain at the top of the semi-circular arc was much smaller than that of the end of the semi-circular arc, which shows that for curved plate dampers, the plastic deformation at the end of the semi-circular arc was large, the stress was also large, and the strain change rate was also large.

(3) The finite element model was established to simulate the loading process of the specimens. Compared with the hysteretic curves obtained in the test, it was found that the two curves were basically the same except for the individual specimens. Because there are many influencing factors in the testing process, the positive and negative hysteretic asymmetry of the test curve will occur when the small displacement is loaded. In case of the large displacement, the hysteretic area of the finite element analysis curve is slightly larger. The mechanical properties of the damper can be obtained through experiments, finite element simulations, and theoretical calculations. It is reasonable to obtain a coefficient beta of 1.78 by numerical and theoretical computations.

The maximum error was within the allowable range. At the same time, the correctness of the finite element model and the theoretical formulae was proved.

In conclusion, the curved plate damper had the characteristics of a simple structure, clear mechanical performance, and good stability.

Author Contributions: Conceptualization, J.Z., A.L. and C.Z.; methodology, J.Z.; software, J.Z.; formal analysis, J.Z.; writing—original draft preparation, J.Z.; Writing—review and editing, C.Z. All authors have read and agreed to the published version of the manuscript.

Funding: The research was financially supported by the Ministry of Science and Technology of China (Grant No. 2017YFC0703603), the National Natural Science Foundation of China (Grant No. 51678322), the Qingdao postdoctoral application research project (Grant No. 2018152), the Taishan Scholar Priority Discipline Talent Group program funded by the Shandong Province, the Cooperative Innovation Center of Engineering Construction and Safety in Shandong Blue Economic Zone, and the first-class discipline project funded by the Education Department of Shandong Province.

Conflicts of Interest: The authors declare no conflicts of interest.

References

1. Housner, G.W.; Bergman, L.A.; Caughey, T.K.; Chassiakos, A.G.; Claus, R.O.; Masri, S.F.; Skelton, R.E.; Soong, T.T.; Spencer, B.F.; Yao, J.T. Structural control: Past, present, and future. *ASCE J. Eng. Mech.* **1997**, *123*, 897–971. [CrossRef]
2. Soong, T.T.; Dargush, G.F. *Passive Energy Dissipation Systems in Structural Engineering*; Wiley: London, UK, 1997.
3. Mahmoodi, P.; Keel, C.J. Performance of viscoelastic structural dampers for the Columbia center building. In *Building Motion in Wind*; ASCE: New York, NY, USA, 1986; pp. 83–106.
4. De Domenico, D.; Ricciardi, G.; Takewaki, I. Design strategies of viscous dampers for seismic protection of building structures: A review. *Soil Dyn. Earthq. Eng.* **2019**, *118*, 144–165. [CrossRef]
5. Zhang, C.W.; Li, L.Y.; Ou, J.P. Swinging motion control of suspended structures: Principles and applications. *Struct. Control Health Monit.* **2010**, *17*, 549–562. [CrossRef]
6. Fu, W.Q.; Zhang, C.W.; Sun, L.; Askari, M.; Samali, B.; Chung, K.L.; Sharafi, P. Experimental investigation of a base isolation system incorporating mr dampers with the high-order single step control algorithm. *Appl. Sci.* **2017**, *7*, 344. [CrossRef]
7. Zhang, C.W.; Ou, J.P.; Zhang, J.Q. Parameter optimization and analysis of a vehicle suspension system controlled by magnetorheological fluid dampers. *Struct. Control Health Monit.* **2006**, *13*, 885–896. [CrossRef]
8. Wei, M.; Sun, L.; Hu, G. Dynamic properties of an axially moving sandwich beam with magnetorheological fluid core. *Adv. Mech. Eng.* **2017**, *9*. [CrossRef]
9. Sun, L.; Huang, W.M. Wet to shrink: An approach to realize negative expansion upon wetting. *Adv. Compos. Mater.* **2009**, *18*, 95–103. [CrossRef]
10. Wei, M.H.; Sun, L.; Zhang, C.; Qi, P.P.; Zhu, J. Shear-thickening performance of suspensions of mixed ceria and silica nanoparticles. *J. Mater. Sci.* **2019**, *54*, 346–355. [CrossRef]
11. Zhang, C.W.; Ou, J.P. Modeling and Dynamical Performance of the Electromagnetic Mass Driver System for Structural Vibration Control. *Eng. Struct.* **2015**, *82*, 93–103. [CrossRef]
12. Zhang, C.W.; Ou, J.P. Control Structure Interaction of Electromagnetic Mass Damper System for Structural Vibration Control. *ASCE J. Eng. Mech.* **2008**, *134*, 428–437. [CrossRef]
13. Zhang, C.W. Control Force Characteristics of Different Control Strategies for the Wind-excited 76-story Benchmark Building Structure. *Adv. Struct. Eng.* **2014**, *17*, 543–559. [CrossRef]
14. Ju, Y.K.; Kim, M.H.; Kim, J.; Kim, S.D. Component tests of buckling-restrained braces with unconstrained length. *Eng. Struct.* **2009**, *31*, 507–516. [CrossRef]
15. Xie, Q. State of the art of buckling-restrained braces in Asia. *J. Constr. Steel Res.* **2005**, *61*, 727–748. [CrossRef]
16. Chen, S.J.; Jhang, C. Experimental study of low-yield-point steel plate shear wall under in-plane load. *J. Constr. Steel Res.* **2011**, *67*, 977–985. [CrossRef]
17. Whittaker, A.S.; Bertero, V.V.; Thompson, C.L.; Alonso, L.J. Seismic testing of steel plate energy dissipation devices. *Earthq. Spectra* **1991**, *7*, 563–604. [CrossRef]

18. Chan, R.W.K.; Albermani, F. Experimental study of steel slit damper for passive energy dissipation. *Eng. Struct.* **2008**, *30*, 1058–1166. [CrossRef]
19. Li, H.N.; Li, G. Experimental study of structure with "dual function" metallic dampers. *Eng. Struct.* **2007**, *29*, 1917–1928. [CrossRef]
20. Whittaker, A.; Bertero, V.; Alonso, J.; Thompson, C. *Earthquake Simulator Testing of Steel Plate Added Damping and Stiffness Elements*; Report No. UCB/EERC-89/02; Earthquake Engineering Research Center, University of California: Berkeley, CA, USA, 1989.
21. Symans, M.D.; Charney, F.A.; Whittaker, A.S.; Constantinou, M.C.; Kircher, C.A.; Johnson, M.W.; McNamara, R.J. Energy dissipation systems for seismic applications: Current practice and recent developments. *J. Struct. Eng.* **2008**, *134*, 3–21. [CrossRef]
22. Bedon, C.; Amadio, C. Passive control systems for the blast enhancement of glazing curtain walls under explosive loads. *Open Civ. Eng.* **2017**, *11*, 396–419. [CrossRef]
23. Tsai, K.C.; Chen, H.W.; Hong, C.P.; Su, Y.F. Design of steel triangular plate energy absorbers for seismic-resistant construction. *Earthq. Spectra* **1993**, *9*, 505–528. [CrossRef]
24. Gray, M.; Christopoulos, C.; Packer, J.; de Oliveira, C. A new brace option for ductile braced frames. *Mod. Steel Constr.* **2012**, *52*, 40–53.
25. Balendra, T.; Lim, E.L.; Liaw, C.Y. Large scale seismic testing of knee brace frame. *J. Struct. Eng.* **1997**, *123*, 11–19. [CrossRef]
26. Mahmoudi, M.; Montazeri, S.; Abad, M.J.S. Seismic performance of steel X-knee-braced frames equipped with shape memory alloy bars. *J. Constr. Steel Res.* **2018**, *147*, 171–186. [CrossRef]
27. Zhang, C.W.; An, D.; Zhu, L. Axial Compressive Behavior Behavior of Steel-Damping Concrete Composite Wall. *Appl. Sci.* **2019**, *9*, 4679. [CrossRef]
28. Zhu, L.M.; Zhang, C.W.; Guan, X.M.; Uy, B.; Sun, L.; Wang, B.L. The multi-axial strength performance of composite structural B-C-W members subjected to shear forces. *Steel Compos. Struct.* **2018**, *27*, 75–87.
29. Lee, C.H.; Ju, Y.K.; Min, J.K.; Lho, S.H.; Kim, S.D. Non-uniform steel strip dampers subjected to cyclic loadings. *Eng. Struct.* **2015**, *99*, 192–204. [CrossRef]
30. Shih, M.H.; Sung, W.P. A model for hysteretic behavior of rhombic low yield strength steel added damping and stiffness. *Comput. Struct.* **2005**, *83*, 895–908. [CrossRef]
31. Kato, S.; Kim, Y.B.; Nakazawa, S.; Ohya, T. Simulation of the cyclic behavior of J-shaped steel hysteresis devices and study on the efficiency for reducing earthquake responses of space structures. *J. Constr. Steel Res.* **2005**, *61*, 1457–1473. [CrossRef]
32. Kato, S.; Kim, Y.B. A finite element parametric study on the mechanical properties of J-shaped steel hysteresis devices. *J. Constr. Steel Res.* **2006**, *62*, 802–811. [CrossRef]
33. Deng, K.; Pan, P.; Su, Y.; Xue, Y. Shape optimization of U-shaped damper for improving its bi-directional performance under cyclic loading. *Eng. Struct.* **2015**, *93*, 27–45. [CrossRef]
34. Aguirre, M.; Sanchez, A.R. Structural seismic damper. *J. Struct. Eng.* **1992**, *118*, 1158–1171. [CrossRef]
35. Tagawa, H.; Gao, J. Evaluation of vibration control system with U-dampers based on quasi-linear motion mechanism. *J. Constr. Steel Res.* **2012**, *70*, 213–225. [CrossRef]
36. Maleki, S.; Bagheri, S. Pipe damper, part I: Experimental and analytical study. *J. Constr. Steel Res.* **2010**, *66*, 1088–1095. [CrossRef]
37. Maleki, S.; Mahjoubi, S. Dual-pipe damper. *J. Constr. Steel Res.* **2013**, *85*, 81–91. [CrossRef]
38. Aghlara, R.; Tahir, M.M. A passive metallic damper with replaceable steel bar compo-nents for earthquake protection of structures. *Eng. Struct.* **2018**, *159*, 185–197. [CrossRef]
39. Motamedi, M.; Nateghi, F. Study on mechanical characteristics of accordion metallic damper. *J. Constr. Steel Res.* **2018**, *142*, 68–77. [CrossRef]
40. Aghlara, R.; Tahir, M.M.; Adnan, A.B. Experimental study of pipe-fuse damper for passive energy dissipation in structures. *J. Constr. Steel Res.* **2018**, *148*, 351–360. [CrossRef]
41. Zibasokhan, H.; Behnamfar, F.; Azhari, M. Experimental study of a new pure bending yielding dissipater. *Bull. Earthq. Eng.* **2019**, *17*, 4389–4410. [CrossRef]
42. Deng, K.; Pan, P.; Wang, C. Development of crawler steel damper for bridges. *J. Constr. Steel Res.* **2013**, *85*, 140–150. [CrossRef]

43. Lee, C.H.; Lho, S.H.; Kim, D.H.; Oh, J.; Ju, Y.K. Hourglass-shaped strip damper subjected to monotonic and cyclic loadings. *Eng. Struct.* **2016**, *119*, 122–134. [CrossRef]
44. Long, Y.; Bao, S. *Structural Mechanics*; Higher Education Press: Beijing, China, 1994.

applied sciences

MDPI

Article

A Multi-Degree of Freedom Tuned Mass Damper Design for Vibration Mitigation of a Suspension Bridge

Fanhao Meng [1,2,3,*], Jiancheng Wan [1], Yongjun Xia [1], Yong Ma [1] and Jingjun Yu [2,*]

1 China Electric Power Research Institute, Beijing 100055, China; wanjiancheng@epri.sgcc.com.cn (J.W.); xiayj@epri.sgcc.com.cn (Y.X.); mayong@epri.sgcc.com.cn (Y.M.)
2 Robotics Institute, Beihang University, Beijing 100191, China
3 Active Structures Laboratory, Université Libre de Bruxelles, 1050 Brussels, Belgium
* Correspondence: Fanhao.Meng@ulb.ac.be (F.M.); jjyu@buaa.edu.cn (J.Y.)

Received: 12 November 2019; Accepted: 31 December 2019; Published: 8 January 2020

Abstract: This paper proposes a synthetic approach to design and implement a two-degree of freedom tuned mass damper (2DOFs TMD), aimed at damping bending and torsional modes of bridge decks (it can also be applied to other types of bridges like cable-stayed bridges to realize the energy dissipation). For verifying the effectiveness of the concept model, we cast the parameter optimization of the 2DOFs TMDs conceptual model as a control problem with decentralized static output feedback for minimizing the response of the bridge deck. For designing the expected modes of the 2DOFs TMDs, the graphical approach was introduced to arrange flexible beams properly according to the exact constraints. Based on the optimized frequency ratios, the dimensions of 2DOF TMDs are determined by the compliance matrix method. Finally, the mitigation effect was illustrated and verified by an experimental test on the suspension bridge mock-up. The results showed that the 2DOFs TMD is an effective structural response mitigation device used to mitigate the response of suspension bridges. It was also observed that based on the proposed synthetic approach, 2DOFs TMD parameters can be effectively designed to realize the target modes control.

Keywords: tuned mass damper; multi-degree of freedom; graphical approach; suspension bridges

1. Introduction

With the tendency to use longer spans, the damping of modern suspension bridges is seriously reduced. Complex vibration problems follow, such as wind-induced vibration, human-induced vibration, cable-structure interactions and flutter instabilities [1–4]. It is admitted that the oversensitivity to dynamic excitation of suspension bridges is associated with the very low structural damping in the global bridge modes [5,6]. Therefore, the dissipation of the vibration energy generated by the dynamic loadings is a central issue in their design. At present, the use of damping systems such as tuned mass damper (TMD) [7], viscous dampers [8,9], or active tendon control [10] is a classical way to alleviate the vibrations in structures. This study aims at the design of multi-degree of freedom TMD for vibration damping of a suspension bridge deck.

Considering their simplicity and effectiveness, tuned mass dampers have been widely used in bridges such as the London Millenium bridge, for damping both lateral and vertical vibrations of the deck. Since Frahm et al. proposed the fundamental theory, TMDs have seen numerous applications in civil engineering [11–13]. Thus, Ben Mekki and Bourquin [14,15] proposed a new semi-active electromagnetic TMD of pendulum type to damp the torsional mode of an evolving bridge mock-up. Their studies showed that the tuned pendulum damper (TPD) is very effective in vibration damping, qualitatively and quantitatively confirming the theoretical predictions. However, the TMDs can only control single mode and easy detune, which limits their further development. For the robustness of vibration control and targeting several vibration modes, the multiple tuned mass dampers (MTMDs) has been proposed, and its performance is more effective as compared to the single TMD. The superior effectiveness of the MTMDs is able to control almost any type of vibration in civil structures [16]. The MTMDs is used to damp suspension bridges for several purposes. In some studies, the MTMDs are used to the suppression of buffeting, flutter or increasing the critical flutter wind speed [17,18]. Other studies consider MTMDs for alleviating pedestrian- and jogger-induced vibration [19–22] or traffic-induced vibration [23–26].

Generally, the weight of a TMD is limited to 1–3% of the structure weight. Hence, as the number of targeted modes increases, a large number of TMDs will increase the burden on the primary system and limit the damping performance (called weight penalty). To avoid such a penalty, in our previous study [27], we proposed to design a two-degree of freedom TMD, where the original mass of TMD is redistributed in such a way that the TMD has a bending mode and a torsional mode. In this design the resonance frequencies and the modal damping of the two modes can be tuned independently. In addition, Zuo and Nayfeh have proposed a multiple degrees of freedom TMD (MDOFs TMD), and experimentally demonstrated that the MDOFs TMD can damp six modes of the primary structure. They also showed that a MDOFs TMD can be used to attain better vibration suppression for single mode vibration of a primary structure [28]. Jang et al. described a novel method for selecting the parameters of a 2DOFs TMD with translational and rotational degrees of freedom [29]. Ma and Yang et al. presented a design of a multi-DOFs TMD to alleviate the dominant mode of the work piece/fixture assembly in milling [30–32].

As the DOF increases, by only selecting the appropriate DOFs and tuned frequencies, the TMDs can reach the best vibration control of the primary structure. Therefore, designing a TMD with expected DOFs and natural frequencies becomes an urgent problem. Unfortunately, due to its complicated structure and easily detuning, the further study on the implementation of MDOFs TMD is rare.

In this paper, we propose a synthetic approach based on both the graphical approach and parameterized compliance for the concrete design of the TMDs with the expected DOFs and we verify their feasibility and performance by numerically and experimentally way on a laboratory suspension bridge mock-up. The paper is organized as follows: Section 2 describes the vibration characteristics of the bridge mock-up and builds a concept model of bridge with two 2DOFs TMDs. Based on the equations of motion, the decentralized control technique is directly used to optimize the stiffness and damping coefficients of the springs and dampers to obtain the optimum frequency ratios in Section 3. Section 4 presents the detailed design process of the 2DOFs TMDs based on the graphical approach and compliance analysis. Section 5 mainly focuses on evaluating the damping performance and verifying the proposed design method. Finally, findings and conclusions of the study are summarized at the end.

2. Formulation of the Bridge–TMD System

Our goal is to use two 2DOFs TMDs to control the first four vibration modes of the suspension bridge simultaneously. The 2DOFs TMD is decoupled in the physical coordinates, their mode shapes follow the physical coordinate of the mock-up, and the corresponding resonance frequencies can be tuned independently to match the desired design. The suspension bridge mock-up and its finite element modelling are detailed described in our previous studies [27,33,34]. Here, this paper only lists the vibration characteristics of the bridge mock-up, as shown in Table 1.

Since the tuning TMDs becomes increasingly complex, we cast the parameter optimization of the 2DOFs TMDs as a control problem with decentralized static output feedback for minimizing the response of the bridge system. This method has been used successfully for a single mode vibration control of a MDOFs TMD by Zuo and Nayfeh [28]. The concept model of the 2DOFs TMDs is to take the springs as local feedback elements of relative displacements and the dampers as local feedback elements of relative velocities, as shown in Figure 1.

Table 1. The numerical and experimental natural frequencies and mode shapes of the bridge mock-up.

Mode	Numerical [Hz]	Experimental [Hz]	Experimental Damping	Numerical Mode Shape	Experimental Mode Shape
1st B	4.5	4.4	0.14%		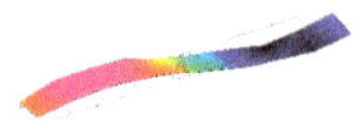
2nd B	6.9	6.4	2.28%	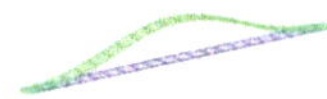	
1st T	9.3	9.4	0.62%	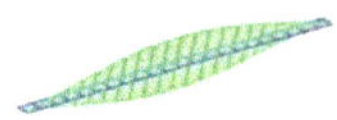	
2nd T	10.2	10.2	0.23%	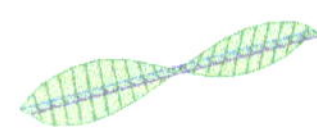	
3rd B	11.2	12.2	0.49%	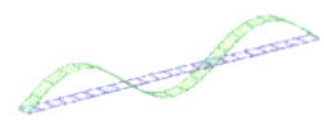	

B stands for bending mode, T stands for torsional mode.

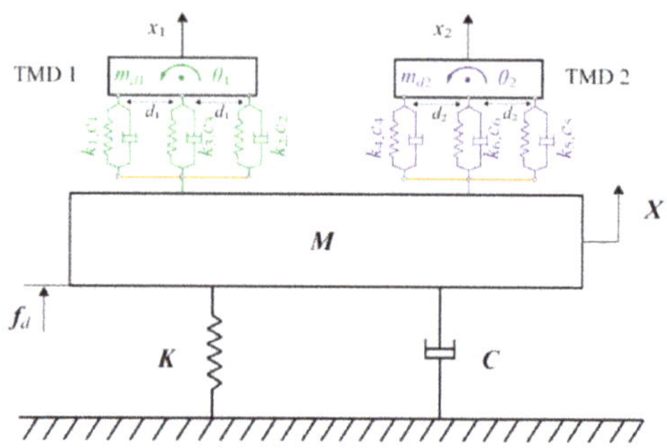

Figure 1. Concept model of the 2DOF TMDs: the bridge system equipped with two TMDs.

In this way, the role of the springs and dampers can be replaced by a control force vector, where the control gain is composed of the spring stiffness and damping coefficients (k_i and c_i for $i = 1, 2, \ldots, 6$). The mass matrix $\boldsymbol{M}_{n\times n}$, stiffness matrix $\boldsymbol{K}_{n\times n}$ and viscous damping matrix $\boldsymbol{C}_{n\times n}$ is extracted from the numerical model of the suspension bridge mock up, respectively. The 2DOFs TMD has two planar degrees of freedom, translation x_1 (x_2) and rotation θ_1 (θ_2). Its mass is m_{d1} (m_{d2}) and the rotational inertia about its center of mass is $I_d = m_d\rho^2$, where ρ is the radius of gyration.

The 2DOFs TMD is connected to the primary system at distances d_1 (d_2) from its center of mass via dashpots and springs. Therefore, the control force vector $[u_1, u_2, \ldots, u_6]$ in this case are given by:

$$u_1 = k_1\left(x_1 - \boldsymbol{B}_2^{\mathrm{T}}\boldsymbol{X} - \theta_1 d_1\right) + c_1\left(\dot{x}_1 - \boldsymbol{B}_2^{\mathrm{T}}\dot{\boldsymbol{X}} - \dot{\theta}_1 d_1\right), \tag{1}$$

$$u_2 = k_2\left(x_1 - \boldsymbol{B}_2^{\mathrm{T}}\boldsymbol{X} + \theta_1 d_1\right) + c_2\left(\dot{x}_1 - \boldsymbol{B}_2^{\mathrm{T}}\dot{\boldsymbol{X}} + \dot{\theta}_1 d_1\right), \tag{2}$$

$$u_3 = k_3\left(x_1 - \boldsymbol{B}_2^{\mathrm{T}}\boldsymbol{X}\right) + c_3\left(\dot{x}_1 - \boldsymbol{B}_2^{\mathrm{T}}\dot{\boldsymbol{X}}\right), \tag{3}$$

$$u_4 = k_4\left(x_2 - \boldsymbol{B}_3^{\mathrm{T}}\boldsymbol{X} - \theta_2 d_2\right) + c_4\left(\dot{x}_2 - \boldsymbol{B}_3^{\mathrm{T}}\dot{\boldsymbol{X}} - \dot{\theta}_2 d_2\right), \tag{4}$$

$$u_5 = k_5\left(x_2 - \boldsymbol{B}_3^{\mathrm{T}}\boldsymbol{X} + \theta_2 d_2\right) + c_5\left(\dot{x}_2 - \boldsymbol{B}_3^{\mathrm{T}}\dot{\boldsymbol{X}} + \dot{\theta}_2 d_2\right), \tag{5}$$

$$u_2 = k_2\left(x_1 - \boldsymbol{B}_2^{\mathrm{T}}\boldsymbol{X} + \theta_1 d_1\right) + c_2\left(\dot{x}_1 - \boldsymbol{B}_2^{\mathrm{T}}\dot{\boldsymbol{X}} + \dot{\theta}_1 d_1\right), \tag{6}$$

where $\boldsymbol{X}$ is the vector of global coordinates of the finite element model, $\boldsymbol{B}_2$ and $\boldsymbol{B}_3$ are the input vector of this two TMDs, respectively.

The equations can govern the vibration of the coupled system can be decomposed into:

$$m_{d1}\ddot{x}_1 = -u_1 - u_2 - u_3, \tag{7}$$

$$m_{d2}\ddot{x}_2 = -u_4 - u_5 - u_6, \tag{8}$$

$$I_{d1}\ddot{\theta}_1 = u_1 d_1 - u_2 d_1, \tag{9}$$

$$I_{d2}\ddot{\theta}_2 = u_4 d_2 - u_5 d_2, \tag{10}$$

The governing equations can then be written as

$$\boldsymbol{M}\ddot{\boldsymbol{X}} + \boldsymbol{C}\dot{\boldsymbol{X}} + \boldsymbol{K}\boldsymbol{X} = \boldsymbol{B}_1 f_d + \boldsymbol{B}_2(u_1 + u_2 + u_3) + \boldsymbol{B}_3(u_4 + u_5 + u_6), \tag{11}$$

where f_d is the external disturbances, $\boldsymbol{B}_1$ is the input vector of the external disturbances. We can express Equations (7)–(11) in matrix form as:

$$\begin{bmatrix} \boldsymbol{M} & 0 & 0 & 0 & 0 \\ 0 & m_{d1} & 0 & 0 & 0 \\ 0 & 0 & m_{d2} & 0 & 0 \\ 0 & 0 & 0 & I_{d1} & 0 \\ 0 & 0 & 0 & 0 & I_{d2} \end{bmatrix} \begin{pmatrix} \ddot{\boldsymbol{X}} \\ \ddot{x}_1 \\ \ddot{x}_2 \\ \ddot{\theta}_1 \\ \ddot{\theta}_2 \end{pmatrix} + \begin{bmatrix} \boldsymbol{C} & 0 & 0 & 0 & 0 \\ 0 & 0 & 0 & 0 & 0 \\ 0 & 0 & 0 & 0 & 0 \\ 0 & 0 & 0 & 0 & 0 \\ 0 & 0 & 0 & 0 & 0 \end{bmatrix} \begin{pmatrix} \dot{\boldsymbol{X}} \\ \dot{x}_1 \\ \dot{x}_2 \\ \dot{\theta}_1 \\ \dot{\theta}_2 \end{pmatrix} + \begin{bmatrix} \boldsymbol{K} & 0 & 0 & 0 & 0 \\ 0 & 0 & 0 & 0 & 0 \\ 0 & 0 & 0 & 0 & 0 \\ 0 & 0 & 0 & 0 & 0 \\ 0 & 0 & 0 & 0 & 0 \end{bmatrix} \begin{pmatrix} \boldsymbol{X} \\ x_1 \\ x_2 \\ \theta_1 \\ \theta_2 \end{pmatrix}$$
$$= \begin{pmatrix} \boldsymbol{B}_1 \\ 0 \\ 0 \\ 0 \\ 0 \end{pmatrix} f_d + \begin{bmatrix} \boldsymbol{B}_2 & \boldsymbol{B}_2 & \boldsymbol{B}_2 & \boldsymbol{B}_3 & \boldsymbol{B}_3 & \boldsymbol{B}_3 \\ -1 & -1 & -1 & 0 & 0 & 0 \\ 0 & 0 & 0 & -1 & -1 & -1 \\ d_1 & -d_1 & 0 & 0 & 0 & 0 \\ 0 & 0 & 0 & d_2 & -d_2 & 0 \end{bmatrix} \begin{pmatrix} u_1 \\ u_2 \\ u_3 \\ u_4 \\ u_5 \\ u_6 \end{pmatrix}, \tag{12}$$

or

$$\boldsymbol{M}_p\ddot{\boldsymbol{p}} + \boldsymbol{C}_p\dot{\boldsymbol{p}} + \boldsymbol{K}_p\boldsymbol{p} = \boldsymbol{B}_f f_d + \boldsymbol{B}_u\boldsymbol{u}, \tag{13}$$

where $\boldsymbol{p} = [\boldsymbol{X}, x_1, x_2, \theta_1, \theta_2]^{\mathrm{T}}$ and $\boldsymbol{u} = [u_1, u_2, \ldots, u_6]$ and T denote the complex conjugate matrix transpose. The matrices $\boldsymbol{M}_p$, $\boldsymbol{C}_p$, $\boldsymbol{K}_p$, $\boldsymbol{B}_f$, and $\boldsymbol{B}_u$ can be obtained from Equation (12) directly.

Defining the state variables of the system as:

$$x = \begin{bmatrix} \boldsymbol{p} \\ \dot{\boldsymbol{p}} \end{bmatrix}, \tag{14}$$

The governing equations are written in first-order form as:

$$\dot{x} = \boldsymbol{A}x + \boldsymbol{B}_{11}w + \boldsymbol{B}_{12}\boldsymbol{u}, \tag{15}$$

where $w = f_d$ and:

$$\boldsymbol{A} = \begin{bmatrix} 0 & \boldsymbol{I} \\ -\boldsymbol{M}_p^{-1} & \boldsymbol{K}_p \end{bmatrix}, \quad \boldsymbol{B}_{11} = \begin{bmatrix} 0 \\ \boldsymbol{M}_p^{-1}\boldsymbol{B}_f \end{bmatrix}, \quad \boldsymbol{B}_{12} = \begin{bmatrix} 0 \\ \boldsymbol{M}_p^{-1}\boldsymbol{B}_u \end{bmatrix}, \tag{16}$$

The cost output can be taken as the absolute or relative displacement, velocity, or acceleration of the primary system, which can be expressed in the form:

$$z = C_1 x + D_{11} w + D_{12} u, \tag{17}$$

For the displacement response of the primary system, the cost output can be written as:

$$z = X = C_1 x, \tag{18}$$

where:

$$C_1 = \begin{bmatrix} I_{n\times n} & O_{n\times 4} & O_{n\times n} & O_{n\times 4} \end{bmatrix}, \tag{19}$$

To complete the state-space description, we rewrite the control force given by Equations (1)–(6) as a static feedback gain F multiplied by the "measurement output" y:

$$u = Fy = \begin{bmatrix} k_1 & c_1 & & & & & & \\ & & k_2 & c_2 & & & & \\ & & & & \cdots & \cdots & & \\ & & & & & & k_6 & c_6 \end{bmatrix} y, \tag{20}$$

where y is given by:

$$\begin{array}{l} y = [x_1 - B_2^{\mathrm{T}} X - \theta_1 d_1, \dot{x}_1 - B_2^{\mathrm{T}} \dot{X} - \dot{\theta}_1 d_1, x_1 - B_2^{\mathrm{T}} X + \theta_1 d_1, \dot{x}_1 - B_2^{\mathrm{T}} \dot{X} + \dot{\theta}_1 d_1, \\ x_1 - B_2^{\mathrm{T}} X, \dot{x}_1 - B_2^{\mathrm{T}} \dot{X}, x_2 - B_3^{\mathrm{T}} X - \theta_2 d_2, \dot{x}_2 - B_3^{\mathrm{T}} \dot{X} - \dot{\theta}_2 d_2, x_2 - B_3^{\mathrm{T}} X + \theta_2 d_2, \\ \dot{x}_2 - B_3^{\mathrm{T}} \dot{X} + \dot{\theta}_2 d_2, x_2 - B_3^{\mathrm{T}} X, \dot{x}_2 - B_3^{\mathrm{T}} \dot{X}]^{\mathrm{T}} = C_2 x + D_{21} w + D_{22} u \end{array}, \tag{21}$$

where C_2 can be obtained from the definition of the state given by Equation (14) and the matrices $D_{21} = 0$ and $D_{22} = 0$. Equations (15), (17), and (21) cast the design of the two 2DOFs TMDs system as a decentralized control problem in the block diagram of Figure 2.

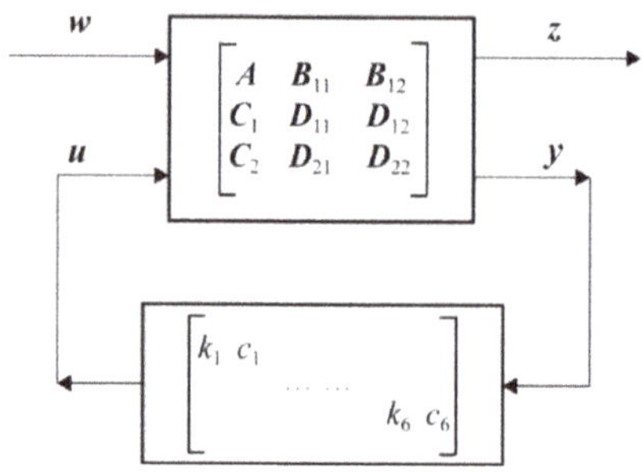

Figure 2. Block diagram of the bridge—2DOFs TMDs system with decentralized control.

3. Numerical Optimization and Simulation

According to above equations, the decentralized control techniques can be directly used to optimize the damping and stiffness coefficients of the dampers and springs to achieve performance (measured by z) under the disturbance w. The minimax numerical method [28] is utilized to minimize the response magnitude of the primary system.

3.1. Optimization Criteria

Traditionally, based on the Den Hartog method, the optimized frequency ratio and TMD damping ratio are aimed to minimize the structural response by the minimization of the structural dynamic magnification function. This classic procedure consists of two separate steps: tuning of the frequency of the damper, and selection of the optimal level of the TMD damping ratio. This classic procedure has been used in our previous research [27]. The goal of this study is to design the parameters (k_i and c_i for $i = 1, 2, \ldots, 6$) in order to determine the optimum frequency ratio ν_j ($j = 1, \ldots, 4$) and the optimum TMD damping ratio ξ_j for minimizing the response of the bridge system.

Due to the fact that the maximum amplitude of the bridge system should be controlled in a reasonable range, herein the damping performance of TMDs is evaluated by the H_∞ criterion. H is the selected FRF value of the bridge system under the excitation f_d. The value range of H_j is constrained by the value range of ω_j ($\omega_j \in [0.7\omega_{sj}, 1.3\omega_{sj}]$). The goal of the optimization is to minimize the maximum value of each value range of H_j. $\chi = [k_1, c_2, \ldots, k_6, c_6]^T$ is selected as the design parameter vector of the TMDs. The optimization problem can be written as:

$$\begin{aligned} &\text{Find}: \quad \chi = [k_1, c_1, k_2, c_2, k_3, c_3, k_4, c_4, k_5, c_5, k_6, c_6]^T \\ &\text{Minimize}: \quad \sum_{j=1}^{4} W_j\left(\max\left|H_j(\chi, \omega_j)\right|\right) \\ &\text{Subject to}: \quad I = \left\{ j \middle| 0.7\omega_{sj} \le \omega_j \le 1.3\omega_{sj},\ \omega_j = \left|eig(A + B_{12}FC_2)\right| \right\} \end{aligned} \tag{22}$$

where j ($j = 1, \ldots, 4$) is the mode number considered and ω_{sj} is the j-th natural frequency of the primary system. ω_j is the evaluation of the eigenvalues, which corresponding to the modal frequencies of the entire system inside the specified frequency band. $W_j = 0.25$, which is the weight coefficient. For each TMD, 2% of the total mass of the structure.

3.2. Numerical Optimization Results

Two TMD devices are used to damp the mode pairs (1B,2T) and (2B,1T); one TMD is placed at the quarter length of the deck (TMD1), the second is located at the center of the deck (TMD2). A detailed description is shown in Section 4.2. A disturbance force f_d is applied at one fixed point of the deck edge and the displacement z is measured at another fixed point, as shown in a small graph of Figure 3. According to Equation (22), the frequency response functions of the primary system with two 2DOFs TMDs are optimized, and the optimum frequency ratios and TMD damping ratios are obtained, as listed in Table 2.

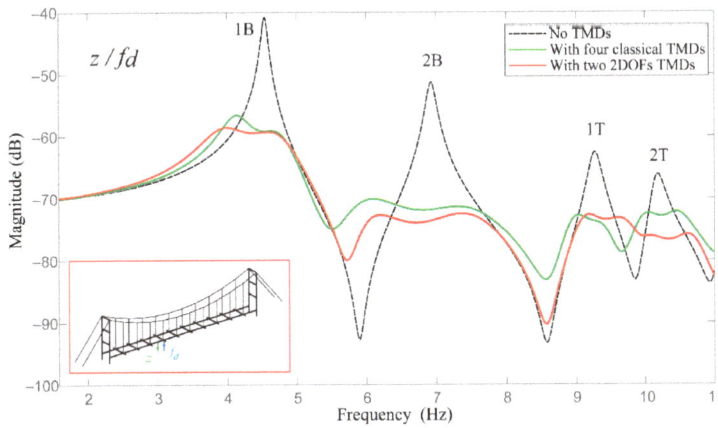

Figure 3. Numerical results: FRFs of the deck without and with the TMDs. The TMDs are targeted for the damping of the mode pair (2B,1T) and (1B,2T).

Table 2. Optimum parameters of 2DOFs TMDs.

TMD	Optimum Frequency Ratios		Optimum Damping Ratios	
	ν_1	ν_2	ξ_{opt1} (%)	ξ_{opt2} (%)
TMD1	0.9422	1.029	13.5	10.7
TMD2	0.9594	1.020	17.6	9.8

Figure 3 shows the frequency responses of the deck z/f_d, when: (i) without any TMD; (ii) the four classical TMDs are targeting to damp the two mode pairs (2B,1T) and (1B,2T), which is achieved by Den Hartog criterion; and (iii) the two 2DOFs TMDs are targeting to damp the two mode pairs (2B,1T) and (1B,2T), which is optimized by that introduced in the present paper. The results indicate that the 2DOFs TMD concept model is effective to suppress both the bending modes and torsional modes of the bridge system at the same time. But, if using the classical configuration of the TMD, we need four TMDs, each of them is tuned on a single mode at the same time: two translation TMDs, with a lumped mass identical to that used in the 2DOFs TMD, and two other TMDs with moment of inertia identical to that of the 2DOFs TMD too. Hence, comparing with the classical configuration of TMD, the two 2DOFs TMDs can reduce the weight penalty.

Figure 4 shows the FRF for different sensor locations, with the same TMD design as in Figure 3. This figure demonstrates the robustness with respect to the FRF used in the TMD design. Figure 5 plots the frequency response curve of the bridge equipped with TMD1, for different values of the TMD1 damping ratio ξ. Here, the TMD1 damping ratio ξ is selected as 10% ξ_{opt}, 25% ξ_{opt}, 50% ξ_{opt} and ξ_{opt}, respectively, and the ξ_{opt} is the optimum damping ratio of TMD1, which is listed in Table 2. From this figure, we see that the dynamic responses of the bridge deck always tend to decrease on increasing the damping ratio of TMD1. Furthermore, the frequency ratio is insensitive to the TMD damping ratio, but if the action is not perfectly resonant, the performance of TMD may decay seriously even though the value of the TMD damping ratio is very high [35]. For TMD structure design, unlike the damping ratio ξ_j which is difficult to quantify, the frequency ratio ν_j is important parameter which can be used to guide the following TMD structural parameters design.

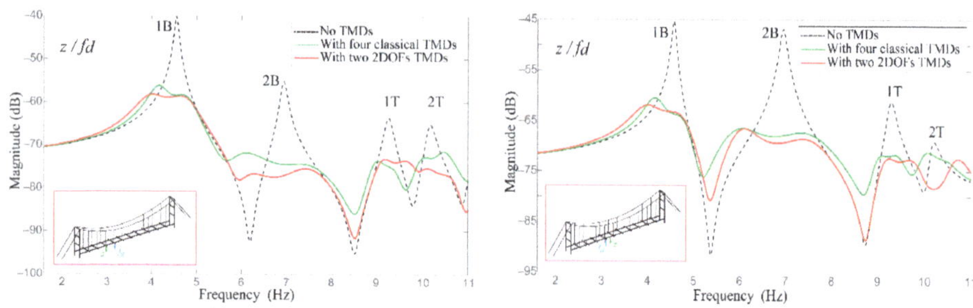

(a) (b)

Figure 4. *Cont.*

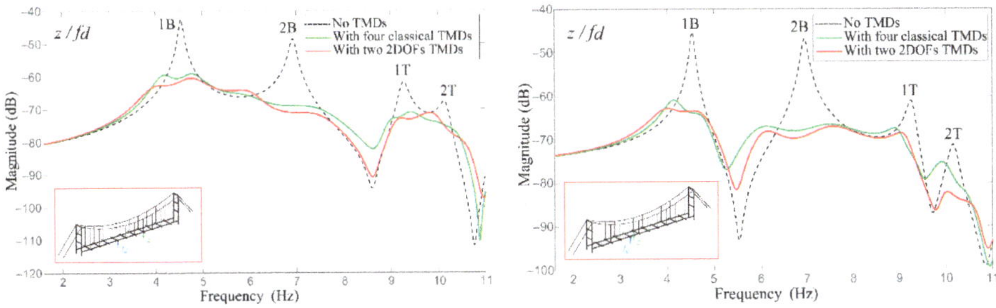

(c) (d)

Figure 4. Numerical results: FRFs of the deck respect to different measuring position (**a–d**). The excitation point is fixed.

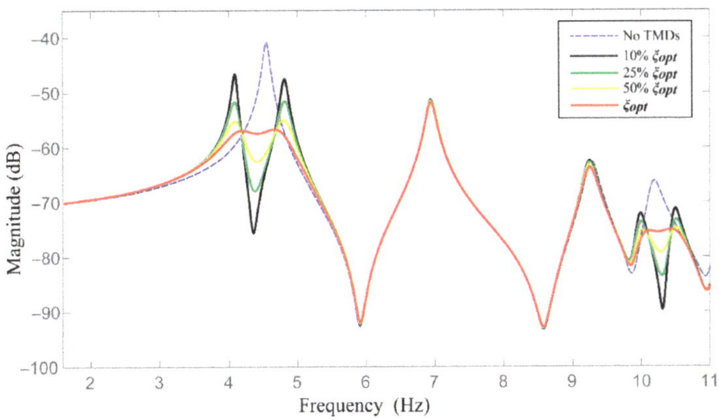

Figure 5. Influence of the damping of TMD1 on the FRF of the system. ξ = 10% ξ_{opt}, 25% ξ_{opt}, 50% ξ_{opt} and ξ_{opt}, respectively.

4. Structural Design of 2DOFs TMD

Due to the lack of an effective theoretical guidance, deterministic structure design of multi-DOFs TMDs is still a challenge in the field of TMD design. The main problems are: (1) kinematic constraints design of the multi-DOFs, it ensures that the TMDs have the expected DOFs; and (2) parametric modeling of the multi-DOFs, that contributes to design the TMD and ensure it has the expected natural frequencies. To solve the above problems, this study presents a synthetic approach based on both the graphical approach [36] and parameterized compliance for the concrete design of the TMD with the expected DOFs.

4.1. Graphical Approach

The screw theory as the theoretical foundation of graphical approach have been widely applied to design and analysis the compliant mechanisms [37,38]. For object behavior design, adding constraints is the most important step to reach the specific motion. Thus, finding the relationship between the constraints and movements is indispensable in TMD design. Currently, the screw theory is the most popular way to describe this relationship. In the screw theory, a unit screw $ is defined by a straight line with an associated pitch and is represented as a pair of three-dimensional vectors:

$$\$ = \begin{bmatrix} s \\ s_0 + hs \end{bmatrix} = \begin{bmatrix} s \\ r \times s + hs \end{bmatrix}, \tag{23}$$

where $s_0 = r \times s$ defines the moment of the screw axis about the origin of a coordinate system, s is a unit vector parallel the screw axis, r is the position vector of any point on the screw axis with respect to the origin of a coordinate system, and h is the pitch of the screw. If h is equal to zero, the screw reduces to a line quantity (Figure 6a):

$$\$_t = \begin{bmatrix} s \\ s_0 \end{bmatrix} = \begin{bmatrix} s \\ r \times s \end{bmatrix}, \tag{24}$$

If h is infinite, the screw reduces to:

$$\$_w = \begin{bmatrix} 0 \\ s \end{bmatrix}, \tag{25}$$

In addition, an infinite-pitch screw can be considered as a line located at infinity, as shown in Figure 6b.

For a better understanding and applicability, the two special cases of unit screw ($\$_t$, $\$_w$) are visualized by geometric patterns in Figure 7. The unit screw of zero pitch ($\$_t$) stands for a pure rotation in freedom space (rotational freedom line) or a unit pure force in static along the line in constraint space (constraint force line). A unit screw of infinite pitch represents a pure translation in freedom space (translational freedom line) or a pure couple in constraint space (constraint couple line). It is worth noting that the rotational freedom line represents the axis of rotational movement, and the constraint couple line stands for the axis of couple imposed on a rigid body.

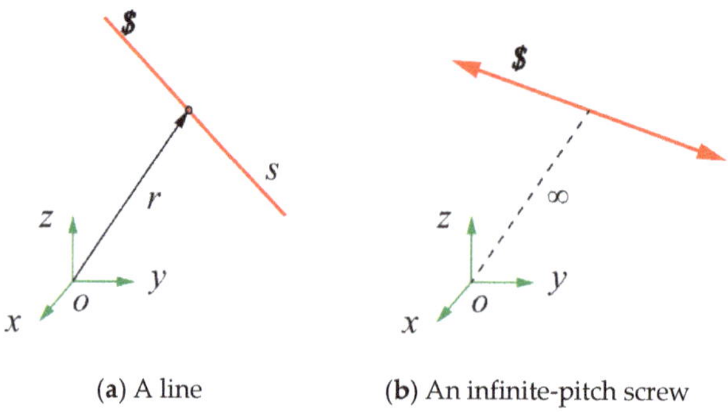

Figure 6. A unit screw.

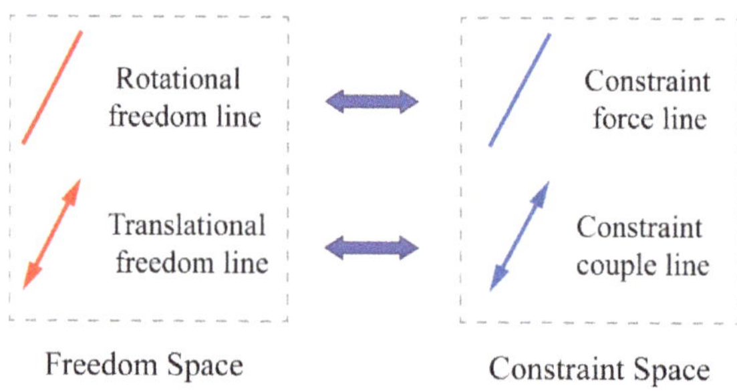

Figure 7. Geometric patterns representing screws.

Based on Maxwell's principles of constraints, the freedoms and constraints in a mechanical system can be defined as:

$$N = 6 - n, \tag{26}$$

where N is the number of DOFs, n is the number of non-redundant constraints. When a rigid body (e.g., TMD) is constrained by several mechanical connections providing n constraints, while N DOFs of the body will remains. In this regard, based on the reciprocal screw theory, n non-redundant constraints form a wrench $\$_1$ in constraint space, and the remained DOFs constitute a twist $\$_2$ in freedom space [39]. Based on the definition, the reciprocity of these two screw systems is expressed as:

$$\begin{aligned} \$_1^T\Delta\$_2 &= s_1\cdot(r_2\times s_2+h_2s_2)+s_2\cdot(r_1\times s_1+h_1s_1) \\ &= (h_1+h_2)(s_1\cdot s_2)+(r_2-r_1)\cdot(s_2\times s_1) \\ &= (h_1+h_2)\cos\alpha_{12}-a_{12}\sin\alpha_{12} \\ &= 0 \end{aligned} \tag{27}$$

where $\Delta=\begin{pmatrix} 0 & I \\ I & 0 \end{pmatrix}$, a_{12} is the normal distance of the two screw axes and α_{12} is the twist angle between the two screws. h_1 and h_2 denote the pitch of $\$_1$ and $\$_2$, respectively.

Thus, according to Equation (27), the relationship between the freedom lines and the constraint lines can be written as a brief form in Table 3.

Table 3. Geometric relationship between the freedom and constraint lines.

Geometric Condition		Freedom Space ($\$_2$)	
		Rotational Freedom Line (h_2 = 0)	**Translational Freedom Line ($h_2=\infty$)**
Constraint space ($\$_1$)	Constraint force line (h_1 = 0)	Coplanar (intersecting or parallel) a_{12} sinα_{12} = 0	Perpendicular α_{12} = 90°
	Constraint couple line ($h_1=\infty$)	Perpendicular α_{12} = 90°	Arbitrary

The 2DOFs TMD with the expected DOFs and mode shapes can be designed by the above geometric relationship.

4.2. Conceptual Design of the TMDs

By studying the characteristics of human-induced vibration, it is found that the first four modes of suspension bridge are easy to be stimulated to produce resonance phenomenon, which is the primary target of vibration reduction. The targeted mode shapes are shown in Figure 8.

The optimal location of a TMD is at the maximum modal displacement. Hence, the center of the deck (point A) is the optimal location for the second bending and the first torsional modes of the deck (2B,1T), while for the first bending and the second torsional modes (1B,2T), it is at the quarter length of the deck (point B). The motion of the TMDs only along the vibration direction of the two pairs can reach the best vibration control. Two TMDs have the same geometric relationship as shown in Figure 9. Therefore, the dimension of the freedom space N is two, and the dimension of the constraint space is four, according to Equation (26).

The translational freedom line and rotational freedom lines are orthogonal in Figure 9a. For the constraint space, the four constraint force lines intersect with the rotational freedom line and are orthogonal with the translational freedom line. Here the constraint couple lines can be ignored due to the fact that four constraint force lines have already formed the expected constraints. Thus, the corresponding constraint space can be divided into two pairs, each pair contains two parallel constraint force lines, and these two pairs are orthogonal (Figure 9b). Then constraint force lines are implemented by flexure elements, each of them can provide single DOF constraint along its axial direction (Figure 9c). Therefore, the exact constraints are formed on the TMDs and the expected DOFs are defined.

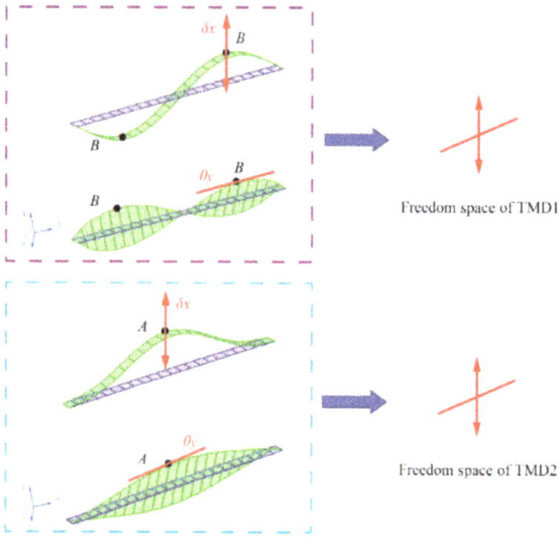

Figure 8. Shapes of the targeted modes and the corresponding optimal location of the TMD: the optimal location of the modes (2B,1T) at point A, while it is point B for modes (1B,2T); the freedom space of the TMDs.

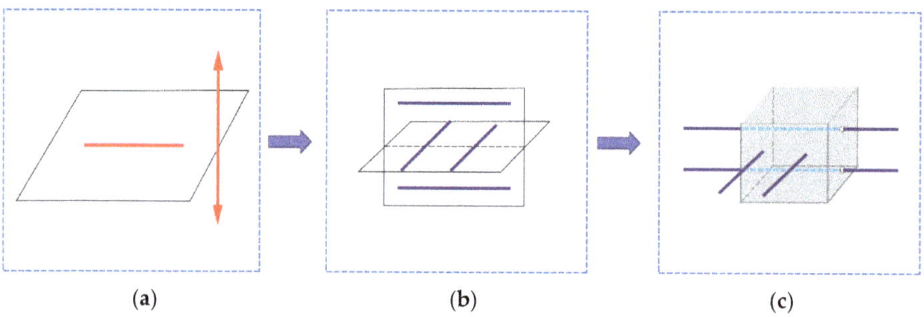

Figure 9. Conceptual design of 2DOFs TMDs: (**a**) freedom space of 2DOFs TMD; (**b**) constraint space of 2DOFs TMD; (**c**) physical model of 2DOFs TMD.

4.3. Parametric Design of 2DOFs TMDs

The conceptual model of TMDs only have two expected DOFs, but in actual design, due to the material and geometric properties, the TMD may have more DOFs than expected. The corresponding redundant modes of TMD may affect the performance of TMD, even lead to TMD failure. This is a key problem that has been perplexing TMD design. In general, the redundant modes which are far from the targeted modes can be ignored, and the empirical design has always been the major tool to achieve this goal. However, as the number of DOF increases, the empirical method gradually fails to realize the complex design of TMDs. Furthermore, the empirical design may cause the increase in time and cost. Therefore, the theoretical guidance has become particularly important in the multi-DOFs TMD design. In this section, we introduce a parameterized compliance approach for parametric design of the 2DOFs TMD.

Figure 10 shows the configuration of the TMD. It can be seen that the TMD is formed by eight slender beams in parallel distributed on mass block, which is transformed from physical models (Figure 9c). Each slender beam is considered as cantilever beam (Figure 10a) with length L, width w, and thickness t. Then, according to the Bernoulli-Euler model, the compliance matrix Cc_p ($p = 1, \ldots, n$) for each slender beam at location coordinate system $Oxyz$ is given as follows:

$$Cc_p = \frac{L_p}{EI_y} diag\left\{ \alpha \quad 1 \quad \frac{1}{\chi\gamma} \quad \frac{L_p^2}{12} \quad \frac{L_p^2}{12}\alpha \quad \frac{L_p^2}{12}\beta \right\}, \tag{28}$$

where:

$$\alpha = \left(\frac{t}{w}\right)^2, \beta_p = \left(\frac{t}{L_p}\right)^2, \chi = \frac{G}{E} = \frac{1}{2(1+\nu)}, \gamma = \frac{J}{I_y}, \tag{29}$$

and

$$I_x = \frac{w^3 t}{12}, I_y = \frac{wt^3}{12}, J = I_x + I_y, \tag{30}$$

where G, E is the shear and Young's modulus, respectively. γ is the ratio of torsion constant over moment of inertia, ν is the Poisson's ratio.

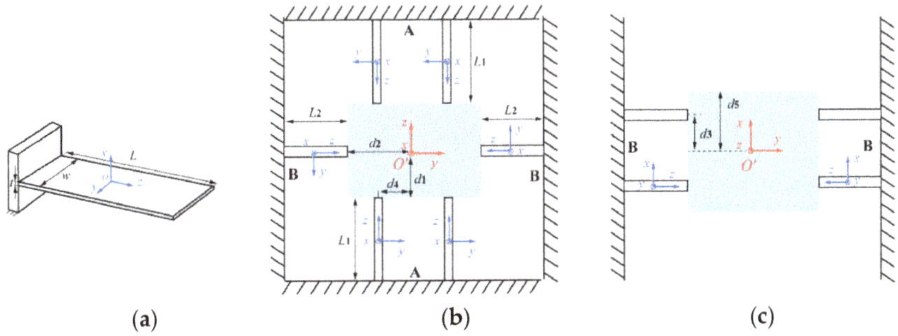

Figure 10. Coordinate systems of the slender beams: (**a**) local coordinate; (**b**) upper view of global coordinate; (**c**) front view of global coordinate.

Further, to simplify the following non-dimensionalization, we choose $t = w$, and according to Equations (29) and (30), $\gamma = 2$. Due to the different length of slender beam (L_1, L_2 and $s = L_1/L_2$), the compliance matrix of two type slender beams can be written as:

$$Cc_A = \frac{L_2}{EI_y}\text{diag}\left[s \quad s \quad \frac{s}{2\chi} \quad \frac{s^3L_2^2}{12} \quad \frac{s^3L_2^2}{12} \quad \frac{sL_2^2}{12}\beta_2 \right], \tag{31}$$

$$Cc_B = \frac{L_2}{EI_y}\text{diag}\left[1 \quad 1 \quad \frac{1}{2\chi} \quad \frac{L_2^2}{12} \quad \frac{L_2^2}{12} \quad \frac{L_2^2}{12}\beta_2 \right], \tag{32}$$

In order to combine the local compliance matrix Cc_p of the eight slender beams, they should be transformed from the local to global coordinate system. The origin O' of the global coordinate system $O'XYZ$ is defined in the centroid of the mass block (Figure 10b). For the parallel flexure mechanism, the global compliance matrix can be given as:

$$C_s = \left(\sum_{p=1}^{m} \left(Ad_p Cc_p Ad_p^T \right)^{-1} \right)^{-1}, \tag{33}$$

where m is the number of slender beams; Ad_p is the adjoint transformation matrix from the pth element to the global system:

$$Ad_p = \begin{bmatrix} R_{x,y,z}(\theta) & 0 \\ TR_{x,y,z}(\theta) & R_{x,y,z}(\theta) \end{bmatrix}, \tag{34}$$

where $\boldsymbol{T}$ is the translation matrix. $\boldsymbol{R}_{x,y,z}(\theta) = \boldsymbol{R}_x(\theta)\boldsymbol{R}_y(\theta)\boldsymbol{R}z(\theta)$, which is the multiplication of rotation matrices. $\boldsymbol{R}_x(\theta)$, $\boldsymbol{R}_y(\theta)$, and $\boldsymbol{R}z(\theta)$ stand for the rotation matrices by an angle θ about the x, y, and z axis, respectively. They are given in Equation (35):

$$\boldsymbol{R}_x(\theta) = \begin{bmatrix} 1 & 0 & 0 \\ 0 & \cos\theta & -\sin\theta \\ 0 & \sin\theta & \cos\theta \end{bmatrix}, \quad \boldsymbol{R}_y(\theta) = \begin{bmatrix} \cos\theta & 0 & \sin\theta \\ 0 & 1 & 0 \\ -\sin\theta & 0 & \cos\theta \end{bmatrix}, \quad \boldsymbol{R}_z(\theta) = \begin{bmatrix} \cos\theta & -\sin\theta & 0 \\ \sin\theta & \cos\theta & 0 \\ 0 & 0 & 1 \end{bmatrix}, \tag{35}$$

For the vector $(\Delta x, \Delta y, \Delta z)^T$ between two position, the translation matrix can be expressed as

$$\boldsymbol{T} = \begin{bmatrix} 0 & -\Delta z & \Delta y \\ \Delta z & 0 & -\Delta x \\ -\Delta y & \Delta x & 0 \end{bmatrix}, \tag{36}$$

Based on Equations (33)–(36), the global compliance matrix of TMD is computed by:

$$\boldsymbol{C}_s = \frac{L_2}{EI_y}\begin{bmatrix} c_{11} & & & & & \\ & c_{22} & & & & \\ & & c_{33} & & & \\ & & & c_{44} & & \\ & & & & c_{55} & \\ & & & & & c_{66} \end{bmatrix}, \tag{37}$$

The principal diagonal elements of $\boldsymbol{C}_s$ are selected as follows:

$$\begin{aligned} c_{11} &= \frac{L_2^2\beta_2 s^3}{16}\left(\frac{1}{\beta_2L_2^2s^3+\beta_2L_2^2s^2+3\beta_2L_2d_1s+3\beta_2L_2d_1s^3+3\beta_2d_1^2+3\beta_2d_2^2s^3+3d_4^2s^2}\right) \\ c_{22} &= \frac{L_2^2s^3}{16}\left(\frac{1}{0.5\chi L_2^2s^3+L_2^2s^2+3L_2d_1s+3d_1^2+3d_3^2s^3}\right) \\ c_{33} &= \frac{L_2^2\beta_2 s^3}{16}\left(\frac{1}{\beta_2L_2^2s^3+0.5\chi\beta_2L_2^2s^2+3\beta_2L_2d_2s^3+3\beta_2d_2^2s^3+3d_3^2s^3+3\beta_2d_4^2}\right) \\ c_{44} &= \frac{L_2^2s^3}{48}\left(\frac{1}{s^3+1}\right) c_{55} = \frac{L_2^2\beta_2s^3}{48}\left(\frac{1}{s^3+\beta_2}\right) \\ c_{66} &= \frac{L_2^2\beta_2s}{16}\left(\frac{1}{\beta_2s+1}\right) \end{aligned}, \tag{38}$$

where c_{11}, c_{22}, and c_{33} are the rotational compliance/stiffness about the x, y and z axis while c_{44}, c_{55}, and c_{66} are the translational compliance/stiffness along the x, y and z axis, respectively. In the end, the natural frequencies of 2DOFs TMDs can be approximate calculated by Equations (39)–(41).

The bending mode:

$$\omega_1 = \frac{1}{2\pi}\sqrt{\frac{k_{44}}{m_d}} = \frac{1}{2\pi}\sqrt{\frac{1}{m_d}\left(\frac{L_2c_{44}}{EI_y}\right)^{-1}}, \tag{39}$$

The torsional mode:

$$\omega_2 = \frac{1}{2\pi}\sqrt{\frac{k_{22}}{I_Y}} = \frac{1}{2\pi}\sqrt{\frac{1}{I_Y}\left(\frac{L_2c_{22}}{EI_y}\right)^{-1}}, \tag{40}$$

and:

$$I_Y = \frac{1}{3}m_d\left(d_1^2 + d_5^2\right), \tag{41}$$

where d_5 is the height of mass block (Figure 10c).

4.4. Results and Discussion

According to the optimum frequency ratios v_i (Table 2) and the target modes of bridge, the expected modes of TMDs are obtained and listed in Table 4.

Table 4. Expected modes of 2DOFs TMDs.

TMD	1st Mode	2nd Mode
	Bending	Torsion
TMD1	4.15 Hz	10.5 Hz
TMD2	6.14 Hz	9.59 Hz

In general, the size of TMD should be smaller and not occupy the space of bridge as much as possible. Thus, considering the processing conditions and the size of deck, we choose $t = w = 1$mm. The mass block is iron and $m_d = 0.18$ kg. The material of slender beams is Acrylonitrile Butadiene Styrene (ABS), and the elastic modulus $E = 2$ GPa, $\chi = 0.37$, the Poisson's ratio v is 0.394. In addition, in order to simplify calculation and TMD design, let $d_4 = d_2$.

The calculation results of 2DOFs TMDs are summarized in Table 5. It is noticed that the three-order natural frequency ω_3 are about 7 times greater than ω_2, and ω_4 is much larger than ω_1 and ω_2. Therefore, the undesired DOFs can be neglected, and the expected 2DOFs TMDs are obtained. According to the dimension parameters, FE model of 2DOFs TMDs are built as shown in Figure 11. The mode shapes of two expected TMDs are shown in Figures 12 and 13, respectively.

Table 5. Calculation results of 2DOFs TMDs.

TMD	Dimension Parameter (m)						Natural Frequency (Hz)			
	L_1	L_2	d_1	d_2	d_3	d_4	ω_1	ω_2	ω_3	ω_4
TMD1	0.045	0.059	0.015	0.0106	0.0075	0.0106	4.16	10.54	74.07	119.7
TMD2	0.0362	0.042	0.0105	0.0105	0.008	0.0105	6.14	9.55	77.7	109.8

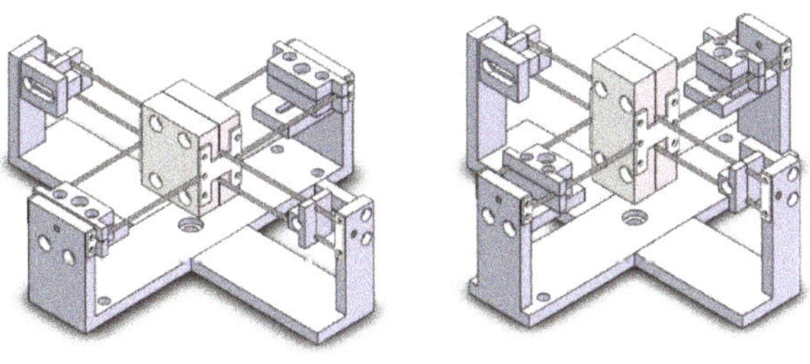

(**a**) (**b**)

Figure 11. CAD view of TMD. (**a**) TMD1; (**b**) TMD2.

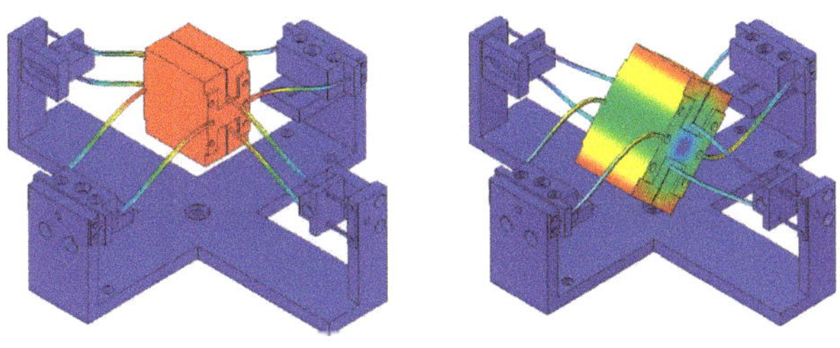

(**a**) (**b**)

Figure 12. Mode shapes of TMD1. (**a**) The 1st mode; (**b**) the 2nd mode.

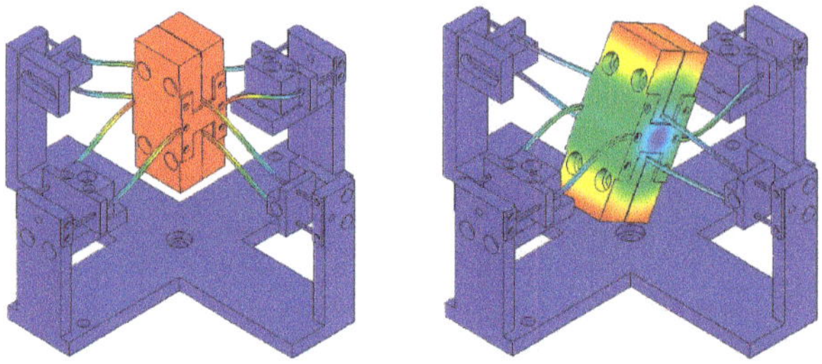

(a) (b)

Figure 13. Mode shapes of TMD2. (**a**) The 1st mode; (**b**) the 2nd mode.

5. Experimental Verification

5.1. Experimental Setup

The laboratory mock-up of the suspension bridge (in Figure 14) is used to investigate the performance of the 2DOFs TMDs. An impact hammer is used to excite the structure. Prior to vibration measurements, the data acquisition system is established, which involves a single-axial accelerometer, positioned to measure vertical accelerations. The position of the accelerometer is illustrated in Figure 14. The output data are obtained by successive hammering all positions on the deck (Figure 15). The modal parameters of the bridge are estimated by hammering method. The natural frequencies and mode shapes of the mock-up are shown in Table 1.

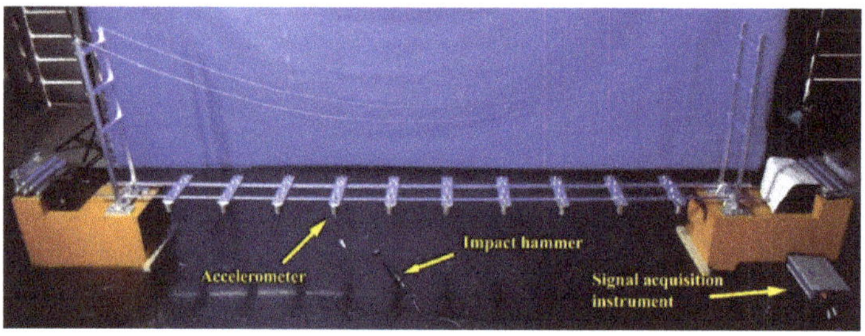

Figure 14. Laboratory mock-up of the suspension bridge.

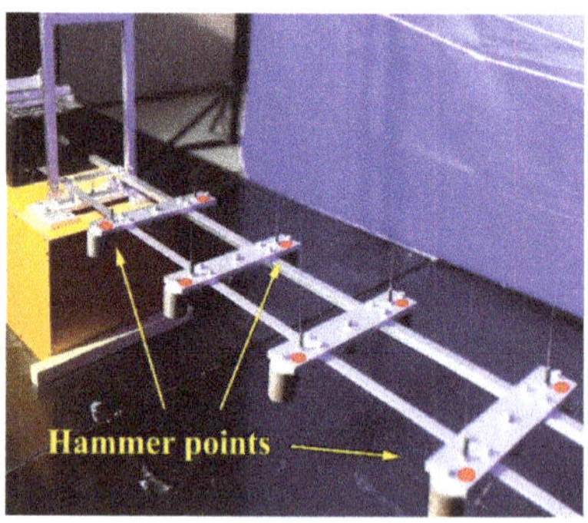

Figure 15. Hammer points on the deck.

5.2. Hammer Tests of TMDs

The structure of 2DOFs TMDs with adjustable natural frequencies is presented in Figure 16. Each TMD consists a mass block, two pedestals, and eight flexible beams with adjusters. For each flexible beam, one side is fixed on the pedestal by bolts and the other side is connected to the mass block. Though changing the positions of the adjuster, the effective length of the flexible beam can be adjusted, then tunable stiffness of the TMD is realized. The adjuster not only improves the tuning ability of TMD, but also compensates for the errors. The errors include the machining error and the calculation error which is caused by ignoring the prestress of flexible beams (Equations (39) and (40)). Thus, the adjuster is an indispensable part of the TMDs.

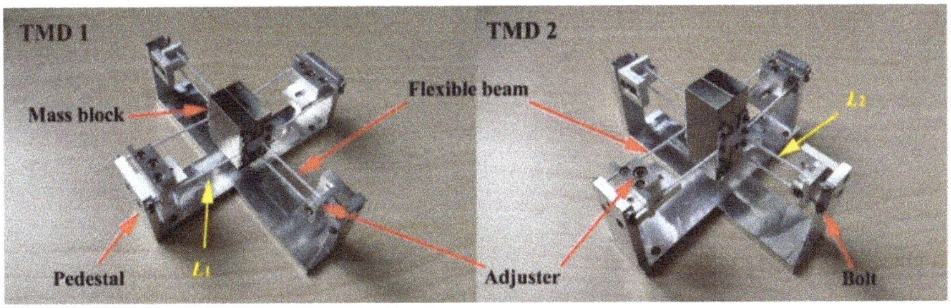

Figure 16. Prototype of the 2 DOFs TMDs.

Based on Equations (39) and (40), the natural frequencies of 2DOFs TMDs are determined by the dimensions of the flexible beams. In order to improve the tuning ability at runtime, the TMDs are designed with tunable L_1 and L_2 which is the effective length of the flexible beams (Figure 16).

In hammer tests, the TMDs are excited by an INV9311impact hammer, and the acceleration responses are recorded by a PSV-500-1D scanning laser vibrometer. The experimental results are shown in Figure 17. The figure shows the experimental acceleration responses of the designed 2DOFs TMDs: (a) when L_1 = 48.5 mm, L_2 = 55.5 mm, the first three order frequency of TMD1 is 4.37 Hz, 10.08 Hz, and 65.81 Hz; (b) when L_1 = 37.5 mm, L_2 = 40 mm, the first three order frequency of TMD2 is 6.21 Hz, 9.47 Hz, and 72.19 Hz. Comparing with the first two order frequency, the third order frequency of TMDs is a relatively large value, which is far beyond the bandwidth (0–15 Hz) we are considering. Therefore, the 2DOFs TMDs meet the design requirements and verify the validity of the theoretical model.

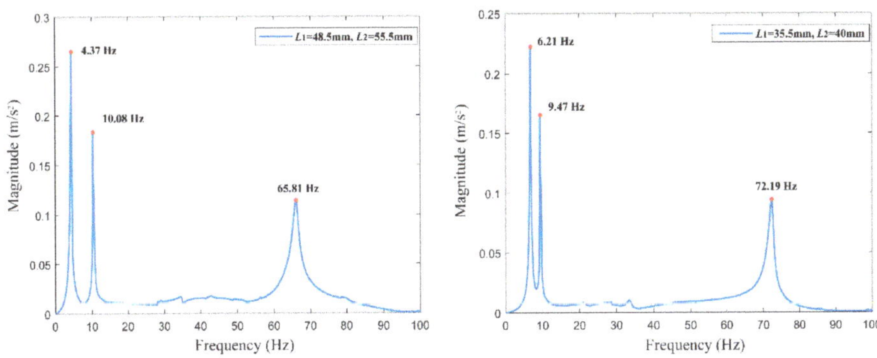

(a) (b)

Figure 17. Experimental acceleration responses of the designed TMDs. (**a**) TMD1; (**b**) TMD2.

5.3. Vibration Suppression of Bridge with TMDs

The designed 2DOFs TMDs are mounted on point A and point B, which has discussed in Section 4.2. Figure 18 shows a close view of the TMD1 (shown in Figure 12); except the different dimension parameters, the two TMDs have the same configuration and can be tuned in the same way. For damping the target modes, the optimum FRF of these two TMDs exhibits two distinct modes at 4.37 Hz, 10.08 Hz, and 6.21 Hz, 9.47 Hz, respectively, as shown in Figure 17.

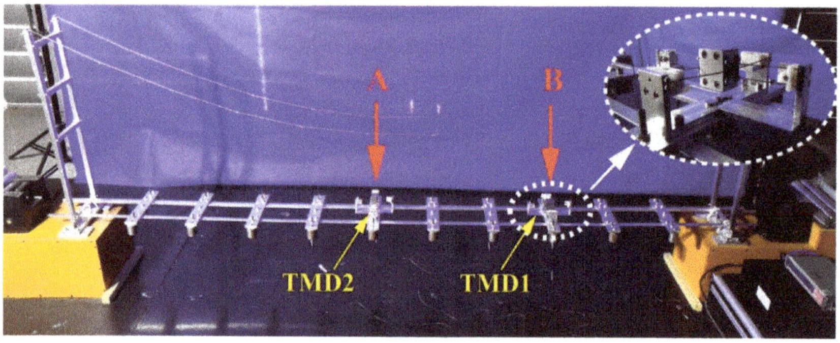

Figure 18. Experimental implementation of the 2DOFs TMDs. The damping is tuned by changing the magnets.

By referring the previous design [27], the damping is introduced by using eddy current damping, where two symmetric powerful magnets are attached on the pedestal below the mass block of the TMD in Figure 19. The damping value can be tuned by manually adjusting the size and number of the magnets. In Figure 19, the replaceable magnets are used to set the translational damping and the rotational damping at the same time. If necessary, another two symmetric magnets which are attached on the pedestal (the position at the red dashed circles in Figure 19) can provide additional damping for the bending mode.

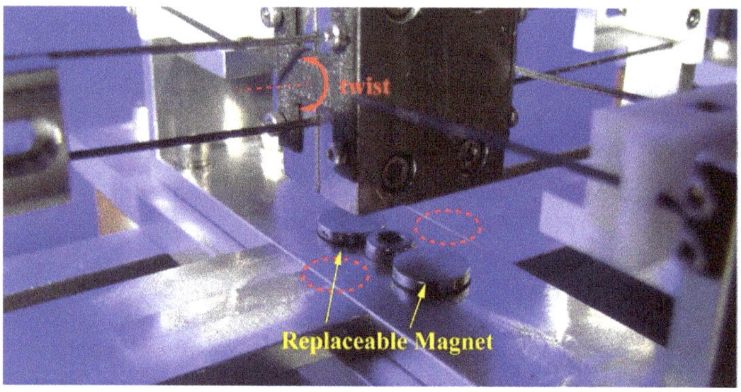

Figure 19. Layout of the symmetric magnets.

The experimental results are shown in Figures 20 and 21. The Figure 20 shows the acceleration responses of the deck: (i) without TMD, (ii) when the structure is equipped with two 2DOFs. The result shows that the 2DOFs TMD is very effective in vibration damping of the bridge. In order to quantitatively confirm the performance of TMDs, Figure 21 shows the FRFs of the deck: (i) without TMD, (ii) when the structure is equipped with two 2DOFs TMDs, targeting the mode pairs (2B,1T) and (1B,2T), respectively. One sees that TMD1 can attain 13.8 dB amplitude reduction of the first bending mode (1B) and 8.8 dB amplitude reduction of the second torsional mode (2T); Meanwhile, a 10.7 dB amplitude reduction of the second bending mode (2B) is observed after using TMD2, and 10 dB amplitude reduction of the first torsional mode (1T).

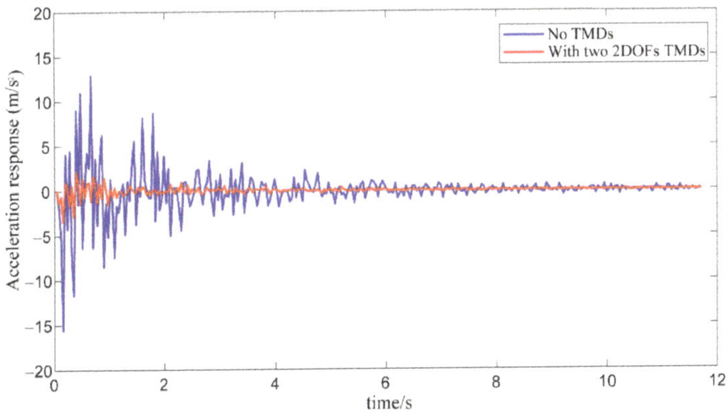

Figure 20. Experimental results: acceleration responses of the deck: blue line stands for without the TMDs; red line stands for with two TMDs.

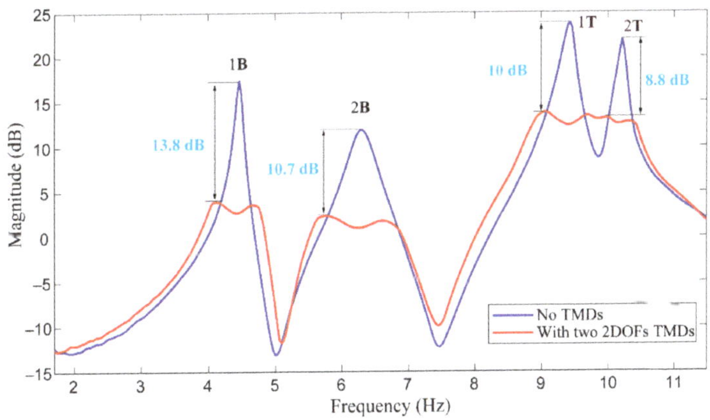

Figure 21. Experimental results: frequency response functions of the deck: blue line stands for without the TMDs; red line stands for with two TMDs of 2 DOF each, targeted for the damping of the mode pairs (2B,1T) and (1B,2T).

The experimental results indicate that the 2DOFs TMDs which designed under the kinematic constraint theory guidance can target the bending and the torsion modes at the same time. Comparing with the classical TMD, the 2DOFs TMD not only achieves vibration reduction, but also avoids increasing the load of the bridge. It is worth mentioning that the final frequency ratios are different from the optimized frequency ratios (Table 4). The optimized frequency ratios are the result of theoretical model calculation. However, the error between the theoretical model and the real structure, as well as the machining and calculation errors, may result in the change of the final frequency ratios. But the role of the optimized frequency ratios in the TMD structural parameter design cannot be neglected.

6. Conclusions

This study proposed a synthetic approach to design and implement 2DOFs TMDs for damping the bending and torsional modes of suspension bridges. For verifying the effectiveness of the concept model, we cast the parameter optimization of the 2DOFs TMDs conceptual model as a control problem with decentralized static output feedback for minimizing the response of the bridge deck. The optimized frequency ratios play a significant role in the TMD structural parameter design. For designing the expected modes of the 2DOFs TMDs, the graphical approach is introduced to arrange flexible beams properly according to the exact constraints. Based on the optimized frequency ratios, the dimensions of TMDs are determined by the compliance matrix method. The proposed design has been simulated and implemented successfully on a suspension bridge mock-up. Based on the presented results and interpretations, the main findings are summarized as follows:

- The synthetic approach based on both the graphical approach and parameterized compliance is an effective way to design the TMD with the expected DOFs (i.e., 1, 2, . . . , 6). It is also an effective complement to the empirical design for the multi-DOFs TMD. Comparing with the empirical design, this synthetic approach can design the expected multi-DOFs TMD without much design experience, which can save time and cost. The disadvantage of this method is that the influence of prestress is not taken into account in the theoretical model. Thus, the adjuster is an indispensable part of TMDs.
- This study verifies the feasibility of the two 2DOFs TMDs in vibration reduction of suspension bridges by numerically and experimentally; comparing with the classical configuration of the TMD, the two 2DOFs TMDs can reduce the weight penalty. The experiment demonstrates the ability of the TMDs for suppressing several vibration modes under laboratory conditions. However, their implementation in a full-scale bridge still needs further research.

Author Contributions: All authors discussed and agreed upon the idea and made scientific contributions. The original idea of the article was provided by F.M. and he also designed the methodology, realized its MATLAB program, and wrote the manuscript; J.W., Y.X. and Y.M. provided with the support of the entire study and revised the manuscript; J.Y. gave suggestions in the theoretical analyses and tests. All authors have read and agreed to the published version of the manuscript.

Funding: This research was funded by the Science and Technology Research Project of State Grid (Research on Key Technologies for Improving Impact Resistance of Suspension Spanning Frame, Grant No. 5442GC180003).

Acknowledgments: The authors thank André Preumont of Active Structures Laboratory from Université Libre de Bruxelles, Brussels, Belgium for his kind guidance and suggestions in the whole research process; and thank the colleagues of FMRL lab from Beihang University, Beijing, China for the realization of the experimental set-up. The reviewers are very much acknowledged.

Conflicts of Interest: The authors declare no conflict of interest.

References

1. Ubertini, F.; Comanducci, G.; Laflamme, S. A parametric study on reliability-based tuned-mass damper design against bridge flutter. *J. Vib. Control.* **2017**, *23*, 1518–1534. [CrossRef]
2. Wang, Z.; Yue, F.; Gao, H. Free Vibration of a Taut Cable with Two Discrete Inertial Mass Dampers. *Appl. Sci.* **2019**, *9*, 3919. [CrossRef]
3. Shi, W.; Wang, L.; Lu, Z.; Zhang, Q. Application of an Artificial Fish Swarm Algorithm in an Optimum Tuned Mass Damper Design for a Pedestrian Bridge. *Appl. Sci.* **2018**, *8*, 175. [CrossRef]
4. Lievens, K.; Lombaert, G.; De Roeck, G.; Van den Broeck, P. Robust design of a TMD for the vibration serviceability of a footbridge. *Eng. Struct.* **2016**, *123*, 408–418. [CrossRef]
5. Caetano, E.; Cunha, Á.; Moutinho, C.; Magalhães, F. Studies for controlling human-induced vibration of the Pedroe Inês footbridge, Portugal. Part 2: Implementation of tuned mass dampers. *Eng. Struct.* **2010**, *32*, 1082–1091. [CrossRef]
6. Van Nimmen, K.; Verbeke, P.; Lombaert, G.; De Roeck, G.; Van den Broeck, P. Numerical and experimental evaluation of the dynamic performance of a footbridge with tuned mass dampers. *J. Bridge Eng.* **2016**, *21*, C4016001. [CrossRef]
7. De Domenico, D.; Ricciardi, G. Improving the dynamic performance of base-isolated structures via tuned mass damper and inerter devices: A comparative study. *Struct. Control Health Monit.* **2018**, *25*, e2234. [CrossRef]

8. De Domenico, D.; Ricciardi, G.; Takewaki, I. Design strategies of viscous dampers for seismic protection of building structures: A review. *Soil Dyn. Earthq. Eng.* **2019**, *118*, 144–165. [CrossRef]
9. Hao, L.F.; Zhang, R.F. Structural safety redundancy-based design method for structure with viscous dampers. *Struct. Eng. Mech.* **2016**, *59*, 821–840. [CrossRef]
10. Preumont, A.; Voltan, M.; Sangiovanni, A.; Mokrani, B.; Alaluf, D. Active tendon control of suspension bridges. *Smart Struct. Syst.* **2016**, *18*, 31–52. [CrossRef]
11. De Domenico, D.; Deastra, P.; Ricciardi, G.; Sims, N.D.; Wagg, D.J. Novel fluid inerter based tuned mass dampers for optimised structural control of base-isolated buildings. *J. Frankl. Inst.* **2018**, *14*, 7626–7649. [CrossRef]
12. Tian, L.; Rong, K.; Zhang, P.; Liu, Y. Vibration Control of a Power Transmission Tower with Pounding Tuned Mass Damper under Multi-Component Seismic Excitations. *Appl. Sci.* **2017**, *7*, 477. [CrossRef]
13. Zhang, R.F.; Zhao, Z.P.; Dai, K.S. Seismic response mitigation of a wind turbine tower using a tuned parallel inerter mass system. *Eng. Struct.* **2019**, *180*, 29–39. [CrossRef]
14. Mekki, O.B.; Bourquin, F.; Maceri, F.; Merliot, E. Unimodal optimal passive electromechanical damping of elastic structures. *Smart Mater. Struct.* **2013**, *22*, 085029. [CrossRef]
15. Mekki, O.B.; Bourquin, F.; Maceri, F.; Van Phu, C.N. An adaptive pendulum for evolving structures. *Struct. Control Health Monit.* **2012**, *19*, 43–54. [CrossRef]
16. Elias, S.; Matsagar, V. Research developments in vibration control of structures using passive tuned mass dampers. *Annu. Rev. Control* **2017**, *44*, 129–156. [CrossRef]
17. Casalotti, A.; Arena, A.; Lacarbonara, W. Mitigation of post-flutter oscillations in suspension bridges by hysteretic tuned mass dampers. *Eng. Struct.* **2014**, *69*, 62–71. [CrossRef]
18. Chen, S.R.; Wu, J. Performance enhancement of bridge infrastructure systems: Long-span bridge, moving trucks and wind with tuned mass dampers. *Eng. Struct.* **2008**, *30*, 3316–3324. [CrossRef]
19. Li, Q.; Fan, J.; Nie, J.; Li, Q.; Chen, Y. Crowd-induced random vibration of footbridge and vibration control using multiple tuned mass dampers. *J. Sound Vib.* **2010**, *329*, 4068–4092. [CrossRef]
20. Daniel, Y.; Lavan, O.; Levy, R. Multiple-tuned mass dampers for multimodal control of pedestrian bridges. *J. Struct. Eng.* **2011**, *138*, 1173–1178. [CrossRef]
21. Tubino, F.; Carassale, L.; Piccardo, G. Human-induced vibrations on two lively footbridges in Milan. *J. Bridge Eng.* **2016**, *21*, C4015002. [CrossRef]
22. Wen, Q.; Hua, X.G.; Chen, Z.Q.; Yang, Y.; Niu, H.W. Control of human-induced vibrations of a curved cable-stayed bridge: Design, implementation, and field validation. *J. Bridge Eng.* **2016**, *21*, 04016028. [CrossRef]
23. Lin, C.C.; Wang, J.F.; Chen, B.L. Train-induced vibration control of high-speed railway bridges equipped with multiple tuned mass dampers. *J. Bridge Eng.* **2005**, *10*, 398–414. [CrossRef]
24. Li, J.; Su, M.; Fan, L. Vibration control of railway bridges under high-speed trains using multiple tuned mass dampers. *J. Bridge Eng.* **2005**, *10*, 312–320. [CrossRef]
25. Luu, M.; Zabel, V.; Könke, C. An optimization method of multi-resonant response of high-speed train bridges using TMDs. *Finite Elem. Anal. Des.* **2012**, *53*, 13–23. [CrossRef]
26. Stăncioiu, D.; Ouyang, H. Structural modification formula and iterative design method using multiple tuned mass dampers for structures subjected to moving loads. *Mech. Syst. Signal Process.* **2012**, *28*, 542–560. [CrossRef]
27. Mokrani, B.; Zhui, T.; Alaluf, D.; Meng, F.H.; Preumont, A. Passive damping of suspension bridges using multi-degree of freedom tuned mass dampers. *Eng. Struct.* **2017**, *153*, 749–756. [CrossRef]
28. Zuo, L.; Nayfeh, S.A. The two-degree-of-freedom tuned-mass damper for suppression of single-mode vibration under random and harmonic excitation. *J. Vib. Acoust.* **2006**, *128*, 56–65. [CrossRef]
29. Jang, S.J.; Brennan, M.J.; Rustighi, E.; Jung, H.J. A simple method for choosing the parameters of a two degree of freedom tuned vibration absorber. *J. Sound Vib.* **2012**, *331*, 4658–4667. [CrossRef]
30. Yang, Y.; Dai, W.; Liu, Q. Design and implementation of two-degree-of-freedom tuned mass damper in milling vibration mitigation. *J. Sound Vib.* **2015**, *335*, 78–88. [CrossRef]
31. Yang, Y.; Dai, W.; Liu, Q. Design and machining application of a two-DOF magnetic tuned mass damper. *Int. J. Adv. Manuf. Technol.* **2017**, *89*, 1635–1643. [CrossRef]
32. Ma, W.S.; Yang, Y.Q.; Yu, J.J. General routine of suppressing single vibration mode by multi-DOF tuned mass damper: Application of three-DOF. *Mech. Syst. Signal Process.* **2019**, *121*, 77–96. [CrossRef]

33. Meng, F.H.; Mokrani, B.; Alaluf, D.; Yu, J.J.; Preumont, A. Damage detection in active suspension bridges: An experimental investigation. *Sensors* **2018**, *18*, 3002. [CrossRef]
34. Meng, F.H.; Yu, J.J.; Alaluf, D.; Mokrani, B.; Preumont, A. Modal flexibility-based damage detection for suspension bridge hangers: A numerical and experimental investigation. *Smart Struct. Syst.* **2019**, *23*, 15–29.
35. Tubino, F.; Piccardo, G. Tuned mass damper optimization for the mitigation of human-induced vibrations of pedestrian bridges. *Meccanica* **2015**, *50*, 809–824. [CrossRef]
36. Yu, J.J.; Li, S.; Pei, X.; Bi, S.; Zong, G. A unified approach to type synthesis of both rigid and flexure parallel mechanisms. *Sci. China Technol. Sci.* **2011**, *54*, 1206–1219. [CrossRef]
37. Hopkins, J.B.; Culpepper, M.L. Synthesis of multi-degree of freedom, parallel flexure system concepts via Freedom and Constraint Topology (FACT)–Part I: Principles. *Precis. Eng.* **2010**, *34*, 259–270. [CrossRef]
38. Yu, J.J.; Li, S.; Su, H.J.; Culpepper, M.L. Screw theory-based methodology for the deterministic type synthesis of flexure mechanisms. *J. Mech. Robot.* **2011**, *3*, 031008. [CrossRef]
39. Jia, M.; Jia, R.P.; Yu, J.J. A Compliance-Based Parameterization Approach for Type Synthesis of Flexure Mechanisms. *J. Mech. Robot.* **2015**, *7*, 031014. [CrossRef]

applied sciences

MDPI

Article

Experimental Campaign of a Low-Cost and Replaceable System for Passive Energy Dissipation in Precast Concrete Structures

Álvaro Mena [1,*], Jorge Franco [2], Daniel Miguel [2], Jesús Mínguez [1], Ana Carla Jiménez [2], Dorys Carmen González [1] and Miguel Ángel Vicente [1]

1 Department of Civil Engineering, University of Burgos, 09001 Burgos, Spain; jminguez@ubu.es (J.M.); dgonzalez@ubu.es (D.C.G.); mvicente@ubu.es (M.Á.V.)

2 ICONKRETE 2012, S.L., 28035 Madrid, Spain; jorge.franco@iconkrete.com (J.F.); daniel.miguel@iconkrete.com (D.M.); ana.jimenez@iconkrete.com (A.C.J.)

* Correspondence: amena@ubu.es; Tel.: +34-947-259423

Received: 29 December 2019; Accepted: 7 February 2020; Published: 11 February 2020

Abstract: This research develops a new low-cost energy dissipation system, capable of being implemented in residential structures in developing countries with high seismic activity, in which the current solutions are not economically viable. These residential structures are entirely made of precast concrete elements (foundations, walls, and slabs). A solution is developed that consists of a new connection between a precast foundation and a structural wall, which is capable of dissipating almost all the seismic energy, and therfore protecting the rest of the building from structural damage. To validate the solution, a testing campaign is carried out, including a first set of "pushover" tests on isolated structural walls, a second set of "pushover" tests on structural frames, and a final set of seismic tests on a real-scale three-storey building. For the first and second set of tests, ductility is analyzed in accordance with ACI 374.2R-13, while for the third one, the dynamic response to a reference earthquake is evaluated. The results reveal that the solution developed shows great ductility and no relevant damage is observed in the rest of the building, except in the low-cost energy dissipation system. Once an earthquake has finished, a precast building implemented with this low-cost energy dissipation system is capable of showing a structural performance level of "immediate occupancy" according to ACI 374.2R-13.

Keywords: seismic test; pushover test; precast concrete structure; shake table

1. Introduction

Precast concrete construction represents a very important percentage of all civil works in the world, given its enormous advantages from a constructive point of view. The reduction of the manufacturing time, the improvement in the quality of the work (due to the improvement of the working conditions), and the reduction of uncertainties related to the geometric and structural deviation of the solution with respect to the project are some of the advantages of this constructive procedure.

Precast concrete is especially efficient in residential structures, which are generally made up of a few groups of different structural elements (columns, slabs, footings, walls, etc.), formed by many identical units. In particular, precast concrete is especially interesting in developing countries, where it can be difficult to find enough skilled labour to perform in situ constructions.

A significant number of these developing countries are in areas of high seismic activity. This is an inconvenience for precast solutions, since it is penalized by international regulations by granting lower reduction coefficients (R) for energy dissipation. This is due to the lower ductility of the connections between elements, that is, the limitation is not due to the precast element by itself, but due to the connections between them, which are usually less ductile than traditional solutions cast in situ.

In these cases, the usual way to address the seismic problem is through the use of seismic isolators, dampers, energy dissipators, etc. However, most of them are very expensive solutions, only suitable for special structural elements (tall buildings or very singular buildings). Therefore, they are not economically viable if massive use is intended in areas with low economic resources [1–8].

Therefore, it is necessary to develop low-cost energy dissipation systems that are capable of being implemented in inexpensive precast concrete buildings without involving an unacceptable increase in the total cost of the building [9–11].

Another common problem regarding the structural behavior of a building that had been subjected to an earthquake was that it was useless after the seismic event and, therefore, it had to be demolished. Regarding the situation of collapse during an earthquake, although it is a breakthrough, the economic cost for the community is still very high. Consequently, it is highly desirable that low-cost energy dissipation systems prevent damage to the structure, and therefore it can be re-occupied under safe conditions once the seismic event has passed.

Research in seismic response of structures, especially if they are made of concrete, requires tests that are usually complex and expensive. On the one hand, the performance of scale tests of concrete structural elements is usually not a viable or reliable option. On the other hand, conducting seismic tests usually requires expensive facilities. Therefore, alternative methodologies have been developed which are easier to implement and obtain, although partially, information on the seismic response of the structure [12,13].

First, there are the quasi-static or cyclic tests, also called "pushover", which consist of the application of a low number of low frequency cyclic loads with increasing amplitude until collapse. This type of test characterizes the ductility of the structure, as well as analyzes very specific regions (connections between elements, singular construction details, etc.) [14–21].

Second, there are pseudo-dynamic tests that are a special type of quasi-static test in which displacements are introduced at some points in the structure. The difference is that these displacements are not known before the test and are calculated during the test using a step-by-step integration software. Although it is essentially a static test, it is a very complex technique to implement, mainly because a sophisticated adaptive control equipment is required [22–25].

Third, there are the tests carried out on a shake table, which introduce a true dynamic excitation in the base of the structure. This is the most realistic technique for the seismic testing of structures, since the displacements (and therefore, the accelerations) are applied at the base and the structure is subjected to the inertial forces. However, it is a very complex test because of all the equipment required. In addition, its interpretation is also difficult, since a large number of structural mechanisms are involved in the seismic response. Therefore, this type of tests is usually carried out at the end of a much more extensive testing campaign [26–33].

This paper shows the design and the laboratory validation tests for a new low-cost energy dissipation system that can be applied in precast concrete structures composed of precast footings, precast structural walls and precast concrete slabs. This energy dissipation system basically consists of a specific connection between the precast footing and the precast structural wall, formed by a set of threaded steel bars that connect both elements. During an earthquake, the steel bars undergo plastic deformation, absorbing most of the energy generated by the earthquake and preventing damage to the rest of the building. The additional advantage of this solution is that steel bars are easily replaced after the seismic event.

For the purpose of this paper, a testing campaign was carried out, based on three phases. First, pushover tests were carried out on isolated structural walls formed by a precast structural wall and a precast footing. The aim of this first phase is to define the ductility of the dissipation system, in accordance with the requirements of the American standard ACI 374.2R-13 [34].

Secondly, pushover tests were carried out on structural frames, composed of two precast structural walls placed over two precast footings and connected with a precast slab. The aim of this second phase is to evaluate the ductility of the system, including the dissipation capacity of both the connection footing wall and the connection wall slab, in accordance with the aforementioned regulation.

Thirdly, seismic tests using a shake table were carried out on a real-scale three-storey precast concrete building, consisting of two precast structural walls placed over two precast footings, two intermediate precast slabs, and a lightweight roof. In this case, the aim of this third phase is to characterize the dynamic response of the entire structure to a reference earthquake and the energy dissipation capacity of the building.

A customized unidirectional shake table was designed and manufactured specifically for the third test phase. This testing facility was capable of appling a horizontal acceleration up to 1·g to the structural elements with a height up to 6 m, a weight up to 40 tons, and a frequency up to 8 Hz.

All the precast concrete elements, as well as the low-cost energy dissipation system were invented, developed, and designed by the Spanish company ICONKRETE 2012, S.L., and therefore this company is the owner of this structural solution and the testing results. The precast elements were manufactured by the company ZENET in its factory in Escalonilla (Toledo, Spain). The test was carried out in the Laboratory of Large Structures of the University of Burgos (Burgos, Spain).

The structure of this paper is as follows: In Section 2, the experimental program is presented; in Section 3, the results of the tests are described and discussed; and finally, in Section 4, the conclusions are shown.

2. Experimental Program

In this section, the three testing sets of specimens are described, i.e., the isolated structural walls, the structural frames, and the real-scale three-storey precast concrete building, as well as the low-cost energy dissipation system and the shake table. Additionally, the testing procedure is described.

2.1. Isolated Structural Wall

As explained before, the isolated structural walls are composed of a precast structural wall and a precast footing, connected through the low-cost energy dissipation system.

The structural wall is a conventional reinforced concrete precast element 3.0 m high, 2.0 m wide, and 16 cm thick. The wall is placed on a reinforced concrete precast footing, 1.1 m wide, 2.0 m deep, and 0.6 m high. In both cases, the concrete quality is C30/37, according to Eurocode 2 [35]. The compressive strength was obtained following the method described in standard EN 12390-3 [36]. The footing includes a longitudinal pocket 36 cm wide and 36 cm high, where the structural wall is placed. Underneath the wall, a 2 cm neoprene band is placed. The structural wall is rigidly connected to the footing through the low-cost energy dissipation system, which is described later (Figure 1).

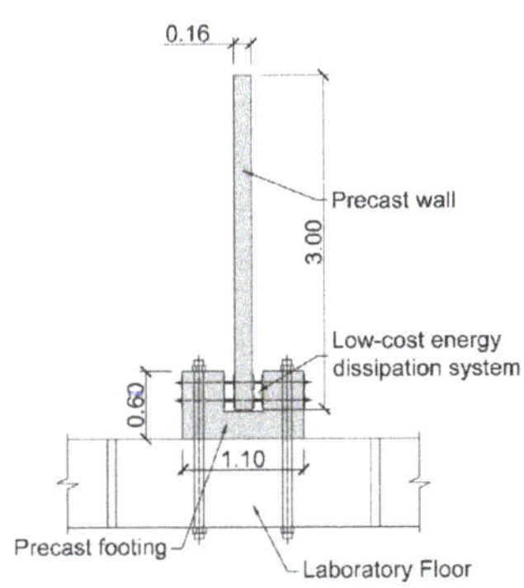

Figure 1. Elevation view of the isolated structural wall.

The structural wall is reinforced on both faces with steel rebars 8 mm in diameter spaced 150 mm in both longitudinal and vertical directions.

2.2. Structural Frame

The structural frame is composed of two isolated structural elements, as described above, connected with a structural concrete slab. The inner distance between the walls is 3.84 m. The slab is composed of a self-supporting precast prestressed concrete slab with a thickness of 8 cm and an upper layer of 14 cm of cast in situ reinforced concrete. The total thickness of the concrete slab is 22 cm. In all cases, the concrete quality is C30/37, according to Eurocode 2 [35].

A "flexible" connection between the walls and the slab was used, which is described next. The self-supporting precast prestressed concrete slab rests on a steel corner profile anchored to the walls using a set of mechanical anchorages. Additionally, a row of conventional steel rebars sew the joint between the wall and the cast-in situ slab. The rebars are L-shaped with a length of 800 and 200 mm, respectively. The long leg of the rebar is placed horizontally inside the cast-in situ reinforced concrete slab, while the short leg is placed vertically inside the wall. The diameter of the rebars are 20 mm, with a spacing of 50 cm. From the structural point of view, this connection is very effective under vertical loads, such as self-weight, dead loads, and vertical live loads. Under horizontal loads, similar to the ones caused by the earthquake, this connection is able to withstand negative bending moments, but not positive bending moments, resulting in a semi-rigid joint. (Figure 2).

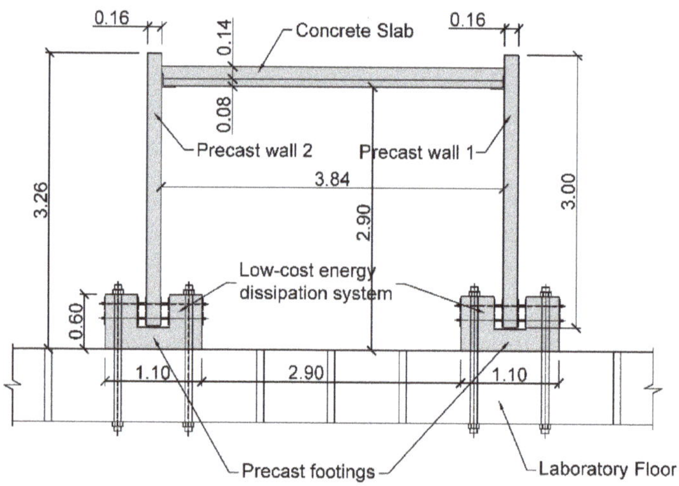

Figure 2. Elevation view of the structural frame.

2.3. Real-Scale Three-Storey Precast Concrete Building

Finally, the real-scale three-storey precast concrete building is composed of two precast concrete walls, two precast footings, two concrete slabs, and a flexible steel roof. The inner distance between the two walls is 2.25 m. In all cases, the concrete quality is C30/37, according to Eurocode 2 [35].

The structural walls are conventional reinforced concrete precast elements 5.62 m high, 2.0 m wide, and 16 cm thick. They are placed on the same reinforced concrete precast footings as described above.

Each of the two concrete slabs is composed of a self-supporting precast prestressed concrete slab with a thickness of 8 cm and an upper layer of 14 cm of cast-in situ reinforced concrete. The connection between the slab and the wall is the same as the one described in the previous subsection. The slabs are located at 2.45 m and 4.20 m high. Finally, the roof is located at the top of the walls. It is a lightweight roof formed by an aluminium sheet which is bolted to two I-beams. (Figure 3).

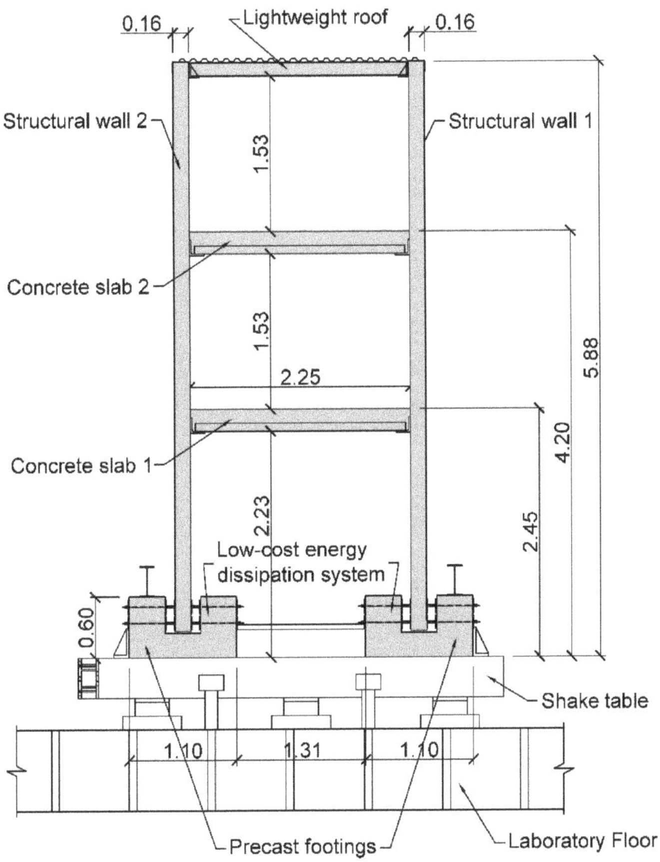

Figure 3. Elevation view of the real-scale three-storey precast concrete building, including the shake table.

2.4. Low-Cost Energy Dissipation System

The low-cost energy dissipation system is a device placed on the connection between the footing and the wall. As previously explained, the footing includes a longitudinal pocket 36 cm wide and 36 cm high, where the structural wall is placed. It consists of a set of threaded bars placed orthogonally to the wall, in such a way that they cross both the footing and the wall. In this case, the system consists of a total of 10 threaded bars, 20 mm diameter, and placed in two rows, spaced 300 mm in both longitudinal and vertical directions. The property class of the steel for the threaded bars is 3.6, according to ISO 898-1 [37].

Since the wall has a thickness of 16 cm, there are two 10 cm gaps between the wall and the footing, one at each side of the wall (Figure 4). Additionally, nuts located at both sides of the wall and the footing are required to fix the wall in its proper position.

When an earthquake occurs, it is expected that most of the energy is consumed in the plastic deformation of the bars, preventing the rest of the structure from damage. The number, position, distribution, and diameter of the bars, as well as the steel quality must be specifically designed to each particular structure, depending on the dimensions of the structure and the location of the building.

One additional advantage of this solution is that bars are easily replaceable once an earthquake has occurred.

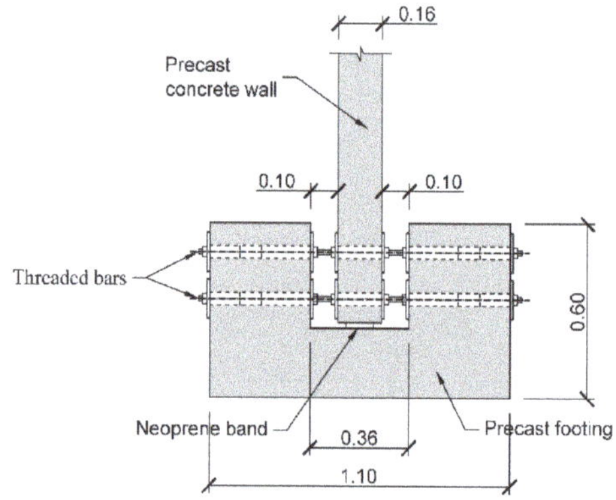

Figure 4. Scheme of the low-cost energy dissipation system.

2.5. Shake Table

A unidirectional shake table was specifically designed and manufactured for this research, in order to fulfill the testing requirements of the real-scale three-storey precast concrete building.

The shake table is composed of a cast steel slab 4.14 m long, 2.87 m wide, and 0.65 m high. It rests on six circular elastomeric bearing pads, reinforced with steel plates. The dimensions of the bearings are 350 mm diameter and 137 mm high.

The seismic loads are applied using a tension-compression MTS 201.70F dynamic actuator (MTS, Eden Prairie, MN, USA), with a capacity of ±1000 kN. The actuator has a load cell MTS 661.31F-01 (MTS, Eden Prairie, MN, USA), with a range of ±1000 kN and an error of below 1% of the range. During the testing, the actuator provided a longitudinal displacement. In order to prevent undesirable transversal movements, four stoppers were placed in both lateral sides of the steel slab (Figures 5 and 6).

Figure 5. Shake table.

Figure 6. Detailed view of the dynamic actuator.

2.6. Testing Procedure

2.6.1. Pushover Tests on Isolated Structural Walls and Structural Frames

The pushover tests carried out on the isolated structural walls and the structural frames were developed according to the American standard ACI374.2R-13 [34]. According to this standard, four levels of structural performance under seismic events are defined, which are "operational", "immediate occupancy", "life safety", and "collapse prevention", moving from least to greatest lateral drift ratio (Figure 7).

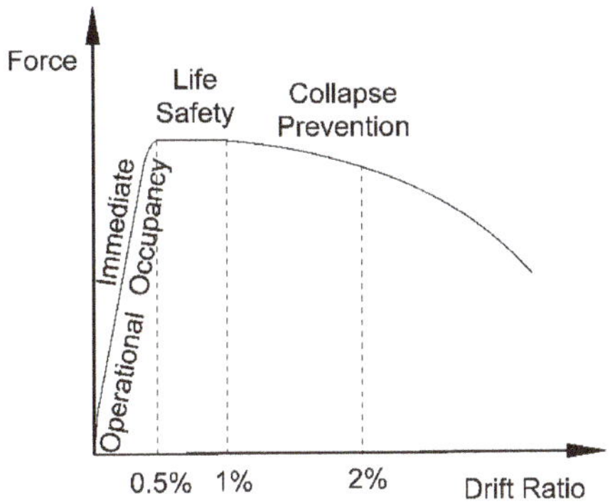

Figure 7. The four levels of structural performance, according to ACI 374.2R-13 [34].

According to the goal of this research, the "immediate occupancy" level is considered for design purposes, which implies that the building can be used once the seismic event has passed. At this level, the stiffness requirements are high, the behavior of the structure must be within the elastic-linear range and, consequently, the seismic loads developed are important. The standard used does not consider other criteria apart from those concerning structural damage. That is the reason why non-structural issues, such as furniture overturning or partition damage, have not been considered in this research. However, it is worth noting that the "immediate occupancy" level implies a very low risk of life-threatening injury as a result of structural damage.

The lateral drift ratio is defined as the quotient between the horizontal displacement of the structure at the loading point and the distance from this point to the centroid of the connection between the wall and the footing. In this case, the maximum allowable lateral drift ratio (see Figure 7) is 0.5%. Additionally, some other criteria must be fulfilled [34]:

1. No plastic behavior is observed in the structure, neither concrete nor steel rebars;
2. Crack width should be below 1.6 mm;
3. No concrete crushing is observed.

Regarding the testing procedure, it consisted of applying a quasi-static horizontal load near the top of the structural element (wall or frame). Displacement controlled reverse cyclic tests were performed. The number of loading cycles for each amplitude of the imposed displacement was kept at two. The value of the amplitude depends on the critical drift φ_y, which is defined as the drift associated with yielding. The first couple of cycles corresponded to a drift equal to $0.5{\cdot}\varphi_y$, the second couple of cycles, to a drift equal to φ_y, the third one, to a drift equal to $2{\cdot}\varphi_y$, and for the remaing couple of cycles the amplitude increased in $1{\cdot}\varphi_y$, i.e., $3{\cdot}\varphi_y$, $4{\cdot}\varphi_y$, $5{\cdot}\varphi_y$, and so on to the conclusion of the test. The test concluded when the maximum load of one cycle was more than 20% lower than the total maximum load of all cycles, according to ACI 374.2R-13 [34].

In the case of the tests on the isolated structural walls, the load was applied at a height of 2.5 m and the critical drift φ_y corresponded to a horizontal displacement of 10.6 mm at the load application point. In the case of the tests on the structural frames, the load was applied at a height of 2.94 m and the critical drift φ_y corresponded to a horizontal displacement of 12.6 mm at the load application point.

The tests on both the isolated structural walls and the structural frames were carried out using a tension-compression MTS 201.70F dynamic actuator (MTS, Eden Prairie, MN, USA), with a capacity of ±1000 kN. The actuator had a load cell MTS 661.31F-01 (MTS, Eden Prairie, MN, USA), with a range of ±1000 kN and an error of below 1% of the range. The tests were displacement controlled. This testing procedure provided greater safety against unexpected collapse and it better meet the requirement of the ACI 374.2R-13 [34].

In order to analyze the structural behavior of the testing specimens, a number of sensors were used, including inclinometers (model PST300, Pewatron AG, Zurich, Switzerland), linear potentiometer displacement transducers (ranged from 50 to 500 mm, Novotechnik, Ostfildern, Germany), and uniaxial strain gauges (150 mm length, Tokyo Sokki Kenkyujo Co., Ltd., Tokyo, Japan). Figures 8 and 9 show the position of the sensors in each of the two sets of testing specimens, i.e., the isolated structural walls and the structural frames.

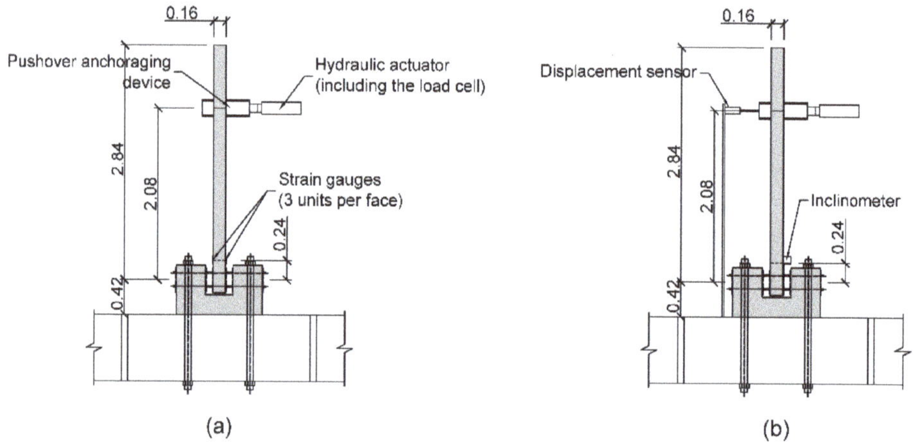

Figure 8. (**a** and **b**) Location of the sensors in the isolated structural walls.

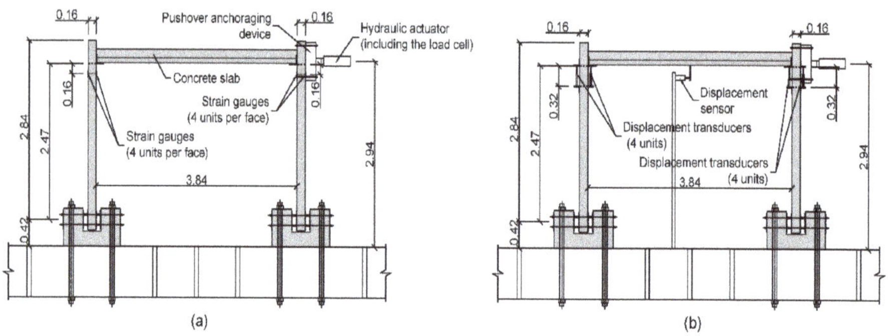

Figure 9. (**a** and **b**) Location of the sensors in the structural frames.

2.6.2. Seismic Tests on Real-Scale Three-Storey Precast Concrete Building

The third phase consists of seismic tests on a real-scale three-storey precast concrete building. In this case, a representative earthquake was reproduced in the laboratory, namely "El Centro" (an earthquake that occurred in the city of El Centro, California, USA in 1979). More specifically, the accelerogram belonging to an orientation of 220° was considered, because this was the most unfavourable in terms on maximum horizontal accelerations. The accelerogram was obtained from the USGS (United States Geological Survey, USA) (Figure 10a).

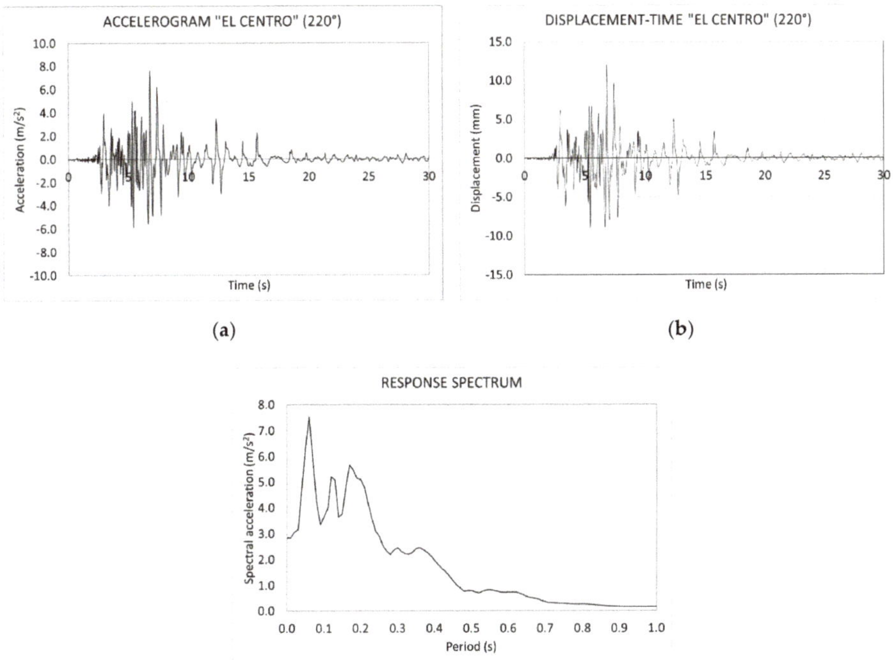

(a) (b) (c)

Figure 10. Earthquake "El Centro". (**a**) Accelerogram; (**b**) diagram of horizontal displacement vs. time; (**c**) response spectrum.

The accelerogram is first transformed in a diagram of horizontal displacement verus time, which is the input signal introduced to the dynamic actuator control equipment (Figure 10b). Additionally, the response spectrum of the earthquake is shown (Figure 10c).

Figure 10c reveals that this earthquake causes the highest acceleration for structures with a natural period of 0.06 s, i.e., for structures with a natural frequency of 16.7 Hz. However, this earthquake is not only very dangerous for rigid structures, but it also provides acceleration values greater than the ground acceleration up to a period of 0.25 s, i.e., a frequency of 4 Hz.

The earthquake was not applied directly on the building, but in a progressive way, similar to foreshocks before the main earthquake. A total of six foreshocks were applied to the building before the main earthquake. To perform it, the ordinate of the seismic signal (i.e., the diagram of horizontal displacement vs. time) was multiplied by a factor. For the first foreshock, the factor was 0.05, i.e., the shape of this earthquake is homothetic to the real "El Centro" earthquake but the displacements are only 5% of the main earthquake. For the rest of the foreshocks, the factors were 0.1, 0.3, 0.5, 0.7, and 0.9 respectively. Finally, the main earthquake was applied.

The sensors used to monitor this test were load cell, accelerometers on the shake table, the intermediate slabs at the top of the walls, and displacement transducers in several positions (Figure 11).

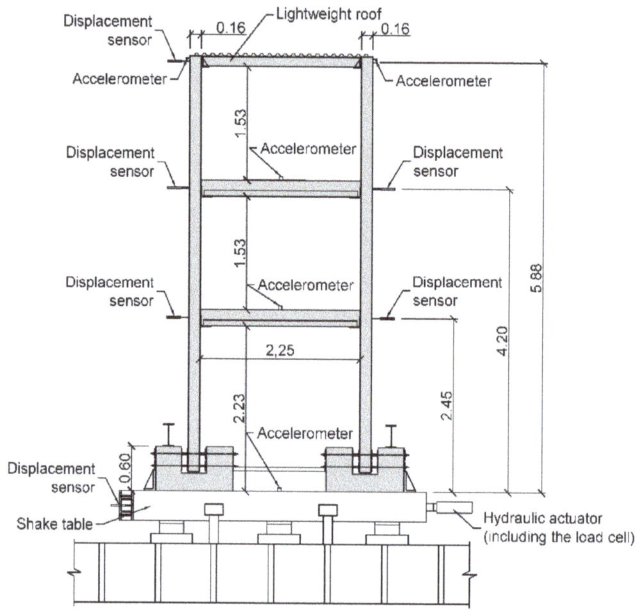

Figure 11. Location of the sensors in the building.

Figure 12 shows a general view of the testing. As can be observer, in addition to the self-weight of the building, a dead load of 1.50 kN/m^2 was included on the intermediate slab. This load was materialized using water tanks.

Figure 12. General view of the real-scale three-storey precast concrete building.

Additionally, a one-cycle impulsive test was performed, before and after the seismic tests, in order to measure the natural frequency and the damping ratio of the building and to compare them with the excitation frequency of the earthquake. Moreover, the comparison of the natural frequency and damping ratio values, before and after the seismic event (including the main earthquakes and the foreshocks), provide useful information about the damage caused by the seismic tests.

In this case, a displacement-time one-cycle square wave signal was applied, with an excitation frequency of 5 Hz and an amplitude of 1 mm.

3. Experimental Results and Discussion

Next, the experimental results of the testings are exposed. In this case, the results of the most representative test of each phase are shown.

3.1. Pushover Tests on Isolated Structural Wall

As previously explained, the aim of this testing phase is to evaluate the ductility of the low-cost energy dissipation system by considering diagrams of horizontal load versus drift, drift versus strain in concrete, and bending moment versus rotation.

3.1.1. Diagram of Horizontal Load versus Drift

Figure 13 shows the diagram of horizontal load versus lateral drift (hysteresis loops). In this case, positive values mean push, and negative values mean pull.

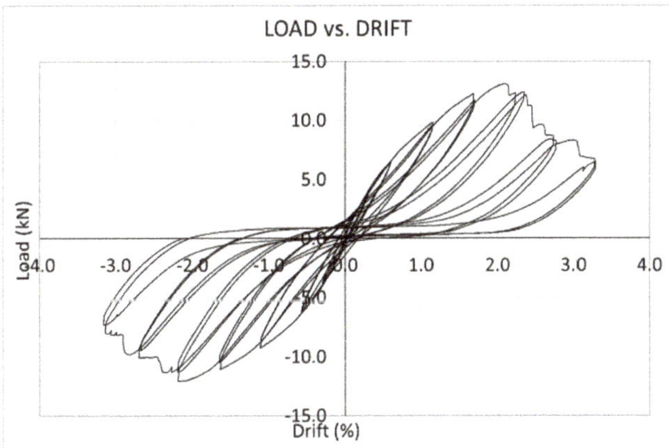

Figure 13. Diagram of load vs. drift. Pushover test on isolated structural wall.

Some interesting conclusions can be obtained from Figure 13. First, a symmetric behavior under push and pull is obsesrved, as expected. Second, it can be observed that the behavior of the specimen is linear elastic up to the critical drift φ_y (which is 0.5% according to ACI374.2R-13 [34]). In consequence, this solution agrees with one of the requirements of the American standard mentioned above.

Once the drift is greater than 0.5%, a progressive plastic behavior is observed, i.e., the specimen begins to dissipate energy at the expense of a higher deformation. The area enclosed by the hysteresis loop is proportional to the energy dissipated during the testing and represents the structural element capacity to mitigate the earthquake effect inelastically.

The maximum loads obtained during the testing were 13.0 kN in the push phase and 12.0 kN in the pull phase. In both cases, these loads correspond to a drift of 2%, which is four times greater than the critical drift. The loads obtained at the critical drift were 6.5 kN in the push phase and 6.2 kN in the pull phase.

3.1.2. Diagrams of Drift versus Strain in Concrete

Figure 14 shows the relationship between the drift of the wall and the average vertical strain of the concrete at the base of the wall, at both the dorsal face (where the actuator is placed) and the frontal face (the opposite side). In this case, a positive value of strain denotes tension, and a negative value denotes compression.

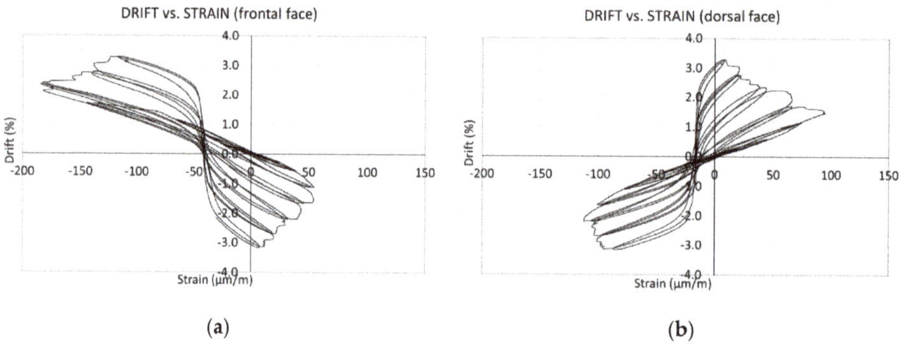

Figure 14. Diagrams of drift vs. vertical strain. Pushover test on isolated structural wall. (**a**) Drift versus strain in the frontal face; (**b**) drift versus strain in the dorsal face.

Figure 14 shows a linear elastic behavior of concrete through the test, since the maximum measured strain is around 200 μm/m under compression and 90 μm/m under tension, which is smaller than maximum elastic strain of concrete (which can be estimated around 1000 μm/m under compression and 100 μm/m under tension). This is particularly true for the critical drift φ_y, where the maximum measured strain is significantly smaller (around 50 μm/m under compression and 10 μm/m under tension). The results satisfy the ACI374.2R-13 [34].

The measured vertical strain values are in accordance with the visual inspections carried out at the end of the tests, where no visible cracks in concrete wall were observed, and, of course, no concrete crushing occured.

The diagrams in Figure 14 show an asymmetric behavior, i.e., compression strains are greater than tension strain. This could be because under tension, small microcracks in concrete occur, relaxing tension stress in concrete (and as a counterpart increasing the tension stress of the reinforcement), resulting in smaller values of tension strain.

This result confirms that the plastic behavior shown by the isolated structural wall is completely caused by the low-cost energy dissipation device. Moreover, once the plastic behavior of the structural wall is observed, a progressive decrease of the maximum vertical strain of concrete occurs. This is because the elastic energy stored in the wall progressively flows to the energy dissipation device, preventing the wall from structural damage.

3.1.3. Diagram of Bending Moment versus Rotation

Figure 15 shows the diagram of bending moment versus rotation of the connection between the wall and the footing. The bending moment is defined as the product of the horizontal force and the vertical distance between the force and the centroid of the low-cost energy dissipation system. The rotation is measured using an inclinometer placed at the level of the centroid of the energy dissipation device (see Figure 8). Positive values mean push, and negative values mean pull.

Figure 15 confirms the findings shown in previous figures. On one side, it is observed that up to the critical drift φ_y, the connection shows a linear elastic behavior. The slopes of the curves are high and quite similar under loading and unloading. The area under the hysteresis loop is small, which means that there is no energy dissipation. Additionally, no lose of stiffness is observed.

Once the structure reaches a drift of 2%, a plastic behavior starts to occur. Then, a progressive decrease of the stiffness of the connection is observed. The area under the hysteresis loop gradually increases, which reveals that there is a progressive energy dissipation. In general, a symmetric behavior of the connection is observed.

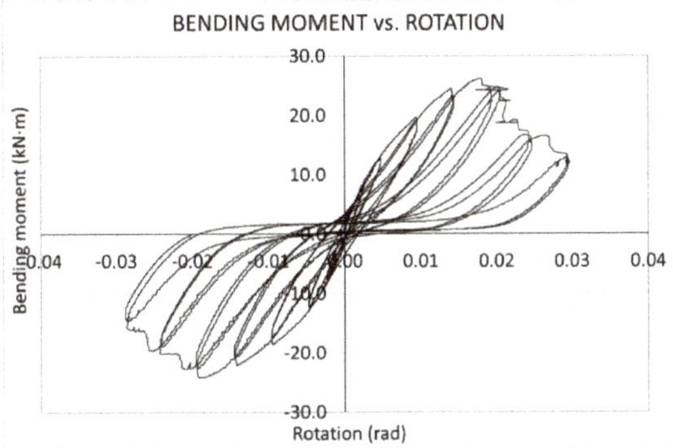

Figure 15. Diagram of bending moment vs. rotation. Pushover test on isolated structural wall.

3.2. *Pushover Tests on Structural Frames*

As previously explained, the aim of this testing phase is to evaluate the ductility of the frame, including the low-cost energy dissipation system, as well as the "flexible" connection between the slab and the wall. To obtain it, the diagrams of load versus drift, drift versus strain in concrete, and bending moment versus rotation are shown.

The tests carried out on the structural frames are not reversal (i.e., push and pull), but they are push and "unpush" (i.e., push the frame up to the maximum displacement of each cycle and return it back to the displacement until zero).

3.2.1. Diagram of Horizontal Force versus Drift

Figure 16 shows the diagram of horizontal load versus lateral drift (hysteresis loops).

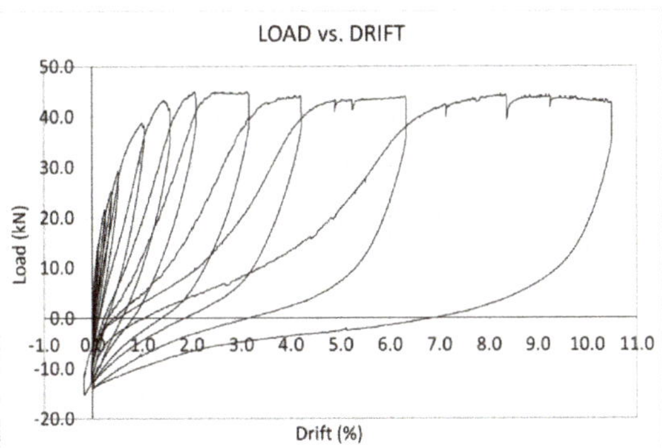

Figure 16. Diagram of load vs drift. Pushover test on structural frame.

Some interesting findings can be observed from Figure 16. First, it is highlighted that a linear-elastic behavior is observed until the critical drift φ_y. This result agrees with the American standard ACI374.2R-13 [34]. Once the critical drift is reached, a progressive plastification of the structure occurs. The area under the hysteresis loop progressively increases, which denotes that an energy dissipation process occurs.

The maximum load is reached for a drift of 3%, i.e., six times the critical draft. Beyond this value, the load does not increase or decrease, but it remains almost constant. However, the area under the hysteresis loops significantly increases. The structural solution shows a huge capacity of energy dissipation without losing structural capacity.

3.2.2. Diagrams of Drift versus Strain in Concrete

Figure 17 shows the relationship between the drift of the walls and the average vertical strain of the walls' concrete, near but below the connection with the slab, at both the internal and the external faces. In this case, a positive value of strain denotes tension, and a negative value denotes compression.

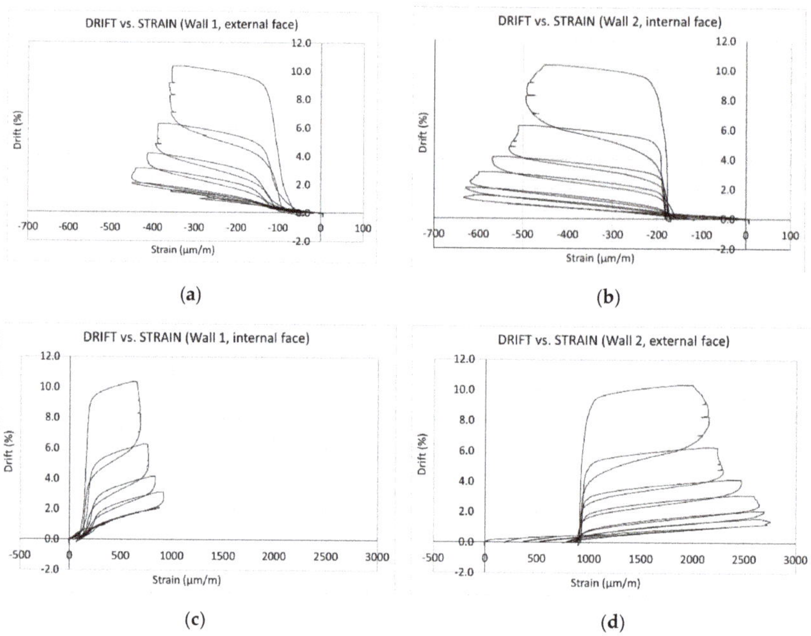

Figure 17. Diagrams of drift vs. vertical strain. Pushover test on structural frame. (**a**) Drift versus strain in the external face of wall 1; (**b**) drift versus strain in the internal face of wall 2; (**c**) drift versus strain in the internal face of wall 1; (**d**) drift versus strain in the external face of wall 2.

Figure 17 reveals the behavior of the connection between the slab and the wall. First, a linear-elastic behavior of the connection, up to a drift of 2%, is observed. In Wall 1, the one in contact with the actuator, compression strain is observed in the exterior face, as well as tension strain in the interior face. On the contrary, in Wall 2, tension strain is observed in the exterior face, as well as compression strain in the interior face. In each loading cycle, the loading and the unloading branches are almost identical and the area under the hysteresis loop is small, which denotes an absence of energy dissipation. During this first phase of the testing, the visual inspections revealed very small horizontal cracks in the walls (especially in the external face of Wall 2 where the tension strain was larger) with a crack width below 0.2 mm, i.e., clearly smaller than 1.6 mm which is the maximum allowable crack width defined by the ACI374.2R-13 [34]. No concrete crushing occurred.

The maximum measured strain values belong a drift of 2%. Beyond this value, there was a progressive decrease in the maximum measured strain, which denotes that the stiffness of the connection between the slab and the wall decreased and a plastic hinge appeared in this connection. Moreover, this plastic hinge showed an asymmetric behavior, i.e., its structural behavior was different when it was subjected to a positive bending (tension in the inner face of the wall and the lower face of the slab) or a negative bending (tension in the outer face of the wall and the upper face of the slab).

Because of the type of loading cycles of the test, the connection between Wall 1 and the slab was always under positive bending while the connection between Wall 2 and the slab was always under negative bending. Wall 1 showed values of strain (both tension and compression) lower than the values observed in Wall 2. This means that the connection shows a stiffness under positive bending smaller than the one under negative bending.

Once the drift was beyond 2%, a clear plastic behavior started to be observed. In each loading cycle, the loading and the unloading branches were different and the area under the hysteresis loop increased with the cycles, which denoted an increased ability of the connection to dissipate energy. In this case, the visual inspections carried out during the testings revealed small horizontal cracks in the walls (especially in the external face of Wall 2 where the tension strain was larger). However, the crack widths were always below 1.6 mm which is the maximum allowable crack width defined by the ACI374.2R-13 [34]. No concrete crushing occurred.

3.2.3. Diagram of Bending Moment versus Rotation

Figure 18 shows the diagrams of bending moment versus rotation of the connections between Walls 1 and 2 and the slab. The bending moment is defined as the product of the horizontal force and the vertical distance between the force and the centroid of the low-cost energy dissipation system. This is, in fact, a "global bending moment" of the frame, and not the real moment of the connection between the wall and the slab. The rotation is defined as the variation of the inner angle between the wall and the slab.

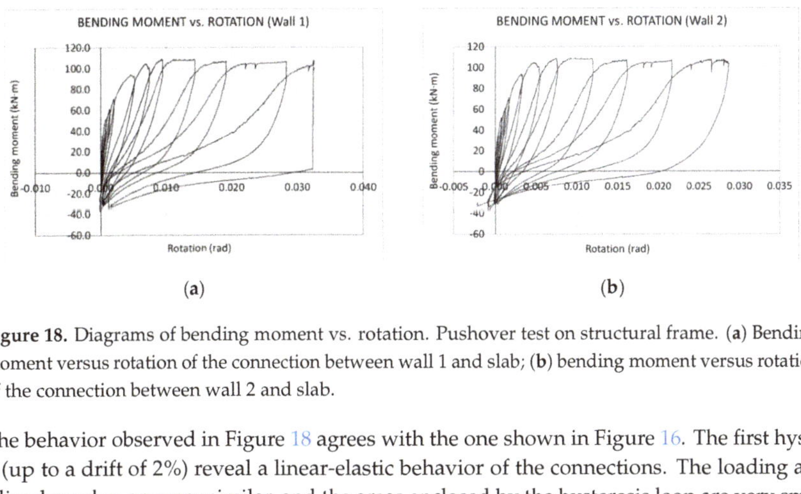

(a) (b)

Figure 18. Diagrams of bending moment vs. rotation. Pushover test on structural frame. (**a**) Bending moment versus rotation of the connection between wall 1 and slab; (**b**) bending moment versus rotation of the connection between wall 2 and slab.

The behavior observed in Figure 18 agrees with the one shown in Figure 16. The first hysteresis cycles (up to a drift of 2%) reveal a linear-elastic behavior of the connections. The loading and the unloading branches are very similar, and the areas enclosed by the hysteresis loop are very small.

Beyond a drift of 2%, the structure begins to show a plastic behavior. In each hysteresis loop, the slope of the curve bending moment versus rotation progressively decreases, and the area enclosed by the hysteresis loop progressively increases. Consequently, the energy dissipation capacity of the connection between the wall and the slab increases. Additionally, the permanent rotation corresponding to null bending moment increases in each cycle, which denotes that the connection suffers damage in each cycle.

An unexpected behavior is observed during the last cycle in the diagram concerning Wall 1. In particular, there is an interruption of the data capture from rotations around 0.03 rad. This can be explained because the measurement range of the transducers used to calculate rotations was exceeded. As a result, the transducers are detached from the concrete surfaces and there is no data collection.

When both connections are compared, it is observed that the one placed in Wall 1 (and consequently subjected to positive bending) shows less stiffness than the one in Wall 2 (subjected to negative bending). This finding agrees with the results of Figure 17. Moreover, the permanent rotation corresponding to null bending moment in the connection of Wall 1 is larger than the one in Wall 2.

3.3. Seismic Tests on Real-Scale Three-Storey Precast Concrete Building

Once the two testing phases have been completed (the first one on isolated structural walls and the second one on structural frames), the seismic tests on a real-scale three-storey precast concrete building were performed. The aim of this third testing phase is to validate the structural solution implemented on a real building subjected to an earthquake. In this case, the structural behavior of the building under the seismic events mainly depends on the connections, both the low-cost energy dissipation systems placed on the connections between the walls and the footings and the flexible connections between the walls and the slabs.

To obtain it, the following parameters are monitored during a real seismic event: Longitudinal displacement of the shake table, longitudinal displacements of the building at the storey levels, and longitudinal accelerations of the shake table and the building. The results shown in the following figures belongs only to the main earthquake and not to the foreshocks.

3.3.1. Longitudinal Displacement of the Shake Table and the Building

Figure 19 shows the diagrams of the longitudinal displacement versus time during the "El Centro" earthquake at the following locations: shake table, Concrete Slab 1, Concrete Slab 2 and lightweight roof.

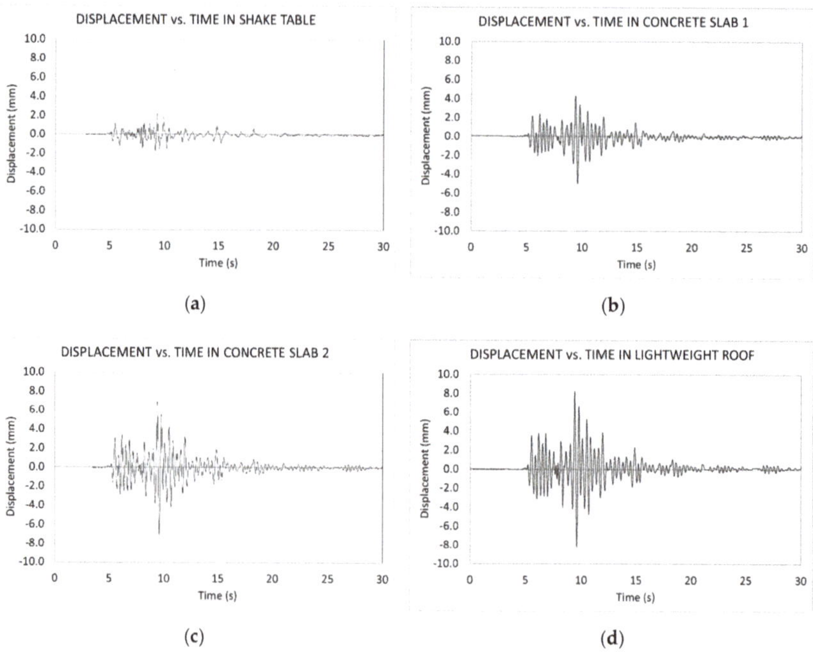

Figure 19. Diagrams of longitudinal displacement vs. time at different locations, from shake table to lightweight roof. (**a**) Displacement versus time in the shake table; (**b**) displacement versus time in concrete slab 1; (**c**) displacement versus time in concrete slab 2; (**d**) displacement versus time in lightweight roof.

Figure 19 reveals some interesting findings. First, a progressive increase of the longitudinal displacement with the height is observed. The measured maximum longitudinal displacements values are 2.18, 4.99, 7.15, and 8.23 milimeters for shake table, Concrete Slab 1, Concrete Slab 2, and flexible roof, respectively. The displacement shows almost a linear trend from shake table to Concrete Slab 2, while the variation is much smaller from Concrete Slab 2 to flexible roof.

Moreover, the visual inspections carried out after the seismic tests revealed that no structural damage is observed in the building (i.e., no cracks in the walls or slabs appeared and, of course, no concrete crushing occurred). This means that the seismic energy was completely dissipated by the connections, i.e., by the low-cost energy dissipation systems placed on the connections between the walls and the footings and the flexible connections between the walls and the slabs. The main aim of the research, which is the design and validation of a low-cost energy dissipation system, as well as the flexible connection between the walls and the slabs, has been reached.

At the end of the seismic event, the residual longitudinal displacements of the both concrete slabs and the flexible roofs are almost zero, which means that the building recovers its original position, that is, the walls recover their upright position.

Finally, it is concluded that the building reached the performance level of "immediate occupancy", according to ACI374.2R-13 [34].

3.3.2. Longitudinal Accelerations of the Shake Table and the Building

Figure 20 shows the diagrams of the longitudinal acceleration versus time during the "El Centro" earthquake at the following locations: shake table, Concrete Slab 1, Concrete Slab 2, and lightweight roof.

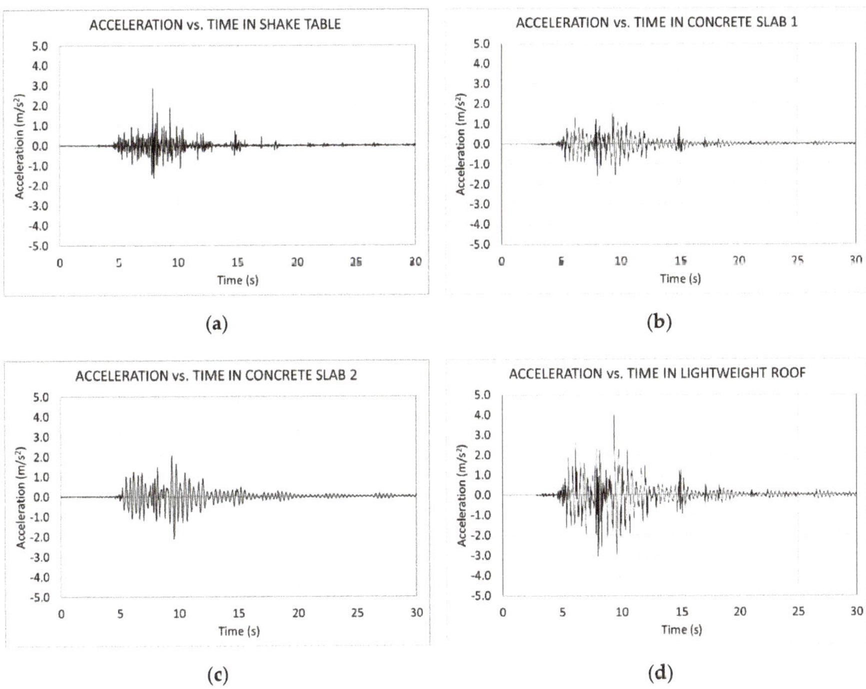

Figure 20. Diagrams of longitudinal acceleration vs. time at different locations, from shake table to lightweight roof. (**a**) Acceleration versus time in shake table; (**b**) acceleration versus time in concrete slab 1; (**c**) acceleration versus time in concrete slab 2; (**d**) acceleration versus time in lightweight roof.

Figure 20 reveals that this structure is especially resistant to the earthquake "El Centro", since the maximum measured longitudinal accelerations of the concrete slabs is smaller than the one on the shake table. The maximum measured longitudinal acceleration at the lightweight roof is a bit larger than the one on the shake table. Specifically, the maximum measured longitudinal acceleration is 2.8 m/s^2 on the shake table, 1.7 m/s^2 on the first concrete slab, 2.1 m/s^2 on the second concrete slab, and 4.0 m/s^2 on the top of the building.

The dominant excitation frequency of the earthquake "El Centro" is around 1.6 Hz (Figure 21), and the natural frequency of the building is around 3.4 Hz before the seismic tests (Figure 22) and 2.6 Hz after them (Figure 23). This large difference between the excitation frequency and the natural frequency implies that the longitudinal accelerations that the earthquake causes in the building are small. Consequently, the horizontal inertial forces are also small, as well as the internal forces caused by the earthquake.

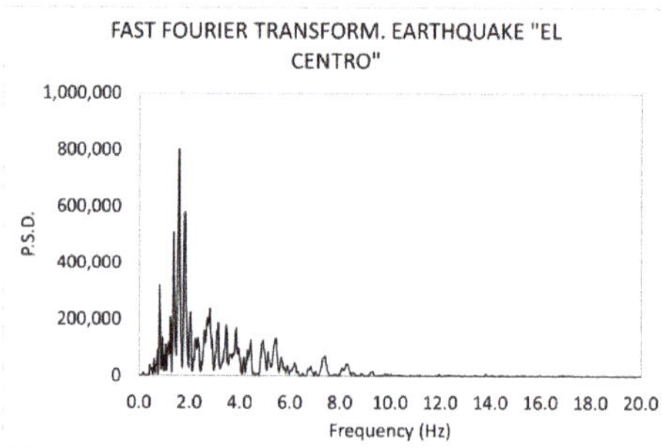

Figure 21. Dominant frequencies of earthquake "El Centro".

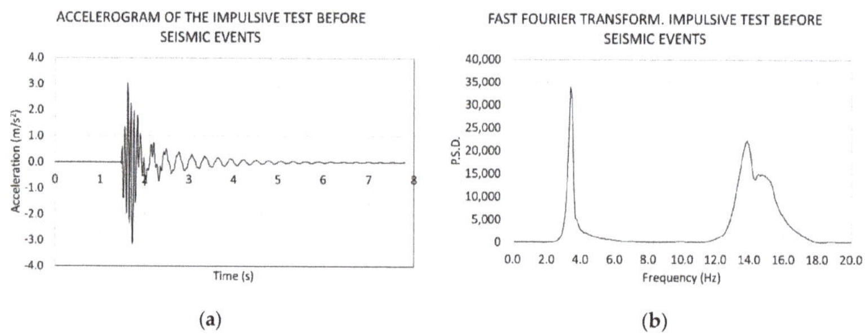

Figure 22. One-cycle impulsive test before seismic events. (**a**) Accelerogram; (**b**) fast Fourier transform.

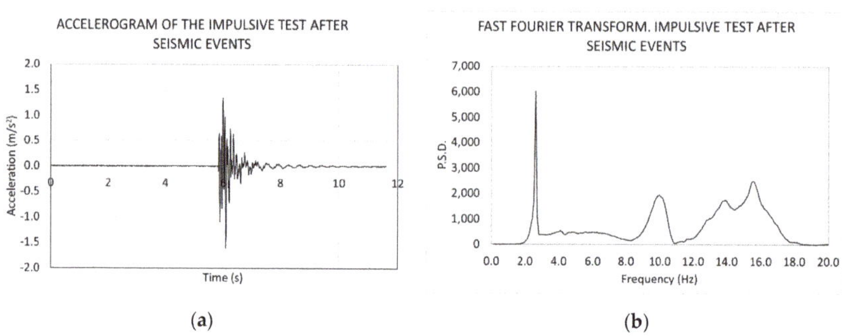

Figure 23. One-cycle impulsive test after seismic events. (**a**) Accelerogram; (**b**) fast Fourier transform.

The seismic event results in a reduction of the natural frequency of the building by 0.8 Hz, that is, 23%. Since no concrete cracks were observed, it is concluded that the damage is completely focused on the connections, both the low-cost energy dissipation system and the flexible connections between slab and walls.

Additionally, the measured damping factor of the building is 4.7% before the seismic tests and 6.2% after them. This increase in the damping factor is also a good indicator of the damage caused by the seismic tests.

When the seismic response of the building is compared to Eurocode 8 [38], it is observed that the measured elastic response spectrum S(T), defined as the ratio between the maximum acceleration of the building and the ground acceleration (i.e., the maximum acceleration of the shake table) is 1.43, which is smaller than the theoretical S(T) provided by this European standard. This means that the standard is conservative, as expected.

Additionally, it is highlighted that the natural frequency of the structure is below 4 Hz before the seismic event (Figure 22b), which is the lowest threshold of the dangerous region of the "El Centro" earthquake (see Figure 10c) and no relevant accelerations are developed during the seismic events. Moreover, the loss of stiffness caused by the seismic events reduces the natural frequency (Figure 23b) of the structure and, consequently, reduces the horizontal accelerations caused by the earthquake, which prevents the structure from aftershock earthquakes.

4. Conclusions

This paper shows the design and the laboratory validation tests of a new low-cost energy dissipation system, for application in precast concrete structures composed of precast footings, precast structural walls, and precast concrete slabs. This energy dissipation system basically consists of a specific connection between the precast footing and the precast structural wall, formed by a set of threaded steel bars that connect both elements. During an earthquake, the steel bars undergo plastic deformation, absorbing most of the energy generated by the earthquake and preventing damage to the rest of the building. The additional advantage of this solution is that steel bars can be easily replaced after the seismic event. Additionally, a flexible connection between walls and slab has been used.

A testing campaign was carried out, including three phases. First, pushover tests were carried out on isolated structural walls formed by one precast structural wall and a precast footing. Second, pushover tests were carried out on structural frames, composed of two precast structural walls placed over two precast footings and connected to a precast slab. Thirdly, seismic tests using a shake table were carried out on a real-scale three-storey precast concrete building, consisting of two precast structural walls placed over two precast footings, two intermediate precast slabs, and a flexible steel roof.

The aim of this structural solution is to fulfill the requirements of the American standard ACI374.2R-13 [34] and more specifically, fulfill the structural performance level of "immediate occupancy", which means that the building can be used without collapse risk once the seismic event has occured.

The pushover tests on isolated structural walls revealed that this solution exhibits a linear-elastic behavior until and beyond the critical drift (which is 0.5%) and no cracks were observed in the structures. The maximum load was reached at a drift of 3%, which was six times greater than the critical drift. Beyond this drift, the structure began to show a plastic behavior. However, no structural damage was observed in the concrete elements, which means that the majority of the energy dissipated by the structural element was through the low-cost energy dissipation device. Additionally, a great ductility of the solution was observed.

The pushover tests on structural frames revealed that the flexible connection between the walls and the slab exhibited an excellent structural behavior. Again, this solution exhibited a linear-elastic behavior until and beyond the critical drift (which is 0.5%) and no cracks were observed in the structures. The maximum load was reached at a drift of 2%, which was four times greater than the critical drift.

Beyond this drift, the structure began to show a plastic behavior. At this moment, the flexible connections worked as an asymmetrical plastic hinge, able to transmit relevant negative bending moments but almost negligible positive bending moments. This reduction in the overall horizontal stiffness of the frame resulted in an increased plastic behavior of the structure and, in consequence, an increased capacity to dissipate seismic energy. In this case, small horizontal cracks in the walls (especially in the external face of the Wall 2 where the tension strain is larger) were observed.

The seismic tests revealed an excellent behavior of the real-scale three-storey precast concrete building. The structure was subjected to a main earthquake and six foreshocks. The earthquake used was "El Centro" (the eathquake that occurred in the city of El Centro, California, USA in 1979). Additionally, two one-cycle impulsive tests were performed, one before the seismic events and the other after the seismic events, in order to measure the dynamic parameters of the building (natural frequency and damping ratio) before and after the seismic events.

The visual inspections carried out after the seismic tests revealed that no structural damage was observed in the building (i.e., no cracks in the walls or slabs appeared and, of course, no concrete crushing occurred). This means that the seismic energy was completely dissipated by the connections, i.e., by the low-cost energy dissipation systems placed on the connections between the walls and the footings and the flexible connections between the walls and the slabs.

At the end of the seismic event, the residual longitudinal displacements of the both concrete slabs and the lightweight roofs were almost zero, i.e., the building recovered its original position.

The impulsive tests revealed that the seismic events caused a decrease in the natural frequency and an increase in the damping ratio, which illustrates the damage given by the seismic tests.

Finally, it is concluded that the building reached the performance level of "immediate occupancy", according to ACI374.2R-13 [34].

Author Contributions: Design of the structural solution, J.F., D.M., and A.C.J.; design of the testing campaign, J.F., D.M., A.C.J., J.M., and M.Á.V.; development and supervision of the tests, Á.M., J.F., D.M., J.M., and D.C.G.; data interpretation, Á.M., J.F., J.M., and M.Á.V.; funding acquisition and project administration, D.M. and A.C.J.; writing, reviewing and editing the paper, Á.M. and M.Á.V. All authors have read and agreed to the published version of the manuscript.

Funding: This project has received funding from the Eurostars-2 joint programme with co-funding from the European Union Horizon 2020 research and innovation programme. The project name is "Novel Seismic-Resistant construction system composed of precast structures" (SEISMPRECAST) and the main partners are ICONKRETE 2012, S.L. and EXERGY, LTD.

Conflicts of Interest: The authors declare no conflict of interest.

References

1. Li, Y.-W.; Li, G.-Q.; Sun, F.-F.; Jiang, J. Experimental study on continuous energy-dissipative steel columns under cyclic loading. *J. Const. Steel Res.* **2018**, *141*, 104–117. [CrossRef]
2. Ismail, M. Seismic isolation of structures, Part II: A case study using the RNC isolator. *HyA* **2018**, *69*, 177–195. [CrossRef]
3. Lobo, P.S.; Almeida, J.; Guerreiro, L. Recentring and control of peak displacements of a RC frame using damping devices. *Soil Dyn. Earth. Eng.* **2017**, *94*, 66–74. [CrossRef]
4. Kramer, A.; Barbosa, A.R.; Sinha, A. Performance of Steel Energy Dissipators Connected to Cross-Laminated Timber Wall Panels Subjected to Tension and Cyclic Loading. *J. Struct. Eng.* **2016**, *142*, E4015013. [CrossRef]
5. Sorace, S.; Terenzi, G.; Mori, C. Passive energy dissipation-based retrofit strategies for R/C frame water towers. *Eng. Struct.* **2016**, *106*, 385–398. [CrossRef]
6. López-Almansa, F.; Segués, E.; Cantalapiedra, I.R. A new steel framing system for seismic protection of timber platform frame buildings. Implementation with hysteretic energy dissipators. *Earthq. Eng. Struct. Dyn.* **2015**, *44*, 1181–1202. [CrossRef]
7. Beziat, A.; Mora, A.; Chase, J.G.; MacRae, G.A.; Rodgers, G.W.; Clifton, C. Performance Analysis of Energy Dissipators and Isolators Placed in Bridges to Prevent Structural Damage in Columns. *J. Earthq. Eng.* **2012**, *16*, 1113–1131. [CrossRef]

8. López, A.; de la Cruz, S.T.; Taylor, C. Experimental study of friction dissipators for seismic protection of building structures. *Earthq. Eng. Eng. Vib.* **2011**, *10*, 475–486. [CrossRef]
9. Symans, M.D.; Charney, F.A.; Whittaker, A.S.; Constantinou, M.C. Energy Dissipation Systems for Seismic Applications: Current Practice and Recent Developments. *J. Struc. Eng.* **2008**, *134*, 3–21. [CrossRef]
10. Sassu, M. The reinforced cut wall (RCW): A low-cost base dissipator for masonry buildings. *Earthq. Spectra* **2006**, *22*, 533–554. [CrossRef]
11. Peng, P.; LiePing, Y.; Wei, S.; HaiYun, C. Engineering practice of seismic isolation and energy dissipation structures in China. *Sci. Ch. Tech. Sci.* **2012**. [CrossRef]
12. Ismail, M. Seismic isolation of structures. Part I: Concept, review and a recent development. *HyA* **2018**, *69*, 147–161. [CrossRef]
13. Villaverde, R. Methods to Assess the Seismic Collapse Capacity of Building Structures: State of the Art. *J. Struct. Eng.* **2007**, *133*, 57–66. [CrossRef]
14. Zhanyu, B.; Jian, G.; Rongyue, Z.; Jianwei, S.; Lee, G.C. Cyclic performance and simplifi ed pushover analyses of precast segmental concrete bridge columns with circular section. *Earthq. Eng. Eng. Vib.* **2016**, *15*, 297–312.
15. Jiang, L.; Zheng, H.; Hu, Y. Seismic behaviour of a steel frame partially infilled with precast reinforced concrete wall. *Adv. Struct. Eng.* **2016**, *19*, 1637–1649. [CrossRef]
16. Ou, Y.-C.; Tsai, M.-S.; Oktavianus, Y.; Chang, K.-C. Cyclic testing of a large tall precast segmental concrete bridge column with a cast-in-place plastic hinge region. *Earthq. Spectra* **2013**, *29*, 1441–1457. [CrossRef]
17. Di Sarno, L.; Manfredi, G. Experimental tests on full-scale RC unretrofitted frame and retrofitted with buckling-restrained braces. *Earthq. Eng. Struct. Dyn.* **2012**, *41*, 315–333. [CrossRef]
18. Guner, S.; Vecchio, F.J. Analysis of Shear-Critical Reinforced Concrete Plane Frame Elements under Cyclic Loading. *J. Struct. Eng.* **2011**, *137*, 834–843. [CrossRef]
19. Ou, Y.-C.; Tsai, M.-S.; Chang, K.-C.; Lee, G.C. Cyclic behavior of precast segmental concrete bridge columns with high performance or conventional steel reinforcing bars as energy dissipation bars. *Earthq. Eng. Struct. Dyn.* **2010**, *39*, 1181–1198. [CrossRef]
20. Braconi, A.; Bursi, O.S.; Fabbrocino, G.; Salvatore, W.; Taucer, F.; Tremblay, R. Seismic performance of a 3D full-scale high-ductile steel–concrete composite moment-resisting frame—Part II: Test results and analytical validation. *Earthq. Eng. Struct. Dyn.* **2008**, *37*, 1635–1655. [CrossRef]
21. Maheri, M.R.; Kousari, R.; Razazan, M. Pushover tests on steel X-braced and knee-braced RC frames. *Eng. Struct.* **2003**, *25*, 1697–1705. [CrossRef]
22. Wenfeng, L.; Tao, W.; Xi, C.; Xiang, Z.; Peng, P. Pseudo-dynamic tests on masonry residential buildings seismically retrofi tted by precast steel reinforced concrete walls. *Earthq. Eng. Eng. Vib.* **2017**, *16*, 587–597.
23. Sucuoglu, H.; Lin, W.-H.W.; Binici, B.; Ezzatfar, P. Pseudo-Dynamic Testing, Performance Assessment, and Modeling of Deficient Reinforced Concrete Frames. *ACI Struct. J.* **2014**, *111*, 1203–1212.
24. Brun, M.; Labbe, P.; Bertrand, D.; Courtois, A. Pseudo-dynamic tests on low-rise shear walls and simplified model based on the structural frequency drift. *Eng. Struct.* **2011**, *33*, 796–812. [CrossRef]
25. Donea, J.; Jones, P.M.; Magonette, G.E.; Verzeletti, G. The Pseudo-dynamic test method. In *Recent Advances in Earthquake Engineering and Structural Dynamics*; Davidovici, V.E., Ed.; Ouest Editions: Nantes, France, 1992; pp. 769–780.
26. Benavent-Climent, A.; Galé-Lamuela, D.; Donaire-Avila, J. Energy capacity and seismic performance of RC waffle-flat plate structures under two components of far-field ground motions: Shake table tests. *Earthq. Eng. Struct Dyn.* **2019**, *48*, 949–969. [CrossRef]
27. Li, C.; Bi, K.; Hao, H. Seismic performances of precast segmental column under bidirectional earthquake motions: Shake table test and numerical evaluation. *Eng. Struct.* **2019**, *187*, 314–328. [CrossRef]
28. Gavridou, S.; Wallace, J.W.; Nagae, T.; Matsumori, T.; Tahara, K.; Fukuyama, K. Shake-Table Test of a Full-Scale 4-Story Precast Concrete Building. I: Overview and Experimental Results. *J. Struct. Eng.* **2017**, *143*. [CrossRef]
29. Chen, M.C.; Pantoli, E.; Wang, X.; Astroza, R.; Ebrahimian, H.; Hutchinson, T.C.; Conte, J.P.; Restrepo, J.I.; Marin, C.; Walsh, K.D.; et al. Full-scale structural and nonstructural building system performance during earthquakes. I: Specimen description, test protocol and structural response. *Earthq. Spectra* **2016**, *32*, 737–770. [CrossRef]
30. Onat, O.; Lourenço, P.B.; Koçak, A. Experimental and numerical analysis of RC structure with two leaf cavity wall subjected to shake table. *Struct. Eng. Mech.* **2015**, *55*, 1037–1053. [CrossRef]

31. Nagae, T.; Ghannoum, W.M.; Kwon, J.; Tahara, K.; Fukuyama, K.; Matsumori, T.; Shiohara, H.; Kabeyasawa, T.; Kono, S.; Nishiyama, M.; et al. Design Implications of Large-Scale Shake-Table Test on Four-Story Reinforced Concrete Building. *ACI Struct. J.* **2015**, *112*, 135–146.
32. Panagiotou, M.; Restrepo, J.I.; Conte, J.P. Shake-Table Test of a Full-Scale 7-Story Building Slice. Phase I: Rectangular Wall. *J. Struct. Eng.* **2011**, *137*, 691–704. [CrossRef]
33. Lee, H.-S.; Woo, S.-W. Seismic performance of a 3-story RC frame in a low-seismicity region. *Eng. Struct.* **2002**, *24*, 719–734. [CrossRef]
34. American Concrete Institute Committee 374. *ACI374.2R-13: Guide for Testing Reinforced Concrete Structural Elements under Slowly Applied Simulated Seismic Loads*; ACI: Farmington Hills, MI, USA, 2013.
35. European Committee for Standardization. *Eurocode 2, Design of Concrete Structures*; European Committee for Standardization: Brussels, Belgium, 2004.
36. European Committee for Standardization. *Testing Hardened Concrete—Part 3: Compressive Strength of Test Specimens*; EN 12390-3; European Committee for Standardization: Brussels, Belgium, 2009.
37. International Organization for Standardization. *ISO 898-1:2013, Mechanical Properties of Fasteners Made of Carbon Steel and Alloy Steel. Part 1: Bolts, Screws and Studs with Specified Property Classes—Coarse Thread and Fine Pitch Thread*; ISO: Geneva, Switzerland, 2013.
38. European Committee for Standardization. *Eurocode 8: Design of Structures for Earthquake Resistance—Part 1: General Rules, Seismic Actions and Rules for Buildings*; European Committee for Standardization: Brussels, Belgium, 2009.

applied sciences

MDPI

Article

Seismic Response Mitigation of Base-Isolated Buildings

Mohammad Hamayoun Stanikzai [1], Said Elias [2,*] and Rajesh Rupakhety [2]

1 Department of Civil and Environmental Engineering, Old Dominion University, Norfolk, VA 23529, USA; mstan006@odu.edu

2 Earthquake Engineering Research Centre, Faculty of Civil and Environmental Engineering, School of Engineering and Natural Sciences, University of Iceland, Austurvegur 2a, 800 Selfoss, Iceland; rajesh@hi.is

* Correspondence: said@hi.is

Received: 30 December 2019; Accepted: 6 February 2020; Published: 12 February 2020

Abstract: Earthquake response mitigation of a base-isolated (BI) building equipped with (i) a single tuned mass damper at the top of the building, (ii) multiple tuned mass dampers (MTMDs) at the top of the building, and (iii) MTMDs distributed on different floors of the building (d-MTMDs) is studied. The shear-type buildings are modeled by considering only one lateral degree of freedom (DOF) at the floor level. Numerical approach of Newmark's integration is adopted for solving the coupled, governing differential equations of motion of 5- and 10-story BI buildings with and without TMD schemes. A set of 40 earthquake ground motions, scaled 80 times to get 3200 ground motions, is used to develop simplified fragility curves in terms of the isolator maximum displacement. Incremental dynamic analysis (IDA) is used to develop simplified fragility curves for the maximum target isolator displacement. It is found that TMDs are efficient in reducing the bearing displacement, top floor acceleration, and base shear of the BI buildings. In addition, it was noticed that TMDs are efficient in reducing the probability of failure of BI building. Further, it is found that the MTMDs placed at the top floor and d-MTMDs on different floors of BI buildings are more efficient in decreasing the probability of failure of the BI building when compared with STMD.

Keywords: Base-Isolated Buildings; bearing displacement; STMD; MTMDs; d-MTMDs; incremental dynamic analysis; earthquake

1. Introduction

Over the last couple of decades, structural vibration control techniques have been popularized for mitigation of dynamic response caused by various environmental actions. Tuned mass dampers (TMDs) are one of the common control methods used for response mitigation of structures under dynamic loadings. Their applications in various situations and loads have been addressed by several researchers [1–7]. As single tuned mass dampers (STMDs) became popular, a more practical solution, for example, by distributing TMD mass over the structure is being investigated in recent times [8–16]. A detailed literature survey on passive TMDs is presented in Elias and Matsagar [17].

Base-isolation (BI) has been one of the most popular and well-established method of seismic response control. This method makes use of special devices such as friction pendulum, lead rubber bearings, etc, to isolate the main structure from the shaking of the ground. Base-isolation system works by making the isolated structure more flexible at the base, thereby reducing acceleration response of the superstructure and therefore base shear force on the structure. As a consequence of added flexibility, displacement demand on the structure gets amplified, and additional damping is provided to keep displacement demand within acceptable limits. Zelleke et al. [18] studied the effectiveness of viscous and visco-elastic dampers on seismic response control of BI buildings. They found that there is a tradeoff between the extent to which acceleration and displacement demand can be controlled by BI system combined with additional damping devices.

Tsai [19] investigated the use of TMD alongside BI and observed that TMD is efficient in response mitigation of BI systems if the dominant period of excitation is longer than the natural period of the structure. Yang et al. [20] studied the efficiency of passive TMDs in response control of BI buildings. Xiang and Nishitani [21] described the effectiveness of optimally designed non-traditional TMD for mitigation of seismic response of BI buildings. Use of TMD with inerter (TMDi) was proposed by De Domenico et al. [22] and De Domenico and Ricciardi [23–25] for diminishing response of BI buildings under earthquakes. Rabiee and Chae [26] reported the effectiveness of MR dampers for response mitigation of BI building under short and long period ground motions. Effectiveness of single TMD (STMD), multiple TMDs (MTMDs), and distributed MTMDs (d-MTMD) on seismic response control of BI buildings was investigated by Stanikzai et al. [27,28]. They noticed that d-MTMDs were more efficient and practical than other schemes.

Past studies on efficiency of TMDs in response mitigation of BI buildings have relied on a limited number of earthquake ground motions. As there is a large uncertainty in the frequency content, amplitude, and duration of ground shaking a structure can experience, a control scheme that is found effective for a certain type of ground motion may not be effective for other ground motions. It is therefore necessary to consider these uncertainties to have a robust understanding of the usefulness of TMDs in response mitigation of BI structures. Therefore, a probabilistic approach, for example, an examination of fragility curves of structures with and without the TMDs, can shed more light on the overall benefits of using TMDs in BI structures. Such analysis of BI buildings equipped with TMDs is lacking in the literature.

Among many other methods, the incremental dynamic analysis (IDA) is one of the popular methods of estimating analytical fragility curves of structures. The idea of incremental dynamic analysis (IDA) was introduced by Bertero [29]. Kennedy et al. [30] proposed the concept of fragility analysis in the field of earthquake engineering. Later, the idea was extended by many researchers including Bazzurro and Cornell [31,32] and Luco and Cornell [33,34]. This method of fragility analysis was further expanded by Vamvatsikos and Cornell [35]. Nowadays, it is broadly employed in seismic risk evaluation of structures. The IDA has been taken by the U.S. Federal Emergency Management Agency (FEMA, [36,37]) standards as a state-of-the-art technique to verify the global collapse capacity of structures. The IDA relies on nonlinear structural analysis using ground motions with increasing intensity, estimating some damage measure (DM) for each ground motion, which is characterized by an intensity measure (IM). Different IMs, for example, peak ground velocity (PGV), peak ground acceleration (PGA), spectral acceleration, etc., can be used to characterize ground motions. Damage measures are related to response parameters such as peak base shear, joint rotation, peak story drift, and bearing displacement. Probabilities of exceeding a specified damage measure for a given excitation intensity level can be estimated if IDA is conducted with an adequate number of ground motions.

This study applies IDA method to estimate seismic fragility of buildings with BI systems and investigates the extent to which such fragility can be reduced using different TMD schemes.

2. Structural Model

Schematic representation of different structures considered here is presented in Figure 1. An idealized N-story BI building is demonstrated in Figure 1a. The masses are lumped at the floor levels, and the floors are deemed to act as a rigid diaphragm. One horizontal translational degree of freedom (DOF) is assigned to each floor. BI structures with different control schemes are shown in Figure 1a–d. In Figure 1b, a TMD is placed at the nth floor, and this scheme is called as BI + STMD. In Figure 1c, MTMDs are placed at the nth floor, and the scheme is called as BI + MTMDs. When the multiple TMDs are placed on various floors, as displayed schematically in Figure 1d, the scheme is called BI + d-MTMDs. For the sake of simplicity, the superstructure is presumed to remain linearly elastic and soil structure interaction (SSI) is not considered.

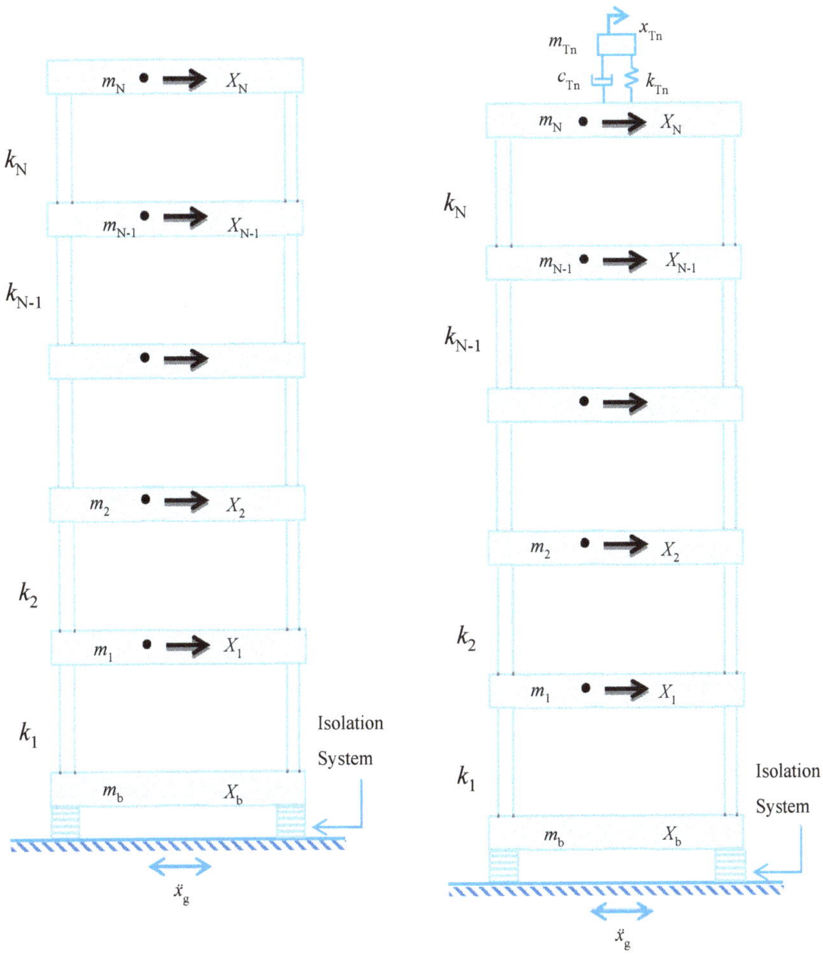

(a) (b)

Figure 1. *Cont.*

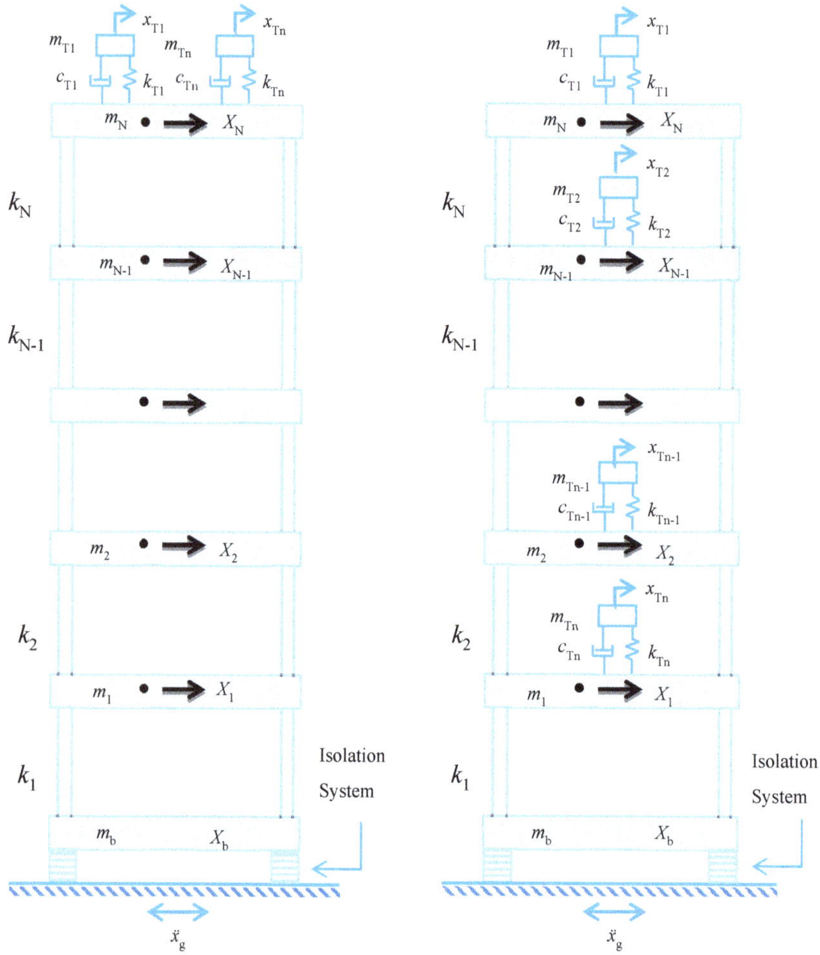

Figure 1. Schematic models of a *N*-story building with (**a**) base-isolation (BI), (**b**) BI + single tuned mass dampers (STMD) at top floor, (**c**) BI + multiple tuned mass dampers (MTMDs) at top floor, and (**d**) BI + distributed multiple tuned mass dampers (d-MTMDs).

The governing equation of motion of the system under ground shaking can be written as

$$[M]\{\ddot{x}(t)\} + [C]\{\dot{x}(t)\} + [K]\{x(t)\} = -[M]\{r\}\{\ddot{x}_g\} \tag{1}$$

where $[M]$ is the mass matrix; $[C]$ is the damping matrix and $[K]$ is the stiffness matrix of the structure; $\{x\} = \{X_1, X_2, \cdots X_N, X_b, \cdots x_{T1}, x_{T2}, \cdots x_{Tn}\}^T$, $\dot{x}$, and $\ddot{x}$ are the unknown relative (floor, isolator, and TMD) displacement, velocity, and acceleration vectors, respectively; $\{\ddot{x}_g\}$ is earthquake ground acceleration; and $\{r\}$ is the vector of influence coefficients. Following Stanikzai et al. [28], the $[M]$, $[C]$, and $[K]$ matrices of the BI buildings installed with TMDs can be written as

$$[M] = \begin{bmatrix} [m_b]_{1\times1} & [M_N]_{1\times N} & [m_n]_{1\times n} \\ [M_N]_{N\times1} & [M_N]_{N\times N} & [0]_{N\times n} \\ [m_n]_{n\times1} & [0]_{n\times N} & [m_n]_{n\times n} \end{bmatrix} \tag{2}$$

$$[C] = \begin{bmatrix} [c_b]_{1\times1} & [0]_{1\times N} & [0]_{1\times n} \\ [0]_{N\times1} & [C_N]_{N\times N} + [c_n]_{N\times N} & -[c_n]_{N\times n} \\ [0]_{n\times1} & -[c_n]_{n\times N} & [c_n]_{n\times n} \end{bmatrix} \tag{3}$$

$$[K] = \begin{bmatrix} [k_b]_{1\times1} & [0]_{1\times N} & [0]_{1\times n} \\ [0]_{N\times1} & [K_N]_{N\times N} + [k_n]_{N\times N} & -[k_n]_{N\times n} \\ [0]_{n\times1} & -[k_n]_{n\times N} & [k_n]_{n\times n} \end{bmatrix} \tag{4}$$

where $[m_b]$ is isolator mass matrix, $[M_N]$ is superstructure mass matrix, and $[m_n]$ is the TMDs schemes mass matrix. $[c_b]$, $[C_N]$, and $[c_n]$ are the corresponding damping matrices. Similarly, $[k_b]$ is isolator stiffness matrix, $[K_N]$ is superstructure stiffness matrix, and $[k_n]$ is the TMDs schemes stiffness matrix. The N story building is isolated by one DOF isolator and equipped by n number of TMDs, that results in the matrices of the order $(N + n + 1) \times (N + n + 1)$.

3. Mathematical Model of Isolator

In this study, a lead rubber bearing (N-Z) isolator is considered. Wen's model (Wen, [38]) is used for characterization of hysteretic behavior of the bearing. Figure 2a shows schematic representation of lead rubber bearings. The restoring force developed in the isolation is given by

$$F_b = c_b\dot{x}_b + \alpha k_b x_b + (1 - \alpha)\, F_y Z \tag{5}$$

where the yield strength of the bearing is denoted by F_y, initial stiffness of the bearing is denoted by k_b, and viscous damping of the bearing is denoted by c_b; α signifies the ratio of post- to pre-yielding stiffness. A non-dimensional hysteretic displacement component is denoted by Z, which satisfies the following non-linear first order differential equation

$$q\dot{Z} = A\dot{x}_b + \beta|\dot{x}_b||Z|Z|^{n_k-1} + \tau\dot{x}_b|Z|^{n_k} \tag{6}$$

where the yield displacement is denoted by q; A, β, τ, and n_k are dimensionless model parameters, often calibrated from experimental tests. The integer parameter n_k controls the smoothness of transition from elastic to plastic deformation. The isolation period (T_b), damping ratio (ξ_b), and normalized yield strength characterizes the N-Z isolation system. Using the post-yielding stiffness (k_p) of the bearing, the isolation period (T_b) and damping ratio (ξ_b) are computed by Equations (7) and (8), respectively.

$$T_b = 2\pi\sqrt{\frac{M}{k_p}} \tag{7}$$

$$\xi_b = \frac{c_b}{2M\omega_b} \tag{8}$$

where ω_b is the isolator frequency and $W = Mg$ is the total weight of the building plus isolator and TMDs, and g is the acceleration due to gravity. Other parameters are maintained constant with $q = 2.5$ cm, $\beta = \tau = 0.5$, $A = 1$, and $n = 2$.

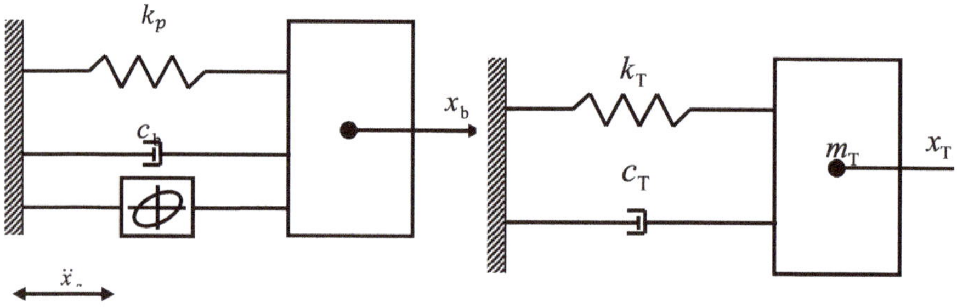

Figure 2. Schematic diagrams for (**a**) N-Z and (**b**) tuned mass dampers (TMD).

4. Mathematical Model of the TMD

Figure 2b shows schematic illustration of a TMD, which is made up of a mass attached to the floor by a spring and a dashpot. The ith TMD is illustrated by its stiffness (k_i), damping constant (c_i), and the mass (m_i). The TMD's mass ratio μ, designed frequency ω_i, tuning frequency ratio f, and damping ratio ξ_i, are given by

$$\omega_i = \sqrt{\frac{k_i}{m_i}} \tag{9}$$

$$\xi_i = \frac{c_i}{2\,\omega_i m_i} \tag{10}$$

$$f_i = \frac{\omega_i}{\omega_b} \tag{11}$$

$$\mu_i = \frac{m_i}{m_s} \tag{12}$$

where m_s represents total mass of the BI building. The proposed formulas by Sadek et al. [39] for optimum tuning frequency ratio $f_{STMD} = \frac{1}{1+\mu_{STMD}}\left[1 - \xi_b\sqrt{\frac{\mu_{STMD}}{1+\mu_{STMD}}}\right]$ and damping ratio $\xi_{STMD} = \frac{\xi_b}{1+\mu_{STMD}} + \sqrt{\frac{\mu_{STMD}}{1+\mu_{STMD}}}$ are considered for STMD.

Figure 1c–d show the main system equipped with n number of TMDs (MTMDs) with various dynamic characteristics. The BI system and each TMD is modeled with a single DOF. In this study, MTMDs and d-MTMDs are designed by considering unequal masses. The designed frequency of ith TMD is stated as

$$\omega_i = \omega_T\left[1 + \left(i - \frac{n+1}{2}\right)\frac{\beta}{n-1}\right] \tag{13}$$

where ω_T is the average frequency of all the MTMDs (i.e., $\omega_T = \sum_{j=1}^{n}\omega_j/n$). Based on the recommendation of De Domenico and Ricciardi [25], the average frequency is taken as 0.78 times the fundamental frequency of the BI building. Non-dimensional frequency spacing parameter β of the MTMDs is defined as

$$\beta = \frac{\omega_n - \omega_1}{\omega_T} \tag{14}$$

For a set of TMD units with equal stiffness (i.e., $k_1 = k_2 = k_3 = \cdots = k_n$) and equal damping ratio (i.e., $\xi_1 = \xi_2 = \xi_3 = \cdots = \xi_n$), the stiffness ($k_j$) is calculated as

$$k_j = \frac{m_n}{\left(1/\omega_1^2 + 1/\omega_2^2 + \cdots + 1/\omega_n^2\right)} \tag{15}$$

where m_n is the total mass of all the TMD(s), and the mass of the ith TMD (m_i) is given by

$$m_i = \frac{k_i}{\omega_i^2} \tag{16}$$

and the damping coefficient of the TMDs is then given by

$$c_i = 2\xi_i m_i \omega_i \tag{17}$$

In this study, n = 3 is assumed for the cases of MTMDs and d-MTMDs; for MTMDs cases, all the TMDs are placed at the top floor, whereas for d-MTMDs, the TMDs are placed at the top three floors.

5. Numerical Study

Numerical study is conducted on response mitigation of 5- and 10-story BI buildings equipped with TMDs at top and at different floor levels under earthquake ground motions. The shear-type buildings are modeled by considering only a lateral DOF at each floor level. The coupled governing differential equation of motion for the system is formulated and solved using Newmark's integration method with linear acceleration assumption between consecutive time steps. For the sake of simplicity, the floor mass and stiffness are kept the same herein. The structural damping matrix is constructed by assuming 2% damping ratio in all the normal modes of vibration. The isolation period and the TMDs mass ratio are assumed to be 2.5 s and 5%, respectively. The total TMD mass is then divided into three TMDs for the cases of MTMDs and d-MTMDs. In addition, the isolation damping ratio, yield displacement, and normalized yield strength (yield strength per unit weight of the structure) are assumed to be 0.05, 5 cm, and 0.05, respectively [40]. The 5- and 10-story buildings have 0.5 s and 1 s fundamental period, respectively, when they are fixed at the base.

Incremental dynamic analysis (IDA) of the above-mentioned buildings is conducted. A set of 40 earthquake ground motions are considered (see Tables 1 and 2). Somerville et al. [41] provide more details of the set of 40 ground motions for a theoretical site in Los Angeles, USA. The set is divided into two categories: (1) design-basis earthquakes (DBE) and (2) maximum considered earthquakes (MCE). The DBE are listed as LA01 to LA20 and the MCE are listed as LA21 to LA40 in Tables 1 and 2, respectively. In order to better understand the nature of the selected DBE and MCE earthquake ground motions, their displacement and acceleration response spectra considering 2% critical damping are shown in Figure 3. It can be seen that that selected list encompasses a wide range of amplitude and frequency content of ground shaking. Each ground motion in the set of 40 are scaled to PGA of 0.025 to 2 g with an increment of 0.025 g (80 times). As a result, a set of 3200 ground motions are obtained, which are used for IDA. Numerical simulation is carried out with MATLAB. The earthquake ground excitations are scaled based on the procedure given in Bhandari et al. [42].

Table 1. List of the design-based earthquakes (DBE)-level earthquake ground motions used in the study and the event details.

SAC Name	Record	Year	Earthquake Magnitude	Distance (km)	Scale Factor	Number of points	Δt (s)	Duration (s)	PGA (g)
LA01	Imperial Valley	1940	6.9	10	2.01	2674	0.02	53.46	0.46
LA02	Imperial Valley	1940	6.9	10	2.01	2674	0.02	53.46	0.68
LA03	Imperial Valley	1979	6.5	4.1	1.01	3939	0.01	39.38	0.39
LA04	Imperial Valley	1979	6.5	4.1	1.01	3939	0.01	39.38	0.49
LA05	Imperial Valley	1979	6.5	1.2	0.84	3909	0.01	39.08	0.3
LA06	Imperial Valley	1979	6.5	1.2	0.84	3909	0.01	39.08	0.23
LA07	Landers	1992	7.3	36	3.2	4000	0.02	79.98	0.42
LA08	Landers	1992	7.3	36	3.2	4000	0.02	79.98	0.43
LA09	Landers	1992	7.3	25	2.17	4000	0.02	79.98	0.52
LA10	Landers	1992	7.3	25	2.17	4000	0.02	79.98	0.36
LA11	Loma Prieta	1989	7	12	1.79	2000	0.02	39.98	0.67
LA12	Loma Prieta	1989	7	12	1.79	2000	0.02	39.98	0.97
LA13	Northridge	1994	6.7	6.7	1.03	3000	0.02	59.98	0.68
LA14	Northridge	1994	6.7	6.7	1.03	3000	0.02	59.98	0.66
LA15	Northridge	1994	6.7	7.5	0.79	2990	0.005	14.945	0.53
LA16	Northridge	1994	6.7	7.5	0.79	2990	0.005	14.945	0.58
LA17	Northridge	1994	6.7	6.4	0.99	3000	0.02	59.98	0.57
LA18	Northridge	1994	6.7	6.4	0.99	3000	0.02	59.98	0.82
LA19	North Palm Springs	1986	6	6.7	2.97	3000	0.02	59.98	1.02
LA20	North Palm Springs	1986	6	6.7	2.97	3000	0.02	59.98	0.99

Table 2. List of the maximum considered earthquakes (MCE)-level earthquake ground motions used in the study and the event details.

ZSAC Name	Record	Year	Earthquake Magnitude	Distance (km)	Scale Factor	Number of Points	Δt (s)	Duration (s)	PGA (g)
LA21	Kobe	1995	6.9	3.4	1.15	3000	0.02	59.98	1.28
LA22	Kobe	1995	6.9	3.4	1.15	3000	0.02	59.98	0.92
LA23	Loma Prieta	1989	7	3.5	0.82	2500	0.01	24.99	0.42
LA24	Loma Prieta	1989	7	3.5	0.82	2500	0.01	24.99	0.47
LA25	Northridge	1994	6.7	7.5	1.29	2990	0.005	14.945	0.87
LA26	Northridge	1994	6.7	7.5	1.29	2990	0.005	14.945	0.94
LA27	Northridge	1994	6.7	6.4	1.61	3000	0.02	59.98	0.93
LA28	Northridge	1994	6.7	6.4	1.61	3000	0.02	59.98	1.33
LA29	Tabas,	1974	7.4	1.2	1.08	2500	0.02	49.98	0.81
LA30	Tabas,	1974	7.4	1.2	1.08	2500	0.02	49.98	0.99
LA31	Elysian Park (simulated)		7.1	17.5	1.43	3000	0.01	29.99	1.3
LA32	Elysian Park (simulated)		7.1	17.5	1.43	3000	0.01	29.99	1.19
LA33	Elysian Park (simulated)		7.1	10.7	0.97	3000	0.01	29.99	0.78
LA34	Elysian Park (simulated)		7.1	10.7	0.97	3000	0.01	29.99	0.68
LA35	Elysian Park (simulated)		7.1	11.2	1.1	3000	0.01	29.99	0.99
LA36	Elysian Park (simulated)		7.1	11.2	1.1	3000	0.01	29.99	1.1
LA37	Palos verdes (simulated)		7.1	1.5	0.9	3000	0.02	59.98	0.71
LA38	Palos verdes (simulated)		7.1	1.5	0.9	3000	0.02	59.98	0.78
LA39	Palos verdes (simulated)		7.1	1.5	0.88	3000	0.02	59.98	0.5
LA40	Palos verdes (simulated)		7.1	1.5	0.88	3000	0.02	59.98	0.63

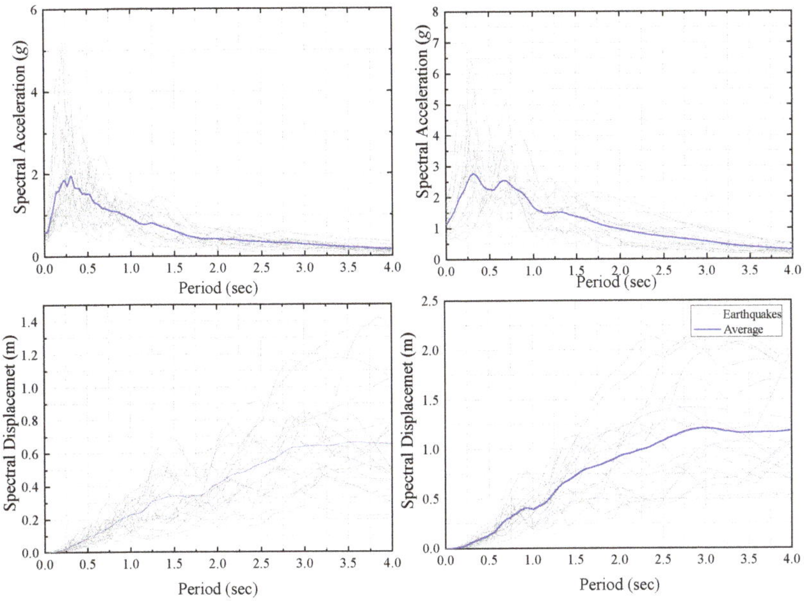

Figure 3. Response spectra of the sets of DBE and MCE earthquake ground motions.

5.1. Performance of TMDs in Response Mitigation

Figure 4 illustrates the time history and Fourier Amplitude Spectra (FAS) of top floor acceleration and bearing displacement of BI and BI equipped by different TMD schemes under LA04 earthquake ground motion. It is observed that the BI building has relatively low acceleration at the top of building, but very large bearing displacement. The TMDs effectively reduce this large displacement demand. The peak bearing displacement of BI, BI + STMD, BI + MTMDs, and d-MTMDs are, respectively, 1.86 m, 1.5 m, 1.18, and 1.12 m. It is observed that the STMD could reduce the displacement by about 20%, whereas, MTMDs and d-MTMDs could reduce displacement by up to 40%. It is also evident that this reduction in bearing displacement is not at the cost of amplified acceleration.

Thereafter, efficacy of the STMD, MTMDs, and d-MTMDs in vibration mitigation of 5- and 10-story BI buildings under DBE and MCE earthquakes are presented in Figures 5 and 6. To identify the reason for effectiveness or vice versa, response of each scheme is arranged based on ratio of T_b/T_f, where T_b is the isolation period and T_f is the dominant period of ground motion [43,44]. The TMD schemes are found to be more effective in controlling displacement response when the isolation period is less than about 3 times the dominant period of ground motion. The d-MTMDs showed marginally more effectiveness for 5-story BI building (see Figure 5). Even if the d-MTMDs show performance similar to STMD and MTMDs, it would be more practical as the TMDs are distributed on different floors, which is better than placing a large mass at the top of the structure. It is to be noted that for acceleration control, all TMD schemes are similar except in few cases. In most cases, MTMDs and d-MTMDs provide higher reduction in displacement. However, STMD showed a superior performance for acceleration response control (see Figures 5 and 6). Overall, the TMD schemes are more effective in controlling response of the 5-story BI building. Hence, it is determined that TMD schemes are effective in reducing response of BI buildings under DBE earthquake excitations, especially when $T_b/T_f \leq 3$.

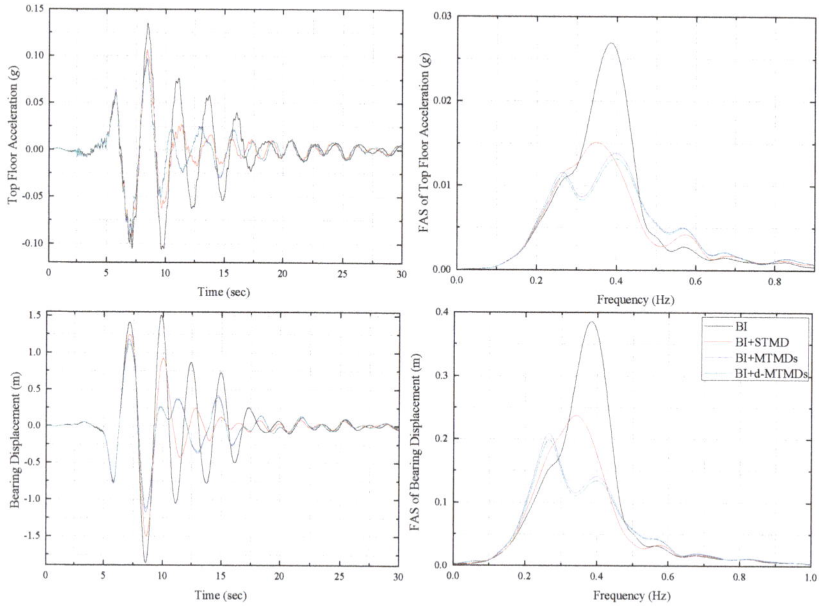

Figure 4. Time history and Fourier Amplitude Spectra (FAS) of top floor acceleration and bearing displacement of 5 story BI and BI equipped by TMD schemes under LA04 earthquake ground motions.

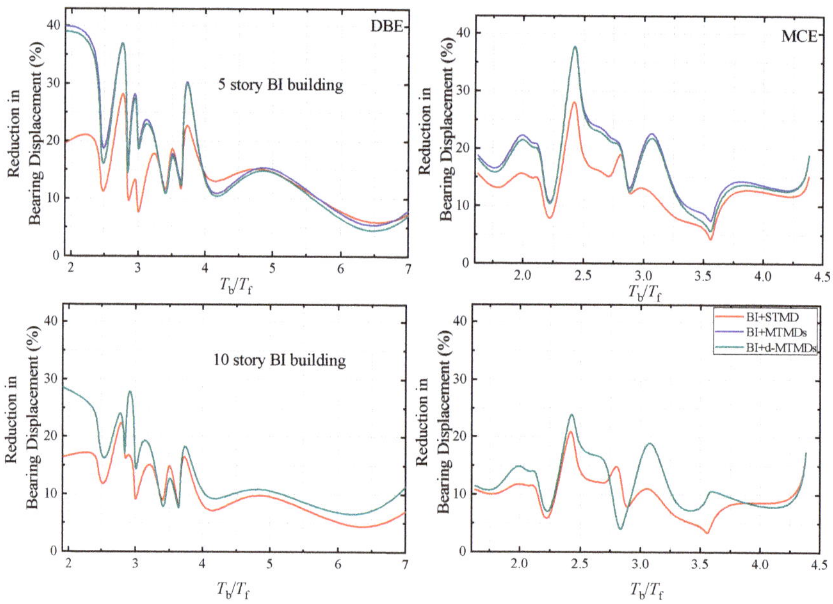

Figure 5. Reduction in average peak bearing displacement of BI and BI equipped by TMD schemes under DBE earthquake ground motions.

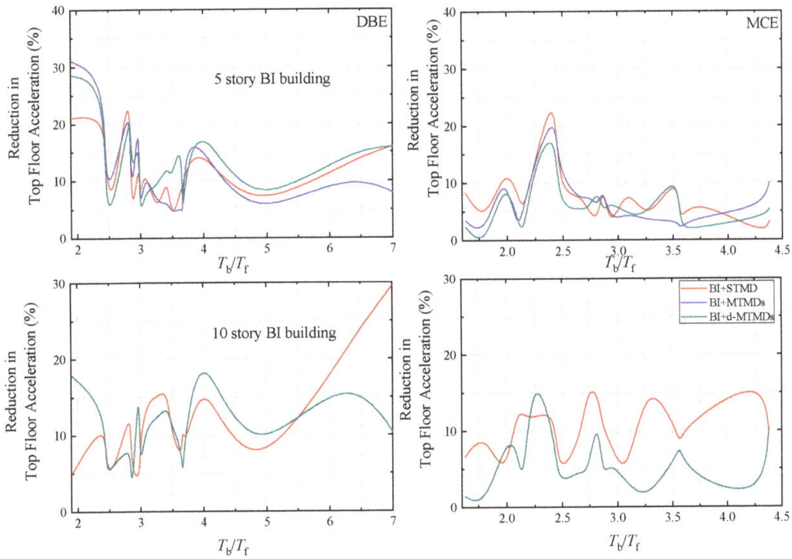

Figure 6. Reduction in average of peak top floor acceleration of BI and BI equipped by TMD schemes under DBE earthquake ground motions.

5.2. *Seismic Effectiveness of Hybrid System*

Probability distribution functions of the BI buildings with and without TMD schemes are presented in this section. Assuming a lognormal model, the probability distribution of bearing displacement, top floor acceleration, and normalized base shear are estimated form the calculated response. Cumulative distribution function (CDF) of response quantities of BI buildings with and without TMD schemes are presented in Figures 7 and 8. Figure 7 shows that the use of TMDs in BI buildings is more effective in mitigation of displacement response than acceleration response and base shear demand. In addition, MTMDs and d-MTMDs are only marginally better than STMDs. Overall, TMDs are found to be effective in decreasing the seismic response (bearing displacement and top floor acceleration) of 5-story BI building. It is noticed that the trend of reduction in displacement and acceleration of the 10-storey building is almost the same (see Figure 8) as that of the 5-story building, but the TMD devices are less effective.

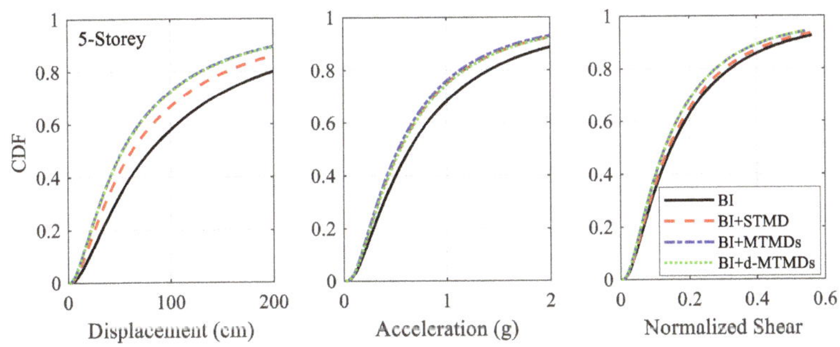

Figure 7. Cumulative probability distribution functions of maximum bearing displacement, peak top floor acceleration, and normalized shear force for 5-story building.

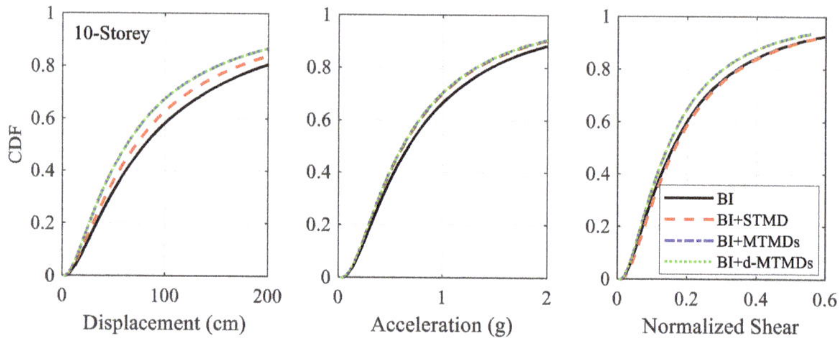

Figure 8. Cumulative probability distribution functions of maximum bearing displacement, peak top floor acceleration, and normalized shear force for 10-story building.

5.3. Seismic Fragility Analysis

Seismic fragility is defined as the probability of exceeding a defined damage state due to a given seismic intensity measure (IM). It happens when the seismic applied load (demand) is greater than the structural capacity. Seismic fragility can be expresses as,

$$\text{Fragility} \approx \text{P}[\text{Demand} > \text{Capacity}] \tag{18}$$

For the fragility analysis of the buildings, following two steps are considered: (1) A failure criteria is defined and (2) a set of seismic intensity measures are selected. Here, the failure criterion is considered as maximum isolator displacement, which ensures the safety of isolation system. In this study, the collapse state of damage is based on limit states proposed by Bhandari et al. (2018). They proposed a limit state for isolator displacement for elastic, elastic–plastic, and plastic states. The maximum target displacement assuming the plastic state is considered 330 mm in this study. Once the target displacement is specified, the analysis of BI buildings equipped with different TMD schemes using the set of 40 earthquakes, each scaled 80 times is carried out. The peak absolute isolator displacement is recorded. The recorded peak response of the BI building with and without TMD schemes for the generated number of ground motions (N_{gen}) are then compared with the corresponding seismic demands. Thereafter, the probability of failure P_f at any certain PGA level is defined as following

$$P_{\text{f}} = \frac{N_{\text{fail}}}{N_{\text{gen}}} \tag{19}$$

where N_{fail} denotes the number of the cases which satisfies Equation (18). Then the process is reiterated for the range of the above-mentioned PGA levels, and the respective fragility curves are achieved assuming a lognormal distribution. The procedure of obtaining the fragility curves for the BI building equipped with the TMD schemes is schematically described in Figure 9. The bearing displacement fragility curves for 5-story BI building equipped with TMDs is depicted in Figure 10.

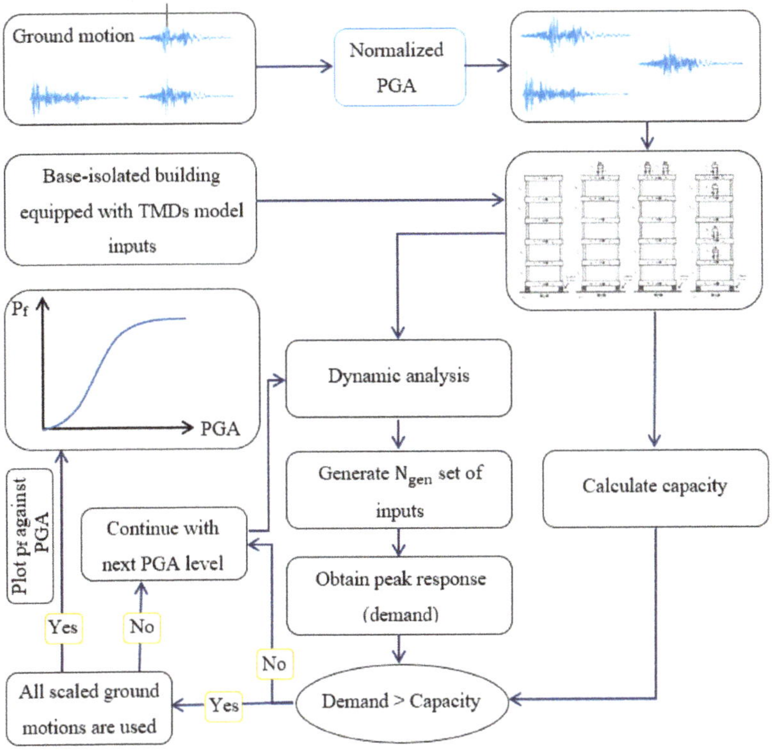

Figure 9. Procedure for seismic fragility analysis of base-isolated building equipped with TMD schemes.

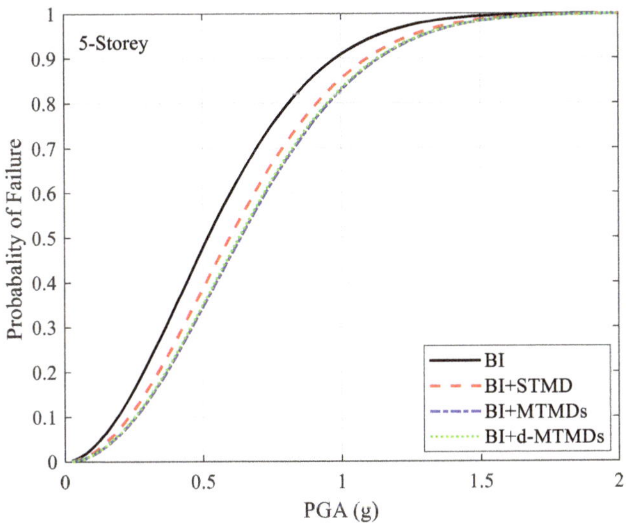

Figure 10. Bearing displacement simplified fragility curves for 5-story base-isolated building equipped with TMDs.

The TMD schemes reduce the probability of the considered damage state by about 10% across a range of PGA. The MTMDs and d-MTMDs are found to perform slightly better than the STMDs.

In addition, similar trend of response reduction was observed for the case of 10-story BI building equipped with TMDs and the results are shown in Figure 11. Since these fragility curves represent uncertainty in ground motion intensity and frequency content, while the structure is assumed to be deterministic, the resulting fragility curves are called as simplified fragility curves in the sense that they might be different from real fragilities of a structure with uncertain parameters. There is future scope for experimental verification of the concept like reference [45].

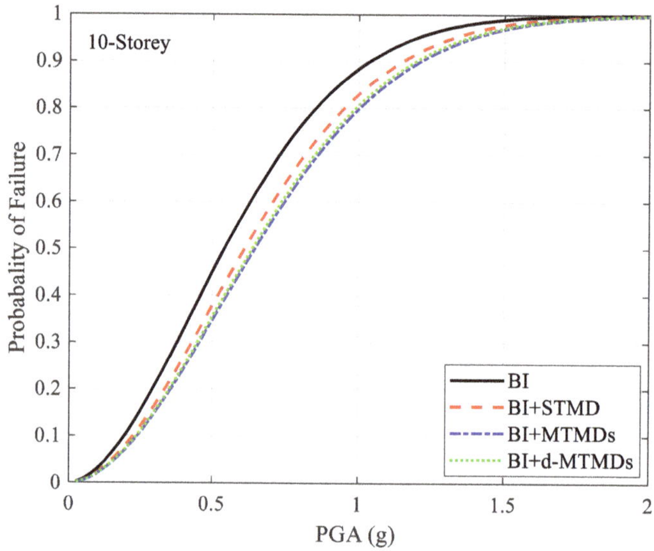

Figure 11. Bearing displacement simplified fragility curves for 10-story base-isolated building equipped with TMDs.

6. Conclusions

Seismic response mitigation of base-isolated (BI) building installed with a single tuned mass damper (STMD), multiple TMDs (MTMDs), and distributed MTMDs (d-MTMDs) are investigated. Numerical analysis of two BI buildings equipped with TMDs is performed using analytical models of 5- and 10-storey buildings. Results were analyzed to investigate effectiveness of different TMD schemes in reducing displacement and acceleration demand of the BI structure. Such effectiveness was investigated in terms of percentage reduction in response as a function of ground motion frequency content, probability distribution functions of response parameters, and fragility curves corresponding to failure of the isolation device due to excessive displacement. The following conclusions can be made from the results of the analyses.

1. TMD schemes are can be used to control bearing displacement of BI buildings subjected to earthquake ground excitations without compromising the control in acceleration response achieved by BI.
2. MTMDs and d-MTMDs are marginally better than STMD in controlling the bearing displacement of the BI building. The effectiveness of MTMDs and d-MTMDs are same as STMD for mitigating top floor acceleration.
3. The TMD schemes reduce the fragility of the structure by about 5% in a wide range of PGA of 0.5g to 1g. For weaker and stronger shaking, the reduction in fragility is not significant. It is to be noted that the TMDs used in this study are not optimized for specific type of ground motions and might experience detuning effects for some ground motions. Designing such TMDs based on

effective period of vibration of a structure for a well-established target displacement obtained from appropriate hazard analysis might provide additional benefits.

Author Contributions: All authors discussed and agreed upon the idea, and made scientific contributions. M.H.S. coded the MATLAB program and wrote initial draft of the paper. S.E. and R.R. improved and revised the initial paper. All authors have read and agreed to the published version of the manuscript.

Funding: It was part of the employment activity no extra funds were given to this study.

Acknowledgments: Authors acknowledge the efforts of reviewers and editors for improving the quality of the paper. First author acknowledges the department of Civil and Environmental Engineering, Old Dominion University, for allowing him to use the work stations and space. Second and third authors acknowledge support from the University of Iceland Research Fund.

Conflicts of Interest: The authors declare no conflict of interest.

References

1. Greco, R.; Marano, G.C. Optimum design of Tuned Mass Dampers by displacement and energy perspectives. *Soil Dyn. Earthq. Eng.* **2013**, *49*, 243–253. [CrossRef]
2. Longarini, N.; Zucca, M. A chimney's seismic assessment by a tuned mass damper. *Eng. Struct.* **2014**, *79*, 290–296. [CrossRef]
3. Reggio, A.; Angelis, M.D. Optimal energy-based seismic design of non-conventional Tuned Mass Damper (TMD) implemented via inter-story isolation. *Earthq. Eng. Struct. Dyn.* **2015**, *44*, 1623–1642. [CrossRef]
4. Lu, Z.; Li, K.; Zhou, Y. Comparative Studies on Structures with a Tuned Mass Damper and a Particle Damper. *J. Aerosp. Eng.* **2018**, *31*, 04018090. [CrossRef]
5. Rezaee, M.; Aly, A.M. Vibration Control in Wind Turbines to Achieve De-sired System-Level Performance under Single and Multiple Hazard Loadings. *Struct. Control Health Monit.* **2018**. [CrossRef]
6. Wang, C.; Shi, W. Optimal Design and Application of a Multiple Tuned Mass Damper System for an In-Service Footbridge. *Sustainability* **2019**, *11*, 2801. [CrossRef]
7. Cao, L.; Li, C. Tuned tandem mass dampers-Inerters with broadband high effectiveness for structures under white noise base excitations. *Struct. Control Health Monit.* **2019**, *26*, e2319. [CrossRef]
8. Elias, S.; Matsagar, V.; Datta, T.K. Effectiveness of distributed tuned mass dampers for multi-mode control of chimney under earthquakes. *Eng. Struct.* **2016**, *124*, 1–16. [CrossRef]
9. Elias, S.; Matsagar, V.; Datta, T.K. Distributed tuned mass dampers for multi-mode control of benchmark building under seismic excitations. *J. Earthq. Eng.* **2019**, *23*, 1137–1172. [CrossRef]
10. Elias, S.; Matsagar, V.; Datta, T.K. Along Wind response control of chimneys with distributed multiple tuned mass dampers. *Struct. Control Health Monit.* **2019**, *26*, e2275. [CrossRef]
11. Elias, S.; Matsagar, V.; Datta, T.K. Dynamic response control of a wind-excited tall building with distributed multiple tuned mass dampers. *Int. J. Struct. Stab. Dyn.* **2019**, *19*, 1950059. [CrossRef]
12. Gill, D.; Elias, S.; Steinbrecher, A.; Schröder, C.; Matsagar, V. Robustness of multi-mode control using tuned mass dampers for seismically excited structures. *Bull. Earthq. Eng.* **2017**, *15*, 5579–5603. [CrossRef]
13. Elias, S. Seismic energy assessment of buildings with tuned vibration absorbers. *Shock Vib.* **2018**, *2018*. [CrossRef]
14. Elias, S. Effect of SSI on vibration control of structures with tuned vibration absorbers. *Shock Vib.* **2019**, *2019*. [CrossRef]
15. Elias, S.; Matsagar, V. Seismic vulnerability of non-linear building with distributed multiple tuned vibration absorbers. *Struct. Infrastruct. Eng.* **2019**, *15*. [CrossRef]
16. Matin, A.; Elias, S.; Matsagar, V. Distributed multiple tuned mass dampers for seismic response control in bridges. *Proc. Inst. Civ. Eng. Struct. Build.* **2019**, 1–18. [CrossRef]
17. Elias, S.; Matsagar, V. Research developments in vibration control of structures using passive tuned mass dampers. *Annu. Rev. Control* **2017**, *44*, 129–156. [CrossRef]
18. Zelleke, D.H.; Elias, S.; Matsagar, V.A.; Jain, A.K. Supplemental Dampers in Base-Isolated Buildings to Mitigate Large Isolator Displacement Under Earthquake Excitations. *Bull. N. Z. Soc. Earthq. Eng.* **2015**, *48*, 100–117. [CrossRef]
19. Tsai, H. The Effect of Tuned-Mass Dampers on the Seismic Response of Base-Isolated Structures. *Int. J. Solid Struct.* **1995**, *32*, 1199–1210. [CrossRef]

20. Yang, J.; Danielianns, A.; Liu, S. Aseismic Hybrid Control Systems for Building Structures. *J. Eng. Mech.* **1991**, *117*, 836–853. [CrossRef]
21. Xiang, P.; Nishitani, A. Optimally Design for More Effective Tuned Mass Damper and Its Application to Base-Isolated Buildings. *Struct. Control Health Monit.* **2014**, *21*, 98–114. [CrossRef]
22. De Domenico, D.; Deastra, P.; Ricciardi, G.; Sims, N.D.; Wagg, D.J. Novel fluid inerter based tuned mass dampers for optimised structural control of base-isolated buildings. *J. Frankl. Inst.* **2018**, *356*, 7626–7649. [CrossRef]
23. De Domenico, D.; Ricciardi, G. Optimal design and seismic performance of tuned mass damper inerter (TMDI) for structures with nonlinear base isolation systems. *Earthq. Eng. Struct. Dyn.* **2018**, *47*, 2539–2560.
24. De Domenico, D.; Ricciardi, G. An enhanced base isolation system equipped with optimal tuned mass damper inerter (TMDI). *Earthq. Eng. Struct. Dyn.* **2018**, *47*, 1169–1192. [CrossRef]
25. De Domenico, D.; Ricciardi, G. Earthquake-resilient design of base isolated buildings with TMD at basement: Application to a case study. *Soil Dyn. Earthq. Eng.* **2018**, *113*, 503–521. [CrossRef]
26. Rabiee, R.; Chae, Y. Adaptive Base-isolation System to Achieve Structural Resiliency under Both Short-and Long-period Earthquake Ground Motions. *J. Intell. Mater. Syst. Struct.* **2019**, *30*, 16–31. [CrossRef]
27. Stanikzai, M.H.; Elias, S.; Matsagar, V.A.; Jain, A.K. Seismic response control of base-isolated buildings using tuned mass damper. *Aust. J. Struct. Eng.* **2019**, 1–12. [CrossRef]
28. Stanikzai, M.H.; Elias, S.; Matsagar, V.A.; Jain, A.K. Seismic Response Control of Base-Isolated Buildings Using Multiple Tuned Mass Dampers. *Struct. Des. Tall Spec. Build.* **2019**, *28*, e1576. [CrossRef]
29. Bertero, V.V. Strength and Deformation Capacities of Buildings under Extreme Environments. *Struct. Eng. Struct. Mech.* **1977**, *53*, 29–79.
30. Kennedy, R.P.; Cornell, C.; Campbell, R.; Kaplan, S.; Perla, H. Probabilistic Seismic Safety Study of An Existing Nuclear Power Plant. *Nuclear Eng. Des.* **1980**, *59*, 315–338. [CrossRef]
31. Bazzurro, P.; Cornell, C.A. Seismic Hazard Analysis for Non-linear Structures. II: Applications. *ASCE J. Struct. Eng.* **1994**, *120*, 3345–3365. [CrossRef]
32. Bazzurro, P.; Cornell, C.A. Seismic Hazard Analysis for Non-linear Structures. I: Methodology. *ASCE J. Struct. Eng.* **1994**, *120*, 3320–3344. [CrossRef]
33. Luco, N.; Cornell, C.A. Effects of random connection fractures on demands and reliability for a 3-story pre-Northridge SMRF structure. In *Proceedings of the 6th U.S. National Conference on Earthquake Engineering*; paper 244; EERI: El Cerrito, CA, USA; Seattle, DC, USA, 1998; pp. 1–12.
34. Luco, N.; Cornell, C.A. Effects of Connection Fractures on SMRF Seismic Drift Demands. *ASCE J. Struct. Eng.* **2000**, *126*, 127–136. [CrossRef]
35. Vamvatsikos, D.; Cornell, C.A. Incremental dynamic analysis. *Earthq. Eng. Struct. Dyn.* **2002**, *31*, 491–514. [CrossRef]
36. FEMA. *Recommended Seismic Design Criteria for New Steel Moment-Frame Buildings*; Report No. FEMA-350, SAC Joint Venture; Federal Emergency Management Agency: Washington, DC, USA, 2000.
37. FEMA. *Recommended Seismic Evaluation and Upgrade Criteria for Existing Welded Steel Moment-Frame Buildings*; Report No. FEMA-351, SAC Joint Venture; Federal Emergency Management Agency: Washington, DC, USA, 2000.
38. Wen, Y.K. Method for random vibration of hysteretic systems. *J. Eng. Mech. Am. Soc. Civ. Eng.* **1976**, *102*, 249–263.
39. Sadek, F.; Mohraz, B.; Taylor, A.W.; Chung, R.M. A Method of Estimating the Parameters of Tuned Vibration Absorbers for Seismic applications. *Earthq. Eng. Struct. Dyn.* **1997**, *26*, 617–635. [CrossRef]
40. Matsagar, V.A.; Jangid, R.S. Influence of isolator characteristics on the response of base-isolated structures. *Eng. Struct.* **2004**, *26*, 1735–1749. [CrossRef]
41. Somerville, P.G.; Smith, M.; Punyamurthula, S.; Sun, J. *Development of Ground Motion Time Histories for Phase 2 of the FEMA/SAC Steel Project*; Report. No. SAC/BD-97/ 04; SAC Joint Venture: Sacramento, CA, USA, 1997.
42. Bhandari, M.; Bharati, S.D.; Sharimali, M.K.; Datta, T.K. Assessment of Proposed Lateral Load Patterns in Pushover Analysis for Base-isolated Frames. *Eng. Struct.* **2018**, *175*, 531–548. [CrossRef]
43. Rathje, E.M.; Faraj, F.; Russell, S.; Bray, J.D. Empirical relationships for frequency content parameters of earthquake ground motions. *Earthq. Spectra* **2004**, *20*, 119–144. [CrossRef]

44. Rupakhety, R.; Sigurdsson, S.U.; Papageorgiou, A.S.; Sigbjörnsson, R. Quantification of ground-motion parameters and response spectra in the near-fault region. *Bull. Earthq. Eng.* **2011**, *9*, 893–930. [CrossRef]
45. Quaglini, V.; Dubini, P.; Furinghetti, M.; Pavese, A. Assessment of Scale Effects in the Experimental Evaluation of the Coefficient of Friction of Sliding Isolators. *J. Earthq. Eng.* **2019**, 1–21. [CrossRef]

applied sciences

MDPI

Article

Numerical Study on Hysteretic Behaviour of Horizontal-Connection and Energy-Dissipation Structures Developed for Prefabricated Shear Walls

Limeng Zhu [1,2], Lingmao Kong [1] and Chunwei Zhang [1,2,*]

[1] School of Civil Engineering, Qingdao University of Technology, Qingdao 266033, China; zhulimeng@qut.edu.cn (L.Z.); konglingmao@qut.edu.cn (L.K.)

[2] Cooperative Innovation Center of Engineering Construction and Safety in Shandong Blue Economic Zone, Qingdao University of Technology, Qingdao 266033, China

* Correspondence: zhangchunwei@qut.edu.cn; Tel.: +86-053285071328

Received: 3 January 2020; Accepted: 7 February 2020; Published: 12 February 2020

Featured Application: This developed "double-step" horizontal-connection and energy-dissipation structure could be used for prefabricated structures such as shear walls and replaceable coupling beams. In its first step, it weakly connects two adjacent shear walls and mainly dissipates the input energy. In its second step, it could strongly integrate two separate adjacent shear walls into one unit to obtain one stronger structure member to resist the collapse of the structural system.

Abstract: This study proposed a developed horizontal-connection and energy-dissipation structure (HES), which could be employed for horizontal connection of prefabricated shear wall structural system. The HES consists of an external replaceable energy dissipation (ED) zone mainly for energy dissipation and an internal stiffness lifting (SL) zone for enhancing the load-bearing capacity. By the predicted displacement threshold control device, the ED zone made in bolted low-yielding steel plates could firstly dissipate the energy and can be replaced after damage, the SL zone could delay the load-bearing and the load-displacement curves of the HES would exhibit "double-step" characteristics. Detailed finite element models are established and validated in software ABAQUS. parametric analysis including aspect ratio, the shape of the steel plate in the ED zone and the displacement threshold in the SL zone, is conducted. It is found that the HES depicts high energy dissipation ability and its bearing capacity could be obtained again after the yielding of the ED zone. The optimized X-shaped steel plate in the ED zone exhibit better performance. The "double-step" design of the HES is a potential way of improving the seismic and anti-collapsing performance of prefabricated shear wall structures against large and super-large earthquakes.

Keywords: energy dissipation; "double-step" characteristics; stiffness lifting; seismic performance; horizontal connection; prefabricated shear wall structural systems

1. Introduction

1.1. Research Status of the Connection for the Prefabricated Shear Wall System

In order to achieve the green and sustainable development and solve the problem of environmental protection and labor shortage, it is particularly significant to develop innovative prefabricated shear wall systems appropriately employed in tall buildings and some special structures [1–3]. The in-cast shear wall system is characterized by great lateral stiffness and bearing capacity. The traditional prefabricated shear wall system is designed according to the in-cast structure standards and its seismic performance fails to meet the requirements of the current seismic design code of buildings in China. With the application of prefabricated structural system especially in highly seismic regions, innovative design theories and systems are to be used to endow this new prefabricated system with high seismic and resilient performance. Therefore, the utilization of some resilient energy dissipation devices in structures is sensible to enhance the performance of prefabricated shear wall systems both in the large and super-large earthquakes [4,5].

The performance of connections between prefabricated shear walls has a significant influence on the seismic behavior of the prefabricated shear wall system. The present connections mainly include cast-in-place bolt connection, casing grouting connection, reserved hole slurry anchoring connection, bolt connection and post-tensioned prestressed connection [6–8]. Shemie [9] proposed a bolt connection between prefabricated panels which makes the connection between the wall panels more flexible and their ductility could be effectively utilized. Zhu et al. [10] conducted a mechanical study on horizontal and vertical joints on fabricated large slab structures, showing that horizontal seams could decrease the lateral stiffness and the shear angle of the vertical joint has a great influence on the distribution of the internal forces. Noel and Soudki [11] performed a reciprocating loading test on prefabricated shear walls and found that the bearing process of the horizontal joint could be defined as three stages which are the elastic stage before slipping, the elastoplastic stage before the damage of horizontal joint and the total slip damage. In the final stage of the slip failure, the strength will drop by 20% and the mortar will be crushed. Sun et al. [12] developed a new-type vertical joint for prefabricated wall and experimental results demonstrated that these connections were strong enough to maintain the global seismic behavior of the prefabricated wall equal to in-cast ones. Smith and Kurama [13] studied the prestressed specimens and found that their strength and initial stiffness are similar to those of cast-in-place specimens. The test piece demonstrated slight damage with a large nonlinear displacement, good self-centering ability but a little decrease in energy dissipation ability. Vaghei et al. [14,15] tested the U-shaped steel channel wall-to-wall connection and found that this type of connection performed better than loop connections. Guo et al. [16] proposed bolt connections for prefabricated wallboard structures and conducted shaking table tests on a 1/2 scale three-story model. The results show that the prefabricated structure system has the characteristics of high stiffness, large bearing capacity, and high collapse margin ratio. Because the current design code regards grouting pile connection as an idealized steel bar, it ignores the restriction of sleeve and composite behavior of components. Son et al. [17] proposed that the sheer force of horizontal connections of members can be resisted by overlapping anchors. The shear behavior of overlapping anchors between prefabricated concrete slabs and reinforced concrete members in simulated tests is analyzed. The results show that the average shear strength of overlapping anchorage connections is 109% of the calculated value. Jiang et al. [18] studied the effect of new bolted connections on the mechanical properties of prestressed concrete shear walls. The results show that the strain of the joints is less than the yield strain, and the joints do not destroy. The mechanical properties of the joints are similar to those of the cast-in-place reinforced concrete shear walls. Therefore, the performance of the connection could significantly influence the structural performance especially in the final stage in the earthquake.

1.2. Research Status of the Shear-Type Metal Damper

Many scholars have conducted extensive research on the behavior of metal dampers used in structural systems. Metal dampers as passive energy dissipation devices commonly serve as non-structural members reciprocating to absorb the input seismic energy and protecting the structural members. This energy dissipation is obtained by plastic deformation in which the structural member is in elastic [19]. Low-yield-point steel has the advantages of low yield strength, large elongation, and good ductility. Its high plastic deformation ability could enhance the structural energy dissipation ability [20]. The shear metal damper proposed by J.M. Kelly was widely used in the damping design of building structures due to simple structure and excellent performance. Whittaker et al. [21] proposed geometrically optimized X-shaped mild steel dampers and triangular soft steel dampers. Zhang and Zhang [22] experimentally researched different ways of weakening the stiffness of the damper in which the shape in the middle has a great influence on the ductility and the flange plates of different shapes have no obvious influence. Abebe et al. [23–25] conducted experimental research and simulation on the hysteretic behavior of low-yield steel shear dampers. Mortezagholi et al. and Zahrai et al. [26,27] proposed a damper with a circular cross-section by geometrically optimized parameter analysis. In order to solve the connection problem between lead blocks and components, Cheng et al. [28] proposed a baffle-type lead damper and its test results demonstrated excellent energy consumption ability. According to the above study, metal dampers could achieve good energy dissipation ability but their failure would result in degradation of the structural stiffness. To some degree, the high rigidity of in-cast structural systems would limit the performance of dampers and the stiffness degradation of the prefabricated structures would result in the collapse especially in large and super-large earthquakes. U-shaped metal yield damper introduced by Jamkhaneh et al. [29] has been tested about its mechanical displacement, lateral strength, elastic stiffness, and energy dissipation ability. Lin et al. [30] developed a detachable buckling restrained shear plate shock absorber. The influence of key design factors, such as the length-width ratio of the slab and the number of internal composite plates, on the seismic performance of the damper, is studied, and the design formulas for calculating the elastic stiffness and ultimate strength of the damper are proposed. Zhu et al. [31] proposed a metal shear plate damper with an optimized shape. The test results show that the metal shear damper has stable energy dissipation capacity and good low cycle fatigue performance. Belleri et al. [32] proposed that the use of passive energy dissipation and re-centering devices could limit the structural damage. Mazza et al. [33,34] successfully proposed a design procedure for the damper braces to attain a designated performance level according to a certain proportion of reinforcement and further developed a new displacement-based design procedure to proportion hysteretic damped braces considering the effect of a structure's seismic degradation. These procedures are verified to be highly effective when being utilized in designing frame structures.

Some gap dampers are proposed by Rawlinson et al. [35] and De Domencio et al. [36] which could be designed to be engaged after an expected displacement and they depict good performance when being utilized in base-isolated structural system. Therefore, an innovative type of wall-to-wall horizontally connecting structure with high energy dissipation and stiffness lifting ability is proposed, which is expected to enhance the seismic and resilient performance of prefabricated shear wall systems.

This study proposes an innovative "double-step" horizontal-connection and energy-dissipation structure (HES) with the character of high energy dissipation and capacity lifting after the decrease. In its first step, it weakly connects two adjacent shear walls and mainly dissipates the input energy. In its second step, it could strongly integrate two respectively working adjacent shear walls into one unit to obtain one stronger structure member to resist the collapse of the structure system. The design procedure of the HES is briefly depicted in Figure 1. First, a shear walls structural system analytical model is built and analyzed in a predicted earthquake is. After that, the shear bearing capacity V_f and the allowable horizontal displacement Δ_{WH} of the shear wall are calculated. The shear threshold displacement D of the HES is determined according to the allowable horizontal displacement Δ_{WH}, by which the input energy during the earthquake is dissipated by the ED zone before yield the shear wall. The shear bearing capacity V_f of shear wall is employed to predict the shear capacity of the SL zone of the HES V_{HSE} to ensure that the HES could strongly integrate two respectively working adjacent shear walls into one unit to obtain one stronger structure member to resist the collapse of the structure system. Numerical analysis is performed to comprehensively study the hysteretic behavior of the HES utilizing the validated finite element models. Their hysteretic load-displacement curves, skeleton curves, shear deformation, and failure mode are discussed in detail and the optimized design methods are suggested.

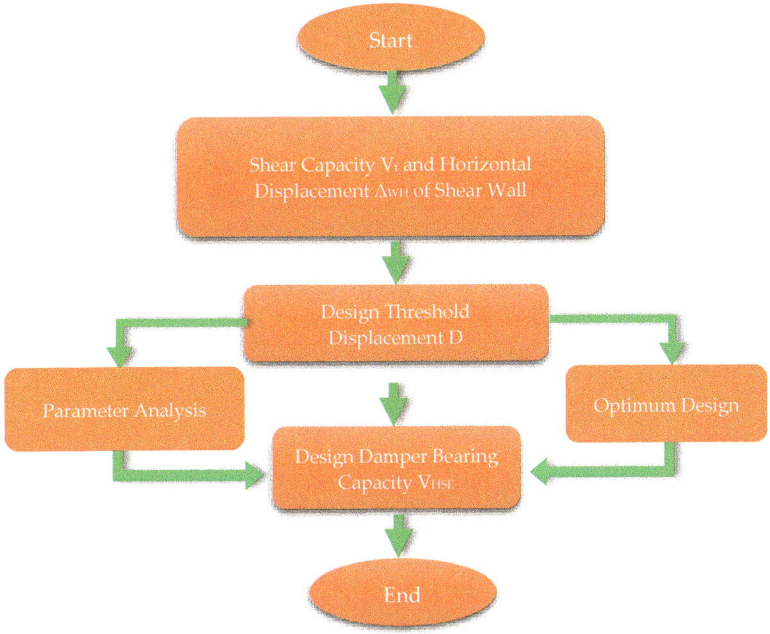

Figure 1. Design flow chart of the horizontal-connection and energy-dissipation structure (HES).

2. The Mechanism of the HES

2.1. General Concepts

The equivalent in-cast connecting method is commonly used in traditional prefabricated shear wall systems which fails to restore the structural function once being damaged in the earthquake. The proposed HES mainly consists of the energy dissipation zone (ED), the stiffness lifting zone (SL) and the horizontal connecting zone (HC) as shown in Figure 2. The HES could horizontally connect two adjacent walls (Figure 2a), the steel plates in the ED zone and SL zone both utilize bolt connection which could be fast replaced after damage in the earthquake and the resilient structural performance is obviously enhanced. The low-yield-strength steel plates are employed in the ED zone mainly dissipating seismic energy. As depicted in Figure 2d–f, the SL zone is composed of the shear stiffness lifting plate, the flange plate, functional bolts, and the buckling restrained plates. The diameter of the functional bolt bar is set to be smaller than that of the circle hole in the shear stiffness lifting plate and this deviated value is the shear displacement threshold. By this threshold control system, the HES exhibits a controlled two-stage mechanical behavior. As depicted in Figure 2, the obvious shear deformation of the HES could be investigated when the in-plane lateral deformation of two adjacent walls are observed in the earthquake. In the first stage, when the shear displacement is smaller than the displacement threshold the functional bolts would not contact the shear stiffness lifting plate and the SL zone has no contribution to the performance of the HES, only the ED zone dissipating the input seismic energy. In the second stage, when the shear displacement is smaller than the displacement threshold, the SL zone would begin to work and the shear stiffness and bearing capacity of the HES increase again. The adjacent walls would be assembled again becoming a strengthening system and the lateral bearing capacity of shear wall system would increase again, which could protect the structure in the large and super-large earthquake.

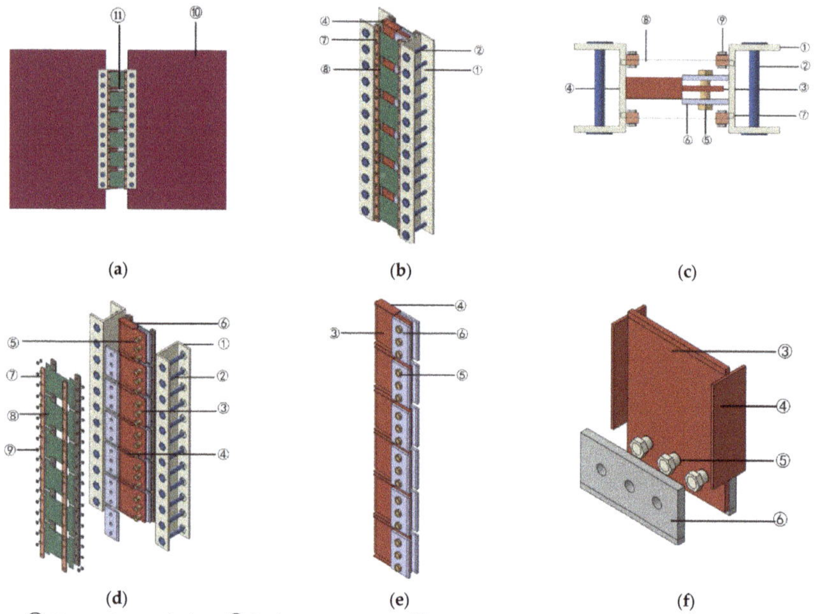

Note: ① Connecting end plate; ② high-strength bolts; ③ the shear stiffness lifting plate; ④ the flange plate; ⑤ functional bolts; ⑥ the buckling restrained plate; ⑦ the backing plateL; ⑧ energy dissipation plate; ⑨ bolts; ⑩ the shear wall; ⑪ the HES

Figure 2. Detailed instruction of the HES system. (**a**) The wall connection, (**b**) assembly model, (**c**) cross-section, (**d**) disassembled model, (**e**) lateral stiffness lifting zone, (**f**) threshold control system.

The shear wall and the HES are assumed to be rigid pieces as shown in Figure 3. The deformation of the HES could be computed approximately by Equation (1) when achieving the horizontal displacement of shear walls.

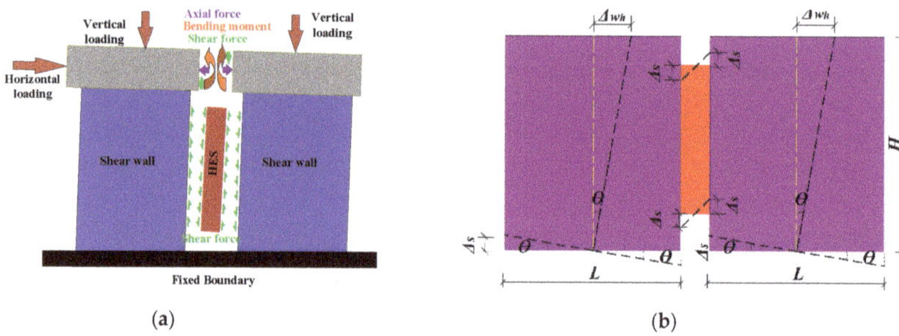

Figure 3. The shear deformation of the HES. (**a**) The boundary condition, (**b**) calculation shear deformation of the HES according to the drift displacement.

The drift ratio is a significant parameter defined in many specifications, which is utilized to evaluate structural performance. Therefore, the limited shear deformation of the HESs could be calculated.

$$\Delta_s = 0.5{\cdot}\theta{\cdot}L = 0.5{\cdot}\Delta_{hw}{\cdot}L/H \quad (1)$$

where, Δ_{hw} is the horizontal displacement of the shear wall, Δ_s is the shear deformation of the shear wall, θ is the drift ratio, L is the width of the shear wall, H is the height of the shear wall.

2.2. The Expected Failure Mode of the HES

The low-yield-point steel plates (LY 100) the yield stress of which is 100 MPa, are utilized in the ED zone. In order to ensure the bearing capacity and shear stiffness of the SL zone, Q345B is utilized the yield stress of which is 345 MPa. The threshold control displacement should be reasonably established to ensure that the ED zone and the SL zone could perform well in sequence. Furthermore, the threshold displacement could be adjusted according to the requirement of the expected structural performance, by changing the hole diameter in the shear stiffness lifting plate. In addition, the shear stiffness and strength of the shear stiffness lifting plate could be effectively enhanced by adding the flange plate on both sides. Simultaneously, the functional bolts are designed to come to failure in the second stage and the other parts of the SL zone are in elastic. When the ED zone and the functional bolts are damaged in the earthquake, they could be fast and easily replaced, largely enhancing the structural resilient performance. Therefore, the ED zone is expected to dissipate the input seismic energy firstly and the functional bolts are expected to fail with high ductility.

In order to clarify the mechanical mechanism of the HES, simplified models are used in finite element modelling analysis. The lateral connection of the HES to the wall is considered to be rigid. The bolted connection of the steel plates in the ED zone is also considered to be a rigid connection. The threshold control displacement of the typical specimen of HES is set to be 3 mm. On the basis of the simplification, this expected failure mode would be validated by numerical analysis.

3. Model Development and Validation

The finite element analysis is an efficient way to predict the performance of testing specimens. The software ABAQUS was employed to simulate the typical hysteretic behavior of the HESs. This numerically modelling method is validated by successfully simulating the cyclic behavior of one low yield strength steel shear plate damper tested by Zhang et al. Using this modelling method, the performance of the HESs are predicted and evaluated.

3.1. Finite Element Model for The HESs

The model and the boundary conditions are depicted in Figure 4. The shell element S4R is used to model the behavior of the steel plates in the ED zone including their buckling. Eight-node-three-dimensional solid element (C3D8R) with reduced integration and hourglass control is utilized to simulate functional bolts, the flange plate and the shear stiffness lifting steel plates. The overall meshed model includes a total of 40,444 elements. A typical surface-to-surface contact with a penalty algorithm is employed between the functional bolts and the shear stiffness lifting plate. A hard contact pressure-over closure relationship is adopted to model the normal contact behavior and the friction coefficient is set to be 0.2 to simulate the tangential slip behavior. The same contact setting is used between the surfaces of buckling restrained steel plates and shear stiffness lifting steel plates. In addition, the geometric nonlinearity is considered to model the intermittent contact behavior in the SL zone. The circular part and the bolts in the SL zone are more elaborately meshed. Considering the computational efficiency and reliability the mesh size of the HES is the same as above which is appropriate for this analysis.

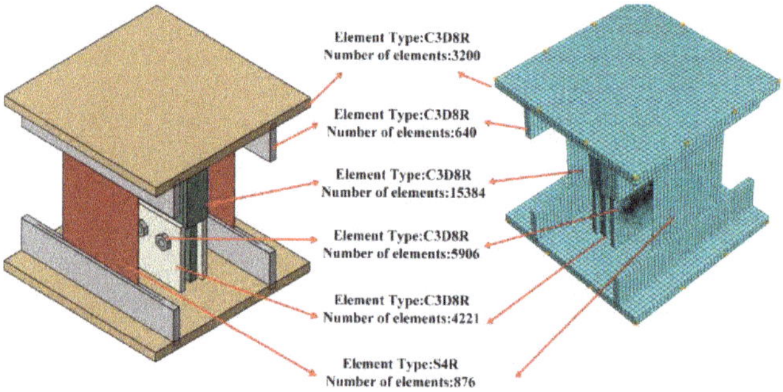

Figure 4. The finite element model for the HES.

3.2. Material Constitutive Models

The von Mises yield criterion and bilinear model are utilized for simulating the behavior of the ED zone made of low yield point steel (LP100) and the SL zone made of Q345B. When the strain ε of the steel sheet is less than the ultimate strain, the actual stress σ of the steel sheet is equal to the elastic modulus multiplied by the strain ε; when the strain ε of the steel sheet is greater than the ultimate strain, the actual stress σ of the steel sheet is equal to the elastic modulus multiplied by the ultimate strain. With the advancement of structural health monitoring technology, high precision strain measurement can be obtained [37,38] to guarantee the quality and reliability of stress calculation under complicated load bearing situation. The elastic modulus of the steel is set to be = 2.05×10^5 MPa, the Poisson's ratio of steel is taken as 0.3 [39,40].

3.3. Validation of the Finite Element Model

3.3.1. Verification

Zhang et al. [41] conducted a cyclic fatigue performance test of a low yield strength steel shear plate damper and the failure mode is depicted in Figure 5b investigating obvious buckling. Using the method above, the finite elemental analysis (Figure 5a) is conducted and the results including the failure mode and the cyclic curve are shown in Figure 5c,d. It can be seen that simulated buckling deformation is consistent with the tested. The simulated curve agrees well with the test curve but a little deviation of the initial stiffness could be observed which might be caused by failing to accurately model the actual loading boundaries. In addition, this finite element model(FEM) accurately predicts the bearing capacity in each loading circle. Therefore, this modelling method could be used to predict the behavior of the HESs and the results could be used to evaluate their performance.

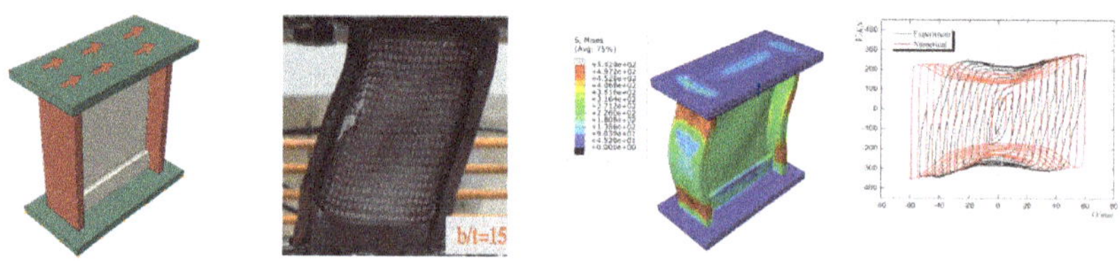

(a) (b) (c) (d)

Figure 5. The validation of the finite element modeling method. (**a**) Finite element model, (**b**) tested failure mode, (**c**) simulated failure mode, (**d**) hysteresis curves.

Xu et al. [39] proposed a metal shear plate damper utilizing the low yield performance of BLY 160 materials for effective energy dissipation and conducted hysteric tests to evaluate its performance.

This metal shear plate damper is composed of four components: shear panels, confined flanges, stiffening ribs and the roof/floor plates for connection (Figure 6b). To validate the simulating method used above, the above modelling method is employed to carry out a finite element analysis of the damper (Figure 6a). The failure mode, the buckling behavior and numerical simulated curves are shown in Figure 6c,d. It is found that the simulated curve has good coincidence with the test curve, but the load-displacement amplitude has some deviation. Some unpredicted slip occurred in the boundary during the experimental loading process, resulting in a deviation between the finite element loading displacement and the test. Therefore, the modelling method could simulate the hysteric behavior of the HESs and could be utilized to evaluate their performance.

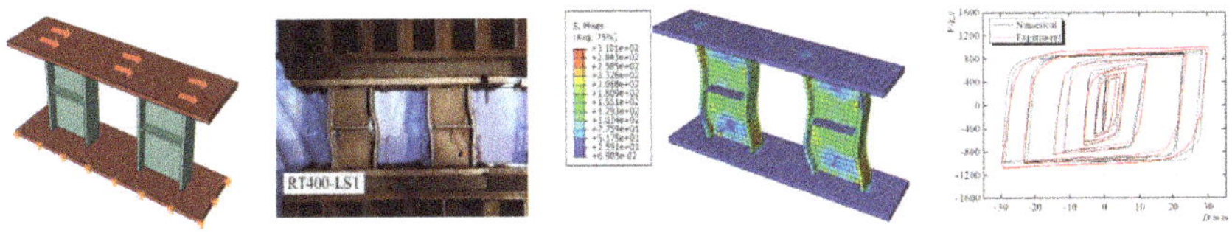

(a) (b) (c) (d)

Figure 6. The validation of the finite element modelling method. (**a**) Finite element model, (**b**) tested failure mode, (**c**) simulated failure mode, (**d**) hysteresis curves.

3.3.2. The Simulation of the Double-Step Performance of the HESs

The detailed geometric parameters are listed in Table 1, Table 2 and Figure 7. The monotonic behavior of specimen HES1 is simulated and the results are shown in Figure 7 and its displacement threshold is set as 3mm. When the shear displacement is applied to 0.24 mm (the drift ratio = 0.12/100), the steel plate in the ED zone begins to yield and the yielding area gradually enlarges from the two ends of the steel plate to the middle part as shown in Figure 8a. With the increment to 3 mm (the drift ratio θ = 1.5/100), the steel plate totally comes into plastic and the shear strength of HES1 comes to the end of its first step, as shown in Figures 8b and 9a. With the development of deformation, the SL zone begin to work and when the horizontal displacement is up to 3.8 mm (the drift ratio = 1.9/100), the middle part of functional bolts yields, the shear stiffness lifting plate is almost in elastic with slight stress concentration at the bolt holes as shown in Figure 8c,d. Finally, the full section of the functional bolts come into plastic and the shear stiffness lifting plate locally yields, as shown in Figure 8e,f. The top and bottom boundaries utilized thick steel plates to simulate the connection to shear walls. When the HES is employed in steel composite shear wall system [42] and steel-damping-concrete composite wall sytems [43] the high strength bolt connection is available.

Table 1. Common geometric parameters for specimens.

	The Typical Dimensions for All HES Specimens (mm)												
	b	H	t	a	h	h_c	t_c	b_a	h_a	t_a	b_f	h_f	t_f
The HESs	330	380	400	65	20	70	10	40	200	56	200	140	10

Table 2. Specific dimensions for HES specimens.

No.	The Geometric Parameters of HES Specimens (mm)													
	b_w	h_w	t_w	b_t	h_t	t_t	R_1	R_2	$D_{(R1-R2)}$	R_3	b_{w1}	h_{w1}	b_{w2}	h_{w2}
HES1	200	200	10	200	330	16	11	8	3	-	-	-	-	-
HES2	200	200	10	200	330	16	11	8	3	-	50	100	-	-
HES3	200	200	12	200	330	16	11	8	3	-	50	100	-	-
HES4	200	200	10	200	330	16	10	8	2	-	-	-	-	-
HES5	200	200	10	200	330	16	13	8	5	-	-	-	-	-
HES6	200	200	10	200	330	16	11	8	3	-	-	-	50	50
HES7	200	200	10	200	330	16	11	8	3	50	-	-	-	-

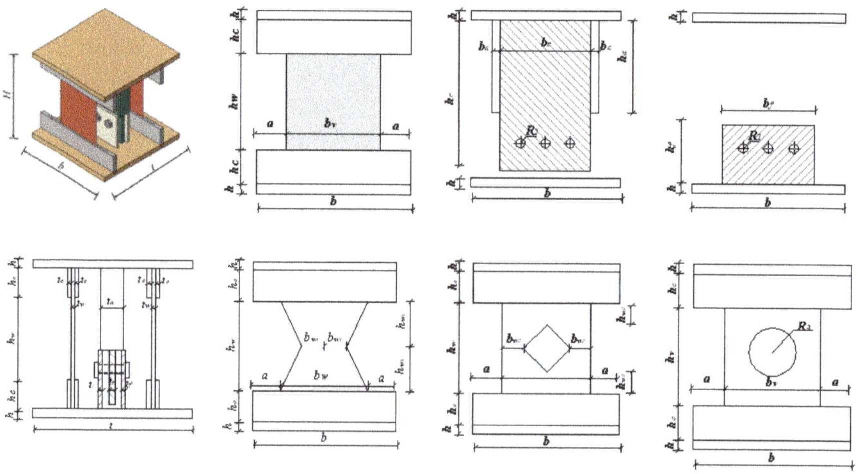

Figure 7. The detailed dimensions of HES specimens. (**a**) 3D model, (**b**) energy dissipation (ED) zone/rectangular steel plates (HES1/HES4/HES5), (**c**) stiffness lifting (SL) zone/the shear stiffness lifting plate, (**d**) SL zone/the shear stiffness lifting plate, (**e**) the top view, (**f**) ED zone/X-shaped steel plates(HES2/HES3), (**g**) ED zone/Rectangular hole (HES6), (**h**) ED zone/Circular hole (HES7).

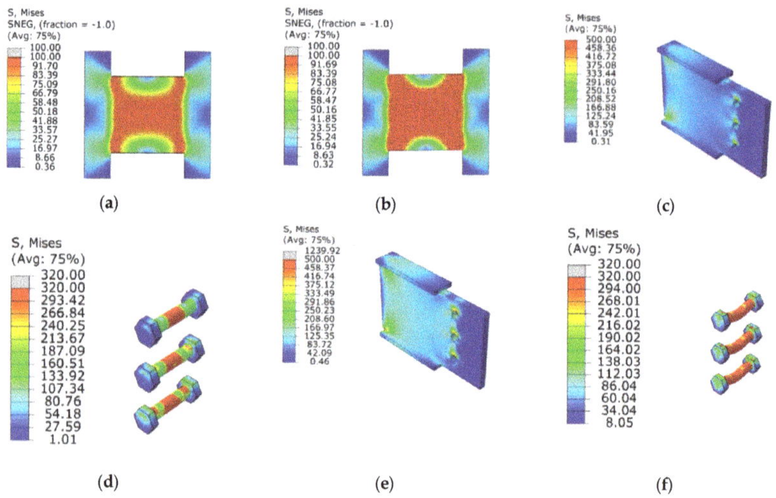

Figure 8. The validation of the finite element modelling method. (**a**) The stress distribution of ED zone ($\theta = 0.12\%$), (**b**) the stress distribution of ED zone ($\theta = 1.5\%$), (**c**) the stress distribution of SL zone ($\theta = 1.9\%$), (**d**) the stress distribution of functional bolts ($\theta = 1.9\%$), (**e**) the stress distribution of SL zone($\theta = 5\%$), (**f**) the stress distribution of functional bolts ($\theta = 5\%$).

The double-step force-displacement curve of HES1 is depicted in Figure 9a and the typical double-step working mechanism of the HESs is shown in Figure 9b. The performance of the HES can be observed with four stages including the completely elastic stage, ED plastic stage, SL elastic stage and functional bolts plastic stage. In the completely elastic stage (the OA line in Figure 9a), steel plates in ED are in elastic and the initial shear stiffness is K1 = 525.19 kN/mm. When coming to the ED plastic stage (the AB line), the input seismic energy is mainly dissipated by the plastic deformation. From point B to point C (the SL elastic stage), the SL zone is almost in elastic and the HES restores the bearing capacity and the shear stiffness with K2 = 166.32 kN/mm. As the shear displacement increases, the bending deformation of the functional bolts gradually develops into the fourth stage (the CD line) and finally the come into yielding as depicted in Figure 9b.

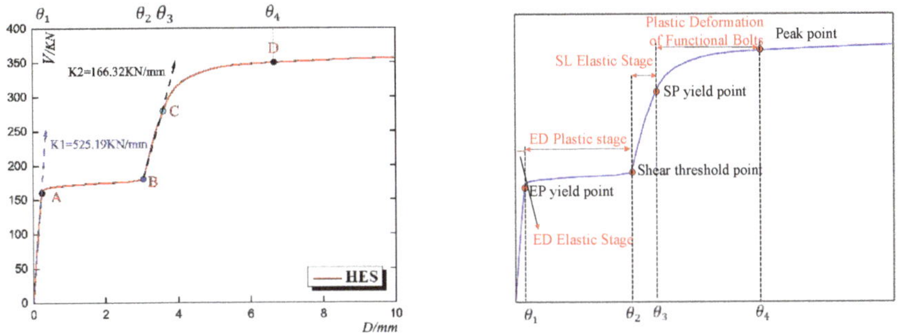

(a) (b)

Figure 9. The typical curve for the HES. (**a**) Monotonic force-displacement curve, (**b**) the failure mode.

3.3.3. The Typical Hysteric Behavior of the HESs

The loading protocol is shown in Figure 10a which is divided into ten stages with two cycles in each stage. In this protocol, two types of calculating the loading displacement are adopted respectively considering the character of the ED zone and the SL zone. In the stages of only the ED zone working, the loading displacement is set according to the yield displacement of the steel plate and the latter loading amplitude is twice the amplitude of the previous loading displacement. In the stages of the ED zone and the SL zone working simultaneously, the loading displacement is set according to the yield displacement of the SL zone and the latter loading amplitude is 1.4 times of the previous loading amplitude. The simulated typical hysteresis curve is depicted in Figure 10b demonstrating the deforming characteristics and energy dissipation performance. Due to the in-plane shear resistance of the steel plate in the ED zone, the HES exhibits a character of the large initial stiffness and the high energy dissipation ability. It can be seen that the cyclic curves are close to rectangular shapes indicating the great energy dissipation ability. When the loading displacement is larger than the designed displacement threshold, the area of the hysteresis curve and the bearing capacity both gradually increase largely. The hysteretic curve exhibits double-step operating characters with both high energy dissipation ability and the shear stiffness re-lifting ability.

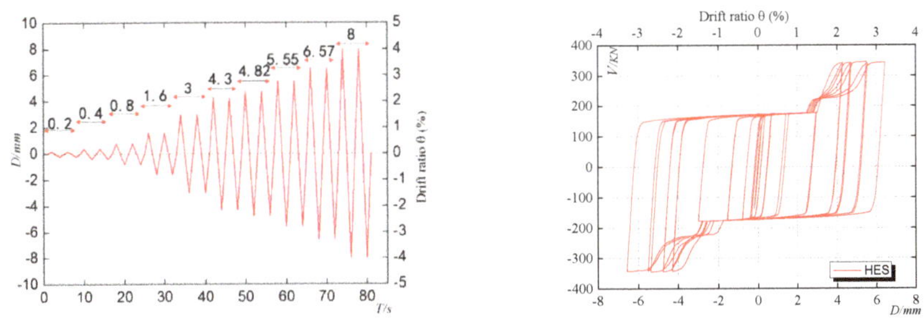

(a) (b)

Figure 10. The typical hysteresis curve of the HESs. (**a**) The loading protocol, (**b**) the hysteretic curve.

4. Parametric Analysis

In order to further optimize the performance of the HES, seven specimens with different threshold displacements, different steel plate width-thickness ratios and different shapes of the steel plates in the ED zone are designed and simulated using software ABAQUS. The detailed geometric parameters are shown in Figure 7, Table 1, and Table 2. The common and specific geometric parameters are respectively listed in Tables 1 and 2. The influence of the shape of the steel plate in the ED zone is investigated by the comparison among specimens HES1, HES2, HES6, and HES7. The influence of the width-thickness ratio is studied by comparing the performance of HES2 and HES3. The threshold displacements of HES2, HES4, and HES5 are changed to investigate their influence.

4.1. Parameter Analysis Under Monotonic Load

4.1.1. The Investigation on the Influence of the Shape of the Steel Plate in the ED Zone

The monotonic behavior of four specimens with different shape types for steel plated in the ED zone including the rectangular shape (HES1), the X-type (HES2), the rectangular shape with one diamond-shaped hole (HES6) and the rectangular shape with one circular hole (HES7) are simulated. Their load-displacement curves, the stress distribution of the steel plates in the ED zone and typical analyzed results are depicted in Figure 11 and Table 3.

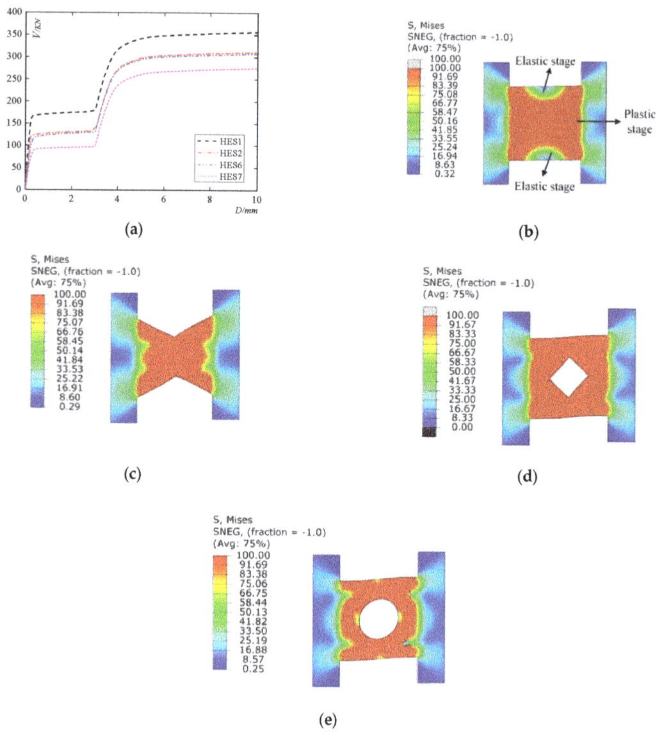

Figure 11. The influence of the shape of the steel plate in the ED zone. (**a**) Load-displacement curves, (**b**) the steel plates in ED zone (HES1), (**c**) the steel plates in ED zone (HES2), (**d**) the steel plates in ED zone (HES6), (**e**) the steel plates in ED zone (HES7).

Table 3. Typical analyzed results.

NO.	The Surface Area (mm^2)	The Initial Stiffness (KN/mm)	Yield Load (KN)	Yield Displacement (mm)
HES1	4.00	525.19	160.05	0.304
HES2	3.00	458.13	116.19	0.253
HES6	3.50	370.01	95.78	0.258
HES7	3.21	319.78	70.63	0.221

It is observed in Figure 11a that the bearing capacity the HES with rectangular steel plates in the ED zone is the largest. Among the four shape-types of the steel plates in the ED zone, the bearing capacity of the rectangular typed specimen is larger than those others but this type of shape failed to fully develop the plastic deformation. Among the three optimized shapes, the X-shaped specimen HES2 with the smallest surface area exhibits the best performance including the initial stiffness and bearing capacity as shown in Table 3.

4.1.2. The Influence of the Width-Thickness Ratio of the Steel Plate in the ED Zone

The monotonic behavior of specimens with optimized X-shaped low-yield-point steel plates (HES2 and HES3) are analyzed and the load-displacement curves are shown in Figure 12a. The thickness of X-shaped low-yield steel plates of specimen HES2 and HES3 are respectively set to be 10 mm and 12 mm which is the only difference between them. With the increase of thickness, the obvious increase of shear capacity (22.5%) can be investigated and the initial shear stiffness is slightly increased.

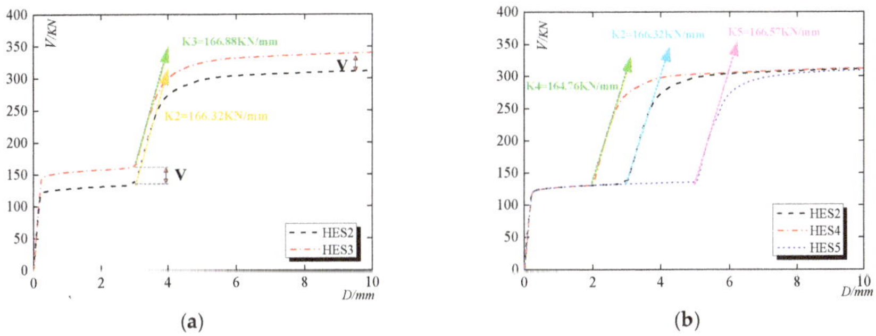

Figure 12. The monotonic load-displacement curves for HES2/HES3/HES4/HES5. (**a**) The influence of the width-thickness ratio, (**b**) the influences of the displacement of threshold.

4.1.3. The Shear Stiffness Lifting Control System

The shear displacement threshold is a significant parameter to decide on which level of shear deformation the SL zone begins to bear loads. The shear displacement threshold ($D_{(R1\text{-}R2)}$) of specimen HES2, HES4 and HES5 are respectively set as 3 mm, 2 mm and 5 mm and the computed load-displacement curves are shown in Figure 12b. It is concluded that this shear stiffness control system endows the HES with double-step character, sufficient energy dissipation ability and the ability to prevent the collapse of the structure in large earthquakes. If the shear displacement threshold is too large, the strength degradation of the steel plate in the ED zone will result in a decrease in the ultimate strength of the HES. Therefore, the shear displacement threshold could be adjusted according to the requirement of performance design.

4.2. Simulated Hysteretic Curves

As depicted in Figure 13, the shapes for the simulated hysteretic curves of the seven specimens are similar to each other and. Due to the shear stiffness lifting control system, two stages of energy dissipation and load-bearing is investigated. When shear displacement is smaller than the threshold displacement, the shape of the hysteretic curves is similar to rectangular and the high energy dissipation performance of the ED zone is obtained. With the increase of the thickness of steel plates, the bearing capacity and the energy dissipation ability are both enhanced. The Bouc–Wen–Baber–Noori model could be adopted to describe the hysteric character and used to analyze the seismic behavior of wall systems. The gap in the SL zone could be further utilized to add some viscous damping material to increase the damping ratio according to the requirement of the wall system.

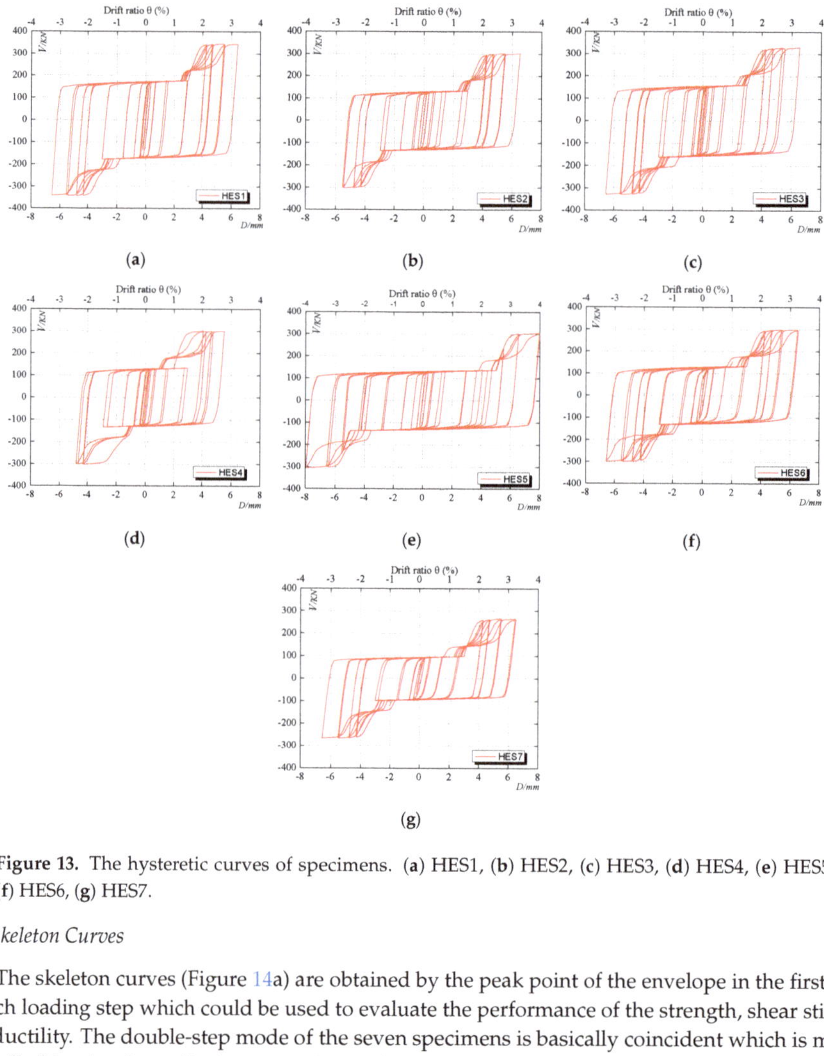

Figure 13. The hysteretic curves of specimens. (**a**) HES1, (**b**) HES2, (**c**) HES3, (**d**) HES4, (**e**) HES5, (**f**) HES6, (**g**) HES7.

4.3. Skeleton Curves

The skeleton curves (Figure 14a) are obtained by the peak point of the envelope in the first cycle of each loading step which could be used to evaluate the performance of the strength, shear stiffness and ductility. The double-step mode of the seven specimens is basically coincident which is mainly controlled by the shear displacement threshold. The comparison HES structural specimens with different threshold displacements, different thicknesses, and shapes of steel plates in the ED zone are respectively shown in Figure 14b–d. It can be seen that the energy-dissipation process of the ED zone is slightly extended as the threshold displacement increases. However, the bearing capacity of the SL zone is not obviously influenced by the threshold displacement. With the increase of thickness, both the bearing capacity of the ED zone and the HES increases. The yield displacement depends on neither the thickness nor the bearing capacity. The X-shaped steel plates with the smallest surface area exhibit the highest bearing capacity in shapes optimized from the rectangular shape which is suggested to be utilized in the ED zone.

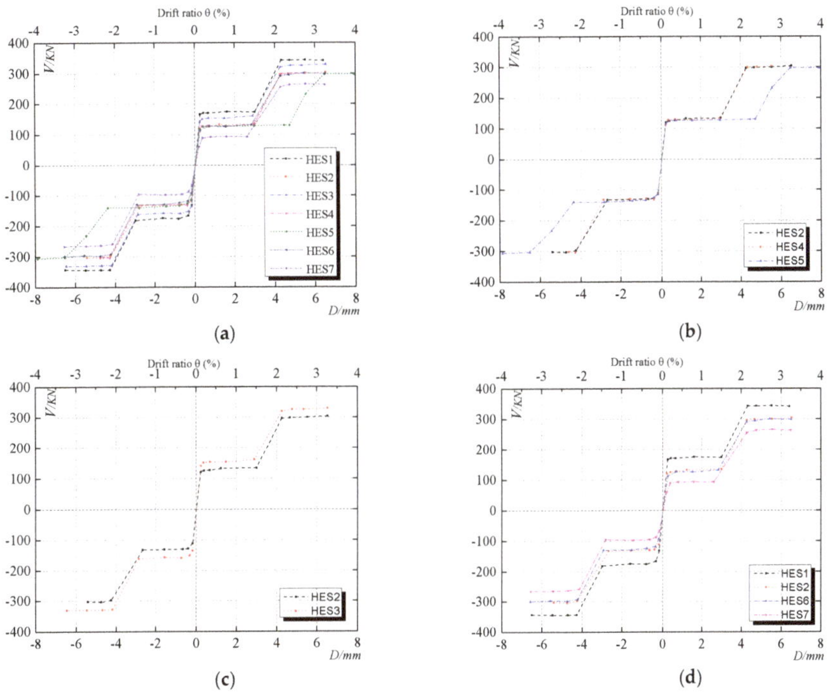

Figure 14. Skeleton curves. (**a**) HES1-HES7, (**b**) HES2/HES4/HES5, (**c**) HES2/HES3, (**d**) HES1/HES2/HES6/HES7.

4.4. The Energy Dissipation Ability of the ED Zone

It can be seen from Figure 15a that, basically, the HES specimens with the rectangular shape of steel plates in the ED zone exhibit larger bearing capacity in the same deformation. But the rectangular steel plate would result in stress concentration and the failure in the bolt connection boundary. The specimen HES2 with optimized X-shape steel plates in the ED zone exhibits the highest energy dissipation capacity compared with that of specimen HES6 and HES7. Because of the thickness increase of steel plates in the ED zone, the energy dissipation capacity of specimen HES3 is larger than that of specimen HES2 as shown in Figure 15b. Figure 15c depicts that the specimens with the larger threshold displacement would dissipate less energy before the SL zone coming to work. When the SL zone begins to bear the load and the plastic deformation of the bolts would increase its energy dissipation ability.

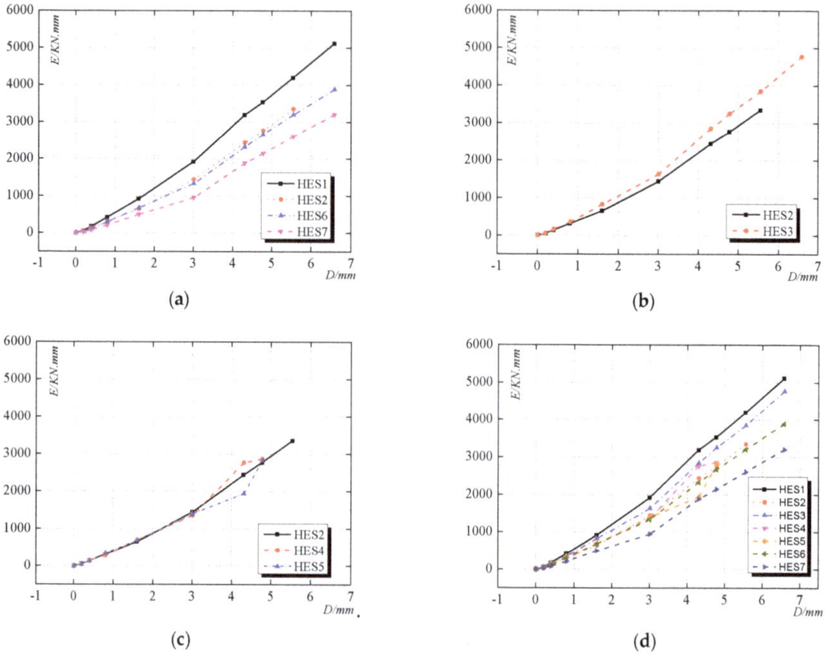

Figure 15. Accumulative energy dissipation curves. (a) HES1/HES2/HES6/HES7, (b) HES2/HES4/HES5, (c) HES2/HES3/HES4, (d) HES1-HES7.

5. Summary and Conclusions

The inherent out-of-plane stiffness and strength of the HES are mainly provided by the steel plates in the ED zone. The strong connections to the two adjacent walls of the HES can also ensure its in plane performance. In the further study a kind of supporting structure will be developed in the SL zone to enhance its out of place performance. This study mainly analyzed the in-plane monotonic and hysteretic behavior of the HESs using software ABAQUS. Seven specimens of the HES are designed with different parameters and the influence of the parameters on their performance is investigated giving some optimized suggestions. The failure mode of the HESs is observed and their typical performance load-displacement is proposed with the character of double-step. Because of the design of the shear stiffness lifting control system, the ED zone would firstly come into plastic dissipating the input seismic energy and the SL zone would come into play when the large deformation occurs in a large and super large earthquake. Therefore, the HES can be used as the horizontal connecting member for the shear wall system and simultaneously enhance its seismic and resilient performance. On the basis of the above simulation and analysis, the following conclusions are obtained.

(1) The proposed shear displacement threshold control system endows the HES with the ability of energy dissipation, stiffness lifting and shear strength lifting by the separate function of the ED zone and the SL zone. The bolt connection in the ED zone and the functional bolts could be easily and rapidly replaced when being damaged in the earthquake, which largely enhances the resilient performance and the recovery capability of the structural system. The threshold can be adjusted according to the requirement of the structural performance, this proposed The HES could be used in prefabricated shear wall system and the performance-based design could be applied.

(2) The rectangular shape for the steel plate in the ED zone exhibits good energy dissipation performance and is easy for construction. According to the parameter analysis of the shape, the

X-shaped steel plate in the ED zone exhibits the best performance and this type is suggested to be utilized in the HESs.

(3) The shear deformation of the HES is caused by the horizontal displacement of the shear walls. When being employed in shear wall structures, the ultimate drift ratio of the HES in this study is about 4%, which could be adjusted to meet the requirement of corresponding horizontal displacement of the shear wall. The ductility coefficient of the steel plate in the energy dissipation zone is about 15 and the use of the low-yield-point steel could effectively enhance the energy dissipation ability in small shear deformation during small earthquakes.

Author Contributions: C.Z. and L.Z. provided the conceptual design of the novel composite wall; L.Z. and L.K. conceived and designed the FE models; L.Z. and L.K. analyzed the data; L.Z. and L.K. wrote the paper; L.Z. and C.Z. revised the paper. All authors have read and agreed to the published version of the manuscript.

Funding: This research was funded by National Natural Science Foundation of China (Grant No. 51708318, 51678322), the International Postdoctoral Exchange Fellowship Program (Grant No. 20190015), and the Ministry of Science and Technology of China (Grant No. 2017YFC0703603).

Acknowledgments: The research is financially and technically supported by the Taishan Scholar Priority Discipline Talent Group program funded by the Shandong Province, the Cooperative Innovation Center of Engineering Construction and Safety in Shandong Blue Economic Zone scheme funded by the Shandong Province and the first-class discipline project funded by the Education Department of Shandong Province.

Conflicts of Interest: The authors declare no conflict of interest.

References

1. Jiang, B.Y.; Li, H.X.; Dong, L.; Wang, Y.; Tao, Y.Q. Cradle-to-Site Carbon Emissions Assessment of Prefabricated Rebar Cages for High-Rise Buildings in China. *Sustainability* **2019**, *11*, 42. [CrossRef]
2. Wu, G.B.; Yang, R.; Li, L.; Bi, X.; Liu, B.S.; Li, S.Y.; Zhou, S.X. Factors influencing the application of prefabricated construction in China: From perspectives of technology promotion and cleaner production. *J. Clean. Prod.* **2019**, *219*, 753–762. [CrossRef]
3. Wu, H.J.; Qian, Q.K.; Straub, A.; Visscher, H. Exploring transaction costs in the prefabricated housing supply chain in China. *J. Clean. Prod.* **2019**, *226*, 550–563. [CrossRef]
4. Kurama, Y.C.; Sritharan, S.; Fleischman, R.B.; Restrepo, J.I.; Henry, R.S.; Cleland, N.M.; Ghosh, S.K.; Bonelli, P. Seismic-Resistant Precast Concrete Structures: State of the Art. *J. Struct. Eng.* **2018**, *144*. [CrossRef]
5. Brunesi, E.; Nascimbene, R. Experimental and numerical investigation of the seismic response of precast wall connections. *Bull. Earthq. Eng.* **2017**, *15*, 5511–5550. [CrossRef]
6. Farhat, M.; Issa, M. Design principles of totally prefabricated counterfort retaining wall system compared with existing cast-in-place concrete structures. *PCI J.* **2017**, *62*, 89–106.
7. Han, C.; Li, Q.N.; Jiang, W.S.; Yin, J.H.; Yan, L. Pseudo-dynamic test of the steel frame—Shear wall with prefabricated floor structure. *Steel Compos. Struct.* **2016**, *20*, 431–445. [CrossRef]
8. Sun, J.; Qiu, H.X.; Xu, J.P. Experimental Verification of Vertical Joints in an Innovative Prefabricated Structural Wall System. *Adv. Struct. Eng.* **2015**, *18*, 1071–1086. [CrossRef]
9. Shemie, M. Bolted connections in large panel system buildings. *PCI J.* **1973**, *18*, 27–33. [CrossRef]
10. Zhu, Y.; Liu, Y.; Rui, C.; Guan, Q.; Shou, G. The Model Test of Large Panel Multistory Building under Horizontal Load. *J. Build. Struct.* **1980**, *1*, 31–46.
11. Noel, M.; Soudki, K. Shear behavior of post-tensioned FRP-reinforced concrete slabs under static and fatigue loading. *Constr. Build. Mater.* **2014**, *69*, 186–195. [CrossRef]
12. Sun, J.; Qiu, H.X.; Jiang, H.B. Lateral load behaviour of a rectangular precast shear wall involving vertical bolted connections. *Adv. Struct. Eng.* **2019**, *22*, 1211–1224. [CrossRef]
13. Smith, B.J.; Kurama, Y.C. Behavior of precast concrete shear walls for seismic regions: Comparison of hybrid and emulative specimens. *J. Struct. Eng.* **2012**, *139*, 1917–1927. [CrossRef]
14. Vaghei, R.; Hejazi, F.; Taheri, H.; Jaafar, M.S.; Ali, A.A.A. A new precast wall connection subjected to monotonic loading. *Comput. Concr.* **2016**, *17*, 1–27. [CrossRef]
15. Vaghei, R.; Hejazi, F.; Firoozi, A.A.; Jaafar, M.S. Performance of Loop Connection in Precast Concrete Walls Subjected to Lateral Loads. *Int. J. Civ. Eng.* **2019**, *17*, 397–426. [CrossRef]

16. Zhao, W.J.; Guo, W.N.; Jin, Q. State of the art research on connection type of vertical components for precast concrete shear wall systems. *Ind. Constr.* **2014**, *44*, 115–121.
17. Son, G.W.; Ha, S.K.; Song, G.T.; Yu, S.Y. Shear Tests on Subassemblies Representing the Multi-anchored Connection between PC Wall Panels and RC Frames. *Ksce J. Civ. Eng.* **2018**, *22*, 5164–5177. [CrossRef]
18. Jiang, S.F.; Lian, S.H.; Zhao, J.; Li, X.; Ma, S.L. Influence of a New Form of Bolted Connection on the Mechanical Behaviors of a PC Shear Wall. *Appl. Sci.* **2018**, *8*, 1381. [CrossRef]
19. Moreschi, L.M.; Singh, M.P. Design of yielding metallic and friction dampers for optimal seismic performance. *Earthq. Eng. Struct. Dyn.* **2003**, *32*, 1291–1311. [CrossRef]
20. Zhang, C.W. Dynamic test and constitutive model of 225MPa low yield point steel material and its energy absorption ability. *Int. J. Protective Struct.* **2011**, *2*, 527–540. [CrossRef]
21. Whittaker, A.S.; Bertero, V.V.; Thompson, C.L. Seismic testing of steel plate energy dissipation devices. *Earthq. Spectra* **1991**, *7*, 563–604. [CrossRef]
22. Zhang, C.F.; Zhang, Z.S.; Shi, J.F. Development of high deformation capacity low yield strength steel shear panel damper. *J. Constr. Steel Res.* **2012**, *75*, 116–130. [CrossRef]
23. Abebe, D.Y.; Jeong, S.J.; Kim, J.W. Analytical Study on Large Deformation in Shear Panel Hysteresis Damper Using low Yield Point Steel[C]. In Proceedings of the New Zealand Society for 'Earthquake Engineering Technical Conference and AGM, Wellington, NZ, USA, 26–28 April 2013; pp. 26–28.
24. Abebe, D.Y.; Choi, J.H.; Jeong, S.J. Effect of Width-to-Thickness Ratio on Large Deformation in Shear Panel Hysteresis Damper Using Low Yield Point Steel[C]. *Appl. Mech. Mater. Trans Tech Publ.* **2014**, *446*, 1460–1465. [CrossRef]
25. Abebe, D.Y.; Jeong, S.J.; Getahune, B.M. Hysteretic characteristics of shear panel damper made of low yield point steel. *Mater. Res. Innov.* **2015**, *19*, S5-902–S5-910. [CrossRef]
26. Mortezagholi, M.H.; Zahrai, S.M. Proposed Relationship for Proper Shear Strength of Elliptical Damper Based on Its Geometrical Parameters. *Int. J. Steel Struct.* **2018**. [CrossRef]
27. Zahrai, S.M.; Mortezagholi, H.M. Cyclic Performance of an Elliptical-Shaped Damper with Shear Diaphragms in Chevron Braced Steel Frames. *J. Earthq. Eng.* **2018**, *22*, 1209–1232. [CrossRef]
28. Cheng, S.L.; Du, S.Y.; Yan, X.S.; Guo, Q.; Xin, Y.J. Experimental study and numerical simulation of clapboard lead damper. *Proc. Inst. Mech. Eng. Part C J. Mech. Eng. Sci.* **2017**, *231*, 1688–1698. [CrossRef]
29. Jamkhaneh, M.E.; Ebrahimi, A.H.; Amiri, M.S. Experimental and Numerical Investigation of Steel Moment Resisting Frame with U-Shaped Metallic Yielding Damper. *Int. J. Steel Struct.* **2019**, *19*, 806–818. [CrossRef]
30. Lin, X.C.; Wu, K.L.; Skalomenos, K.A.; Lu, L.; Zhao, S.L. Development of a buckling-restrained shear panel damper with demountable steel-concrete composite restrainers. *Soil Dyn. Earthq. Eng.* **2019**, *118*, 221–230. [CrossRef]
31. Zhu, B.J.; Wang, T.; Zhang, L.X. Quasi-static test of assembled steel shear panel dampers with optimized shapes. *Eng. Struct.* **2018**, *172*, 346–357. [CrossRef]
32. Belleri, A.; Marini, A.; Riva, P.; Nascimbene, R. Dissipating and re-centring devices for portal-frame precast structures. *Eng. Struct.* **2017**, *150*, 736–745. [CrossRef]
33. Mazza, F.; Vulcano, A. Displacement-based design procedure of damped braces for the seismic retrofitting of r.c. framed buildings. *Bull. Earthq. Eng.* **2014**, *13*, 2121–2143. [CrossRef]
34. Mazza, F. A simplified retrofitting method based on seismic damage of a SDOF system equivalent to a damped braced building. *Eng. Struct.* **2019**, *200*, 109712. [CrossRef]
35. Rawlinson, T.A.; Marshall, J.D.; Ryan, K.L.; Zargar, H. Development and experimental evaluation of a passive gap damper device to prevent pounding in base-isolated structures. *Earthq. Eng. Struct. Dyn.* **2015**, *44*, 1661–1675. [CrossRef]
36. De Domenico, D.; Gandelli, E.; Quaglini, V. Effective base isolation combining low-friction curved surface sliders and hysteretic gap dampers. *Soil Dyn. Earthq. Eng.* **2020**, *130*, 105989. [CrossRef]
37. Sun, L.; Li, C.; Li, J.; Zhang, C.; Ding, X. Strain Transfer Analysis of a Clamped Fiber Bragg Grating Sensor. *Appl. Sci.* **2017**, *7*, 188. [CrossRef]
38. Sun, L.; Hao, H.; Zhang, B.; Ren, X.; Li, J. Strain Transfer Analysis of Embedded Fiber Bragg Grating Strain Sensor. *Journal of Testing and Evaluation* **2016**, *44*, 2312–2320. [CrossRef]
39. Xu, L.Y.; Nie, X.; Fan, J.S. Cyclic behaviour of low-yield-point steel shear panel dampers. *Eng. Struct.* **2016**, *126*, 391–404. [CrossRef]

40. Li, D.X.; Uy, R.A.; Wang, J. Behaviour and design of high-strength steel beam-to-column joints. *Steel Compos. Struct.* **2019**, *31*, 303–317. [CrossRef]
41. Zhang, C.; Zhu, T.; Wang, L. Ultra-low cycle fatigue performance evaluation of the miniaturized low yield strength steel shear panel damper. *J. Constr. Steel Res.* **2017**, *135*, 277–284. [CrossRef]
42. Zhang, C.; An, D.; Zhu, L. Axial Compressive Behavior of Steel-Damping-Concrete Composite Wall. *Appl. Sci.* **2019**, *9*, 4679. [CrossRef]
43. Zhu, L.; Zhang, C.; Guan, X.; Uy, B.; Sun, L.; Wang, B. The multi-axial strength performance of composite structural B-C-W members subjected to shear forces. *Steel and Composite Structures* **2018**, *27*, 75–87. [CrossRef]

applied sciences

MDPI

Article

Study of Lead Rubber Bearings for Vibration Reduction in High-Tech Factories

Shen-Haw Ju *, Cheng-Chun Yuantien and Wen-Ko Hsieh

Department of Civil Engineering, National Cheng Kung University, Tainan 70101, Taiwan; sunny21208@gmail.com (C.-C.Y.); eric3320jh@gmail.com (W.-K.H.)

* Correspondence: juju@mail.ncku.edu.tw

Received: 21 January 2020; Accepted: 20 February 2020; Published: 22 February 2020

Abstract: This paper studies the seismic and micro vibrations of the high-tech factory with and without lead rubber bearings (LRBs) using the three-dimensional (3D) finite element analysis. The soil-structure interaction is included using the p-y, t-z, and Q-z nonlinear soil springs, while the time-history analysis is performed under seismic, wind, or moving crane loads. The finite element results indicate that the moving crane does not change the major ambient vibrations of the factory with and without LRBs. For a normal design of LRBs, the high-tech factory with LRBs can decrease the seismic base shear efficiently but will have a much larger wind-induced vibration than that without LRBs, especially for the reinforced concrete level. Because micro-vibration is a major concern for high-tech factories, one should use LRBs with a large initial stiffness to resist wind loads, and use a small final LRB stiffness to reduce the seismic load of high-tech factories. This situation may make it difficult to obtain a suitable LRB, but it is an opportunity to reduce the seismic response without increasing the micro-vibration of high-tech factories.

Keywords: earthquake; high-tech factory; lead rubber bearing; moving crane; soil structure interaction; vibration; wind load

1. Introduction

The lead rubber bearing (LRB) has the advantage of increasing a building's natural period, which is away from the seismic period range, to avoid the amplification caused by earthquakes, so it is ideal to reduce seismic loads using LRBs for high-tech factories. However, the LRB may increase environmental vibrations induced by moving vehicles and wind loads, which will damage the high-tech production. In the literature, the issue of LRBs for high-tech factories is rarely studied. However, LRBs have many references in building and bridge research and testing. Turkington et al. [1,2] demonstrated the bridge isolation design process that can be applied to all earthquakes and used the numerical simulation of LRB bridges to obtain the long-term periodic displacement and effective damping, due to LRBs, which can improve the seismic capacity of general bridges. Fujita et al. [3] conducted a base isolation test for a building and found that LRBs can effectively reduce the building response. Salic et al. [4] used LRB numerical simulation on eight-layer structures to propose that the structure increases the natural period to avoid the shortest period of earthquake damage. Kalpakidis et al. [5] proposed a theory that predicts the dependence of feature intensity and energy time to predict the behavior of LRBs to simplify the analysis. Kalpakidis and Constantinou [6] proposed the necessary conditions for reducing the LRB scaling test and the need to consider the temperature rise of the lead core. Islam et al. [7] made a multi-layer building foundation combined with finite element simulations of LRBs, suggesting that this isolation technology has the ability to survive buildings under strong earthquakes. Li et al. [8] studied the rational yield ratio of isolation system for buildings, considering the influences of total heights, yield ratios, and seismically isolated schemes, and the rational range of the yield ratio is recommended to be 2%–3%.

In a number of references, correlation studies on the effects of LRB parameters are used to understand the best design parameters. Warn et al. [9] studied the relationship between lateral displacement and vertical stiffness of LRB and found that the vertical stiffness decreases with the increase of lateral displacement. Weisman and Warn [10] conducted experiments and numerical simulations to understand the relationship between LRB critical loads and lateral displacements and found that the critical loads decrease with the increase of lateral displacements. Al-Kutti and Islam [11] proposed that LRB systems with higher characteristic strength and relatively less isolation periods behave better to reduce structural offset, and LRBs with lower characteristic strength and a high isolation period can control the basic shear, providing a small acceleration and low inertia. Several references investigated the biaxial interactions of LRBs, which is convenient for understanding interaction effects. Nagarajaiah et al. [12] considered the formula proposed by Park to simulate the biaxial interaction of LRB. Huang et al. [13] proposed a two-way simulation formula for LRBs and made some experiments to compare the uniaxial and biaxial effects. Abe et al. [14] conducted a biaxial test on LRBs to understand the effect of the torsional coupling effect. It is suggested that the two-axis interaction cannot be ignored. Falborski and Jankowski [15] used the experiment to verify the effectiveness of an isolation system made of polymeric bearings in reducing structural vibrations and demonstrated that the application of this bearing can significantly reduce the lateral acceleration.

Although the application of LRBs is quite mature, there is very little or probably no research that focuses on high-tech factories directly. The reason is because the equipment used to produce high-tech productions requires strict micro-vibration standards, but it is unclear whether micro-vibration will increase significantly when LRBs are installed in high-tech factories. This study thus investigates both the seismic and ambient vibrations, due to the LRB installed in the high-tech factory, using the finite element method, while the ambient vibrations are induced by the wind load and moving crane.

2. Finite Element Modeling of Lead Rubber Bearings

The LRB, as shown in Figure 1, is a single or multiple lead core built into laminated rubber to reduce structural horizontal vibration during earthquakes. Because the laminated rubber has high vertical stiffness, low horizontal stiffness, and high recovery and lead metal has low yield stress, combining the characteristics of the two makes the LRB a good vibration isolation device.

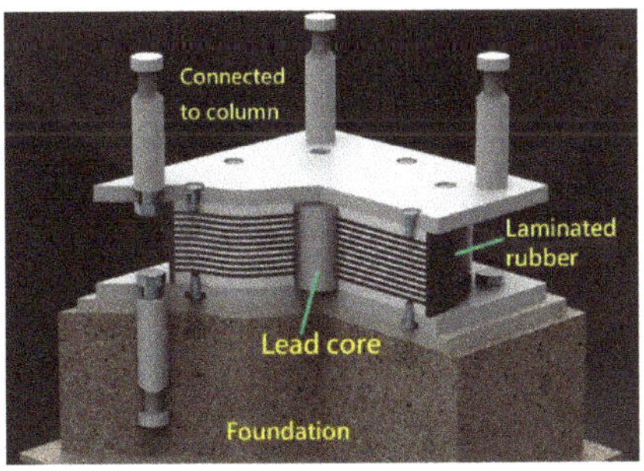

Figure 1. Illustration of the lead rubber bearing (LRB) containing a lead core and laminated rubber.

Nagarajaiah et al. [12] proposed a two-way LRB model as below:

$$\{P\} = \left\{ \begin{array}{c} P_x \\ P_y \end{array} \right\} = \alpha \frac{F_y}{Y}\{U\} + (1-\alpha)F_y\{Z\} = k_d\{U\} + (1-\alpha)F_y\{Z\} \tag{1}$$

$$Y\left\{ \begin{array}{c} \dot{Z}_x \\ \dot{Z}_y \end{array} \right\} = \left(A[I] - \left[\begin{array}{cc} z_x^2\left(\gamma Sign\left(\dot{U}_x Z_x\right) + \beta\right) & Z_x Z_y\left(\gamma Sign\left(\dot{U}_y Z_y\right) + \beta\right) \\ Z_x Z_y\left(\gamma Sign\left(\dot{U}_x Z_x\right) + \beta\right) & z_y^2\left(\gamma Sign\left(\dot{U}_y Z_y\right) + \beta\right) \end{array} \right] \right) \left\{ \begin{array}{c} \dot{U}_x \\ \dot{U}_y \end{array} \right\} \tag{2}$$

where $\{P\} = [P_x, P_y]^T$ is the LRB force vector, $\{U\} = [U_x, U_y]^T$ is the in-plane displacement vector between the LRB two sides, α is the ratio of the final LRB stiffness over the initial LRB stiffness ($\alpha = k_d/k_e$, $k_e = F_y/Y$ = initial LRB stiffness), F_y is the LRB yielding force, Y is the LRB deflection at the yielding force, k_d is the yielding LRB stiffness, $\{Z\} = [Z_x, Z_y]^T$ is the LRB nonlinear variable, γ, β, and A are dimensionless parameters to control the shape of the hysteresis loop used in the two-way theory, where $A/((\gamma + \beta)=1$, and [I] is a unit vector. Equation (2) is nonlinear, and the Newton–Raphson method can be used to find {Z} using $\{\dot{U}\}$ obtained from the finite element analysis. The details can be found in [16]. The finite element stiffness matrix of the shear force contribution is:

$$K_{shear} = \begin{bmatrix} K_{11} & K_{12} \\ K_{21} & K_{22} \end{bmatrix} \tag{3}$$

$$\begin{aligned} \text{Where, } K_{11} &= \frac{\partial P_x}{\partial U_x} = \alpha\left(\frac{F_y}{Y}\right) + (1-\alpha)F_y\frac{\partial Z_x}{\partial U_x},\ K_{12} = \frac{\partial P_x}{\partial U_y} = (1-\alpha)F_y\frac{\partial Z_x}{\partial U_y} \\ K_{21} &= \frac{\partial P_y}{\partial U_x} = (1-\alpha)F_y\frac{\partial Z_y}{\partial U_x},\ \text{and } K_{22} = \frac{\partial P_y}{\partial U_y} = \alpha\left(\frac{F_y}{Y}\right) + (1-\alpha)F_y\frac{\partial Z_y}{\partial U_y} \end{aligned} \tag{4}$$

where $\frac{\partial Z_x}{\partial U_x}$, $\frac{\partial Z_x}{\partial U_y}$, $\frac{\partial Z_y}{\partial U_x}$, and $\frac{\partial Z_y}{\partial U_y}$, can be found in [16]. Equations (3) and (4) produce an unsymmetrical global stiffness matrix, which may cause the double requirement of computer memory and time. Thus, one can empirically set $K_{21} = K_{12} = (K_{12} + K_{21})/2$, and use the Newton–Raphson method to obtain the solution with small unbalance forces. The solution is still accurate, since equation (1) is used to find the LRB internal force vector without errors, but the Newton–Raphson iterations may increase when one direction is loading and the other is unloading. For the vertical direction of the LRB stiffness (K_v), a linear spring is used. The original LRB hysteretic curve under low speed loads can be modified as the functions of the wave frequency, wave speed, and axial load [17,18]. For simplicity, we used the original LRB model for finite element analyses.

Laboratory experiments were conducted to find the LRB characteristics at a vertical compressive force of 6300 kN and a maximum horizontal displacement of 0.149 m. The experimental results are shown in Figure 2, plotted as the dot line, where the LRB calibrated material properties are $K_e = K_d/\alpha$ = 1.9×10^5 kN/m (initial stiffness), $K_v = 5 \times 10^7$ kN/m, F_y = 370 kN, α = 0.0288, β = 0.1, and γ = 0.9 based on equation (1). Finite element analysis using the proposed LRB element mentioned above was then performed to find the hysteresis curve, as shown in the black line in Figure 2. This figure indicates a good agreement between the finite element analysis and experimental result.

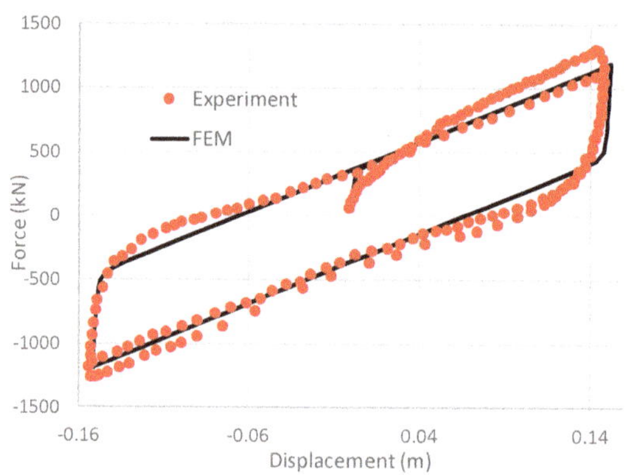

Figure 2. Experimental and finite element results for the LRB hysteresis curve.

3. High-Tech Factory and Finite Element Model

3.1. Illustration of the Structure of the High-Tech Factory

Before explaining the studied factory, we first briefly introduce the micro vibration standards in the high-tech industry. Gordon [19] recommended the vibration criterion (VC) for high-tech factories, where five levels include VC-E to VC-A under the velocity vibration at the floor slab from 42 dB to 66 dB with the increment of 6 dB, where the dB calculation can be found in the references [19,20]. The studied high-tech factory located in southern Taiwan is a three-story building mainly used for producing photovoltaic panels, where the first level is the VC-C reinforced concrete (RC) structure, the second level is the VC-B steel structure, and the third level is the VC-A steel structure. Intensive RC columns are used in the first level to avoid environmental vibration, while large span truss frames are used in the second and third levels to achieve greater production space. It is noted that the studied factory has no currently installed LRBs, and we use it to perform the seismic and micro vibration analyses with and without LRBs. Figure 3 shows the two typical frames in the X and Y direction. In the RC level, the column span is 6 m in the X direction and 5.2 m in the Y direction, where there are 71 and 27 column lines in the X and Y directions, respectively. As shown in Figure 3, the RC columns connected to the steel frame have a big square size of 1.5 m, and others are 0.6 m. For the two steel levels, the column span is 12 m in the X direction and 32 m in the Y direction, while the section properties are shown in Table 1. The thickness of the RC slab is 0.725, 0.55, and 0.45 m for the first to third level, respectively, and the main purpose of thick slabs is to reduce ambient vibration. For the two steel levels, the properties of steel sections are listed in Table 1, where columns are the box section and others are the H-shape section. This high-tech factory used pile foundations of 28 m length to avoid excess environmental vibration, while the reversed circulation piles with two different sections were constructed, and one is the diameter of 1.5 m connected to the big columns and the others are the diameter of 0.6 m connected to other columns. The soil profile contains 10 m inorganic clays of medium plasticity (undrained shear strength S_u = 50 kPa), 5 m sandy soil (submerge internal frictional angle of sand ϕ = 33°), 10 m clay of hard plasticity (S_u = 150 kPa), and the rest is very hard sand (ϕ = 37°). We used the axial forces from columns to select appropriate LRBs, where two types of LRBs were used at the top of piles. As shown in Figure 3, the first type, named LRB1, was used to connect with the big columns, and the second type, named LRB2, was used to connect to other columns, where Table 2 shows the material properties of the two types of LRBs.

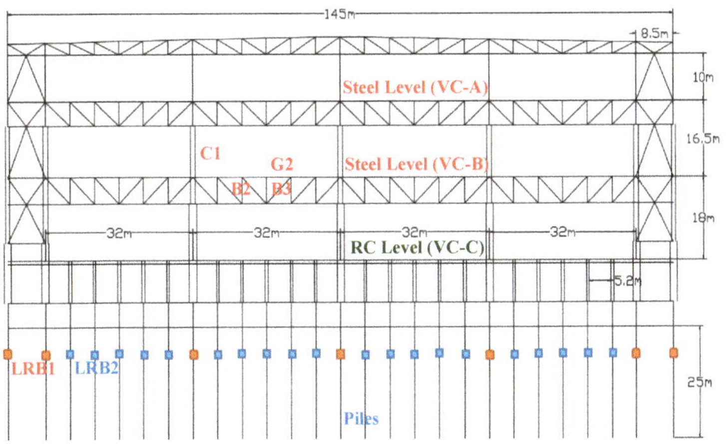

(a) X-direction section.

Figure 3. *Cont.*

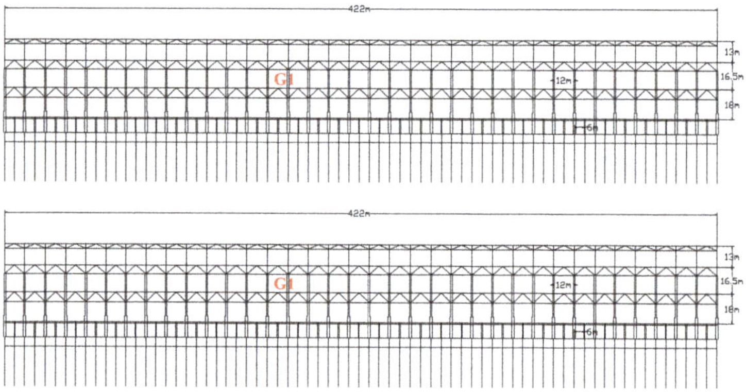

(b) Y-direction section

Figure 3. Two typical frames of the high-tech factories.

Table 1. The steel sections in the second and third levels of the high-tech factory, as shown in Figure 3 (A572 steel with Fy = 345 MPa).

Member	Axis	Label	Member Dimensions (mm) d×bf×tw×tf
Column	ALL	C1	BOX 900 × 900 × 35 × 35
Braces	X	B1	RH 400 × 400 × 13 × 21
Braces	Y	B2	RH 414 × 405 × 18 × 28
Braces	Y	B3	RH 428 × 407 × 20 × 35
Girder	X	G1	RH 588 × 300 × 12 × 20
Girder	Y	G2	BH 375 × 200 × 10 × 25

Table 2. LRB material properties (= 0.1, and = 0.9 based on Equation (1)).

Name	Model/Parameter	$K_e(kN/m)$	F_y(kN)	α	$K_v(kN/m)$
LRB1	EIRL-G4-1000-170	1.9732×10^4	196.0850	0.0769	4.660×10^6
LRB2	EIRL-G4-700-130	1.4058×10^4	114.8299	0.0769	3.259×10^6

3.2. Finite Element Model

The finite element program from reference [21] was used in the finite element analysis, where the LRB element mentioned in Section 2 has been added into this program. The three-dimensional (3D) finite element mesh is shown in Figure 4 with the total number of degrees of freedom of 1,849,662 and 695,643 elements, where the high-tech factory, warrior slabs, crane, and rail system are included. Although the finite element is complicated, the major part of the mesh is modelled using 2-node 3D beam elements, such as beams, columns, piles, and crane rails of the factory, and the end released moments of beam elements are used to model truss members. Waffle slabs are simulated using 2-node 3D beam elements with 0.75 m interval, 0.4 m width, and 0.75, 0.55, and 0.45 m depth on the first, second, and third floor slabs, respectively, where the 0.18 m rigid zone at two beam ends is set. The slabs at the truss bottom on the second and third steel stories are modeled using 4-node plate elements with a thickness of 0.15 m. The soil-structure interaction is modelled using the API p-y, t-z, and Q-z nonlinear soil spring elements [22], where one end of these elements are connected to the beam element nodes of piles, and the other nodes are applied to the time-history seismic displacements for the earthquake load. If LRBs are included, the LRB elements mentioned in Section 2 are generated between foundation beams and the top of piles, as shown in Figure 3. The Rayleigh damping was used in the finite element analysis, where the mass damping equals 0.3/s and the stiffness damping equals 0.0003 s, which gives approximately 4% damping ratio at frequencies of 0.6 and 40 Hz, respectively.

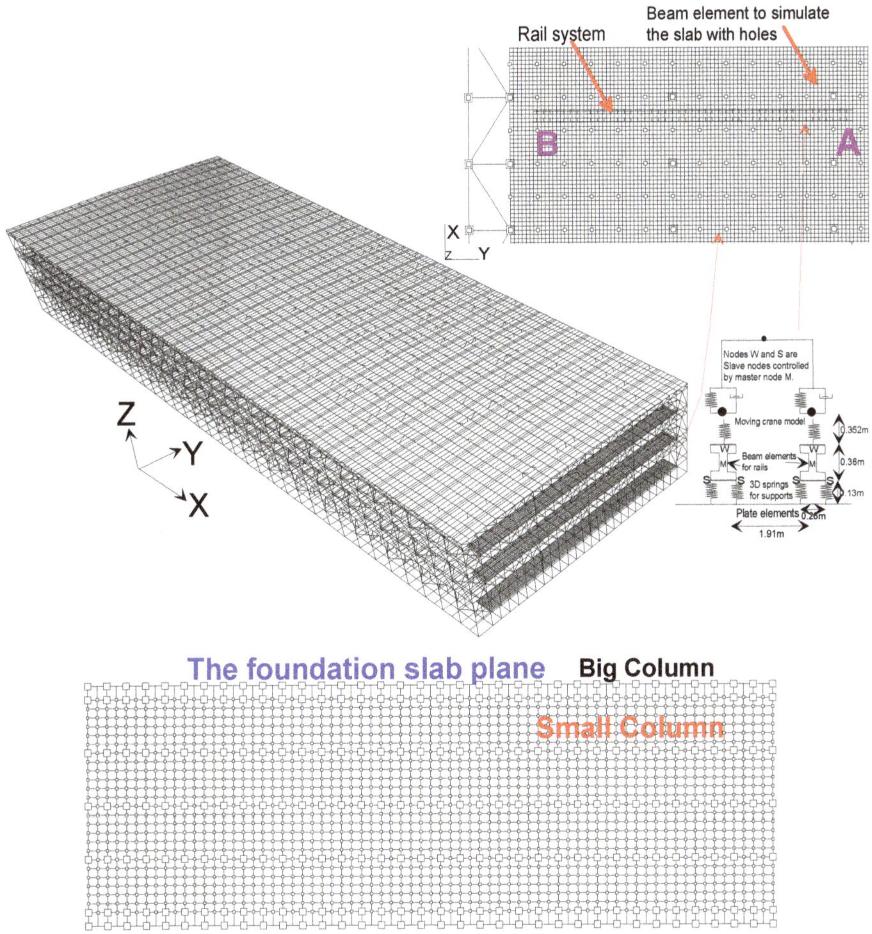

Figure 4. Three-dimensional (3D) Finite element mesh containing the high-tech factory, slabs, rails, and crane (The big and small columns are connected to LRB1 and LRBs, respectively, as shown in Table 2).

The rail and crane system on the second level, as shown in Figure 4, contains two steel rails with the properties of the axial area of 0.17×10^{-2} m^2, I_x of 0.19E-4 m^4, and I_y of 0.6×10^{-4} m^4. The 2-node 3D beam element is used to simulate rails supported by the 1.3 m interval springs with the stiffness of 4.8×10^5 kN/m and the damping of 10 kN-s/m between rails and slabs. Two slave nodes, labeled as node S in Figure 4, are controlled by the master node at the beam center at each support section, while a number of slave nodes W at the rail top are set for the route of moving wheel elements. Thus, the crane finite element model can be moved on the rails which are connected to the slab of the high-tech factory. The crane, as shown in Figure 4, is generated using a beam, spring-damper, and moving wheel elements [20] with the mass of five tons. Except the API soil spring, LRB, and moving wheel elements, other finite elements are linear elastic. The consistent mass method, Newmark's integration method with the average acceleration, and the Newton–Raphson method were used to solve this nonlinear problem with a time step of 0.005 s and a simulation of 20,000 time steps for wind loads and 10,000 time steps for other loads. The finite element analysis contains two stages, where the first stage is the static analysis under the dead weight load using a step, and the second stage is the time-history analysis using 10,000 or 20,000 time steps. It is noted that a comparison against sensors' measurements under both wind-induced and crane-induced vibrations was reported in [20,23] to validate the accuracy of the finite element analysis.

3.3. Illustration of Seismic Loads

The artificial earthquake generation software Simqke [24] was used to generate the time-history seismic acceleration using the spectrum from IBC 2006 [25], as shown in Figure 5. The peak ground accelerations (PGA) of 0.25, 0.28, 0.32, 0.36, and 0.40, respectively, were used for five seismic loads in the global X direction with T_s (Figure 5) of 0.6 s, where one group is shown in Figure 6. Moreover, the important parameter T_s representing the dominant frequency of seismic loads, as shown in Figure 5, was set to 0.4, 0.5, 0.6, 0.7, 0.8, 0.9, 1.0, 1.2, and 1.4 s, respectively, with the PGA of 0.32 g for nine seismic loads in the global X direction. For the other two directions, 70% and 30% of that PGA in the global Y and Z directions, respectively. This three-direction seismic accelerations are applied on the ground surface. We used ten soil layers with the interval of 5 m for the SHAKE 91 [26] input data. The SHAKE 91 program is then used to generate the acceleration field in each soil layer. Finally, the integration to obtain the displacement fields, which are applied to the node of each p-y, t-z, and Q-z curve elements for the seismic simulation.

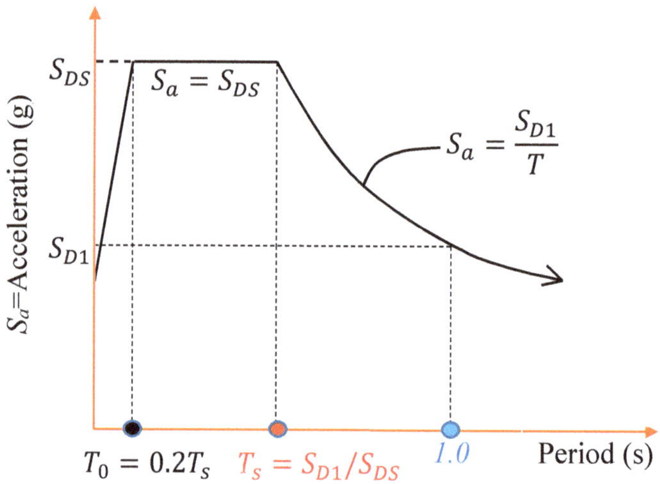

Figure 5. Seismic response spectrum according to IBC 2006 [25].

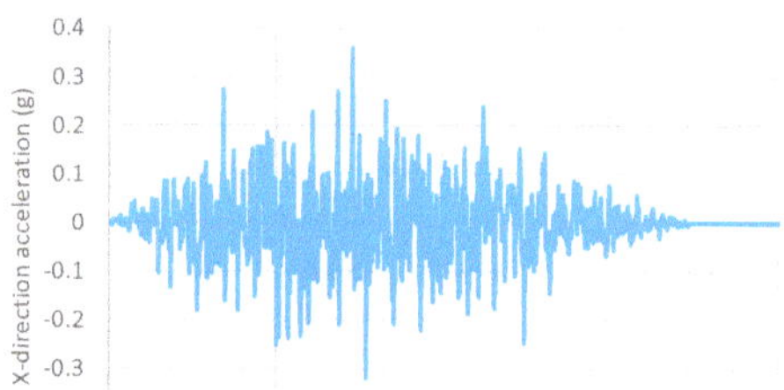

Figure 6. *Cont.*

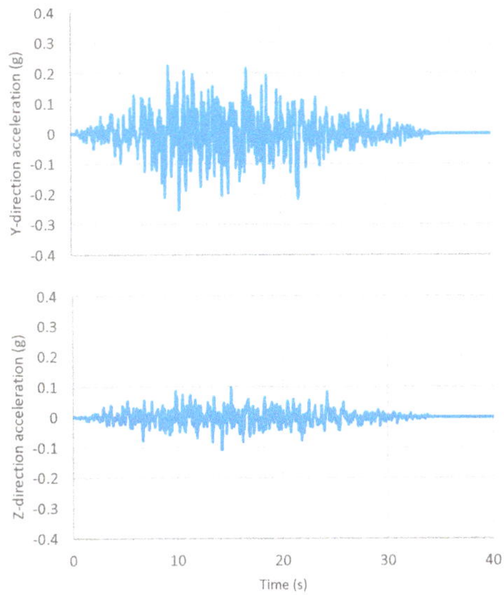

Figure 6. Artificial time-history seismic acceleration for T_s = 0.6 and peak ground accelerations (PGAs) of 0.36, 0.252, and 0.108 g in the local X, Y, and Z directions.

4. Parametric Study Using LRBs in High-Tech Factories

4.1. Earthquake Effect

The base shears in the X and Y directions of the factory with and without LRBs are shown in Figure 7, where the PGA of the applied seismic load is 0.32 g, and T_s is 0.7 and 1.4 s for two cases, respectively. The base shear is determined from the summation of the shear forces at the top of all the piles, and it represents the total seismic loads changing with time for the superstructure of the high-tech factory. To simplify the time-dependent base shears in the X and Y directions, we first find the magnitude ($S(t)$) of the two-direction base shears using the following equation:

$$S(t) = \sqrt{S_x(t)^2 + S_y(t)^2} \tag{5}$$

where $S_x(t)$ and $S_y(t)$, as shown in Figure 7, are the time-dependent base shears in X and Y directions, respectively. Then, we obtain the maximum base shear ($Smax$) of all the time steps during the finite element analysis. Finally, we define the base shear ratio ($R = Smax_{LRB}/ Smax_{NO\text{-}LRB}$) as the maximum base shear of the structure with LRBs ($Smax_{LRB}$) over that without LRBs ($Smax_{NO\text{-}LRB}$)), and this ratio can be used to understand the efficiency of the LRB used to structures during earthquakes. Figure 8 shows this base shear ratio changing with PGA under T_s of 0.6 s, and Figure 9 shows that changing with T_s under the PGA of 0.32 g. These figures indicate the following features:

(1) Figure 8 shows that when PGA increases, the base shear ratio increases slightly. However, for the worse case, the ratio for the PGA of 0.4 g is still small, which means that the LRB can effectively reduce the seismic load regardless of the magnitude of earthquakes. Figure 9 shows that when the dominant period of the earthquake increases, the base shear ratio increases to a noted extent. For long period seismic loads, such as near fault earthquakes, this situation can lead to LRB disadvantages.

(2) Usually, the high-tech factory requires thick floor slabs, big long trusses, and dense RC columns to reduce ambient vibration, but this arrangement will largely increase the building mass that causes large seismic loads during earthquakes. The high-tech factory with LRBs can decrease over 50% of the seismic base shear under $T_s \leq 1.0$, which means that the high-tech factory can resist larger earthquakes

using LRBs for not very long periods of seismic loads. The comparison of base shears, shown in Figure 7a,b, between the factory with and without LRBs indicates the above conclusion, where the time-history base shears of the factory with LRBs are much smaller than those without LRB.

(3) For the earthquake with a very long dominant period, such as 1.4 s, the LRB efficiency to reduce the factory base shear may decrease a little, since the natural period of the factory, due to the full yield of the LRB, can approach the earthquake with a long dominant period. However, earthquakes with this long dominant period often occur in significantly soft soil, and the design of LRBs for the high-tech factory may avoid this condition. Nevertheless, the simulation results indicate that the LRB efficiency for the earthquake with a long dominant period is still in the acceptable range, as shown in Figure 7c,d.

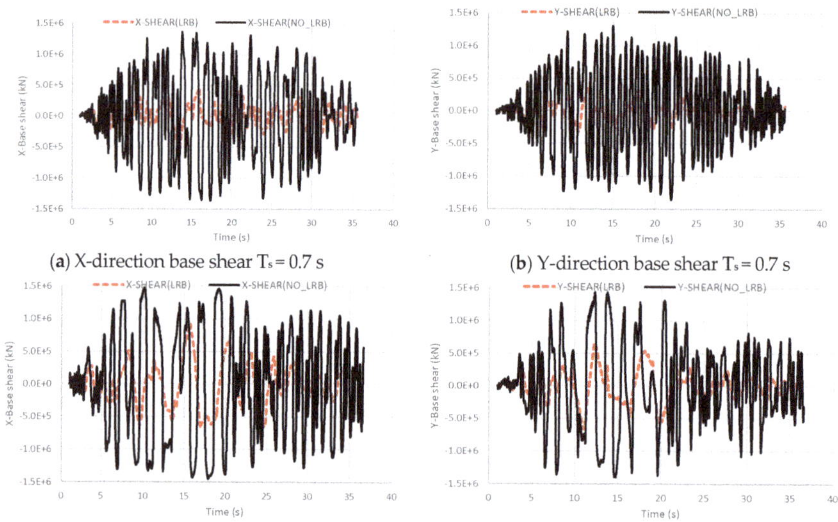

(**a**) X-direction base shear $T_s = 0.7$ s (**b**) Y-direction base shear $T_s = 0.7$ s

(**c**) X-direction base shear $T_s = 1.4$ s (**d**) Y-direction base shear $T_s = 1.4$ s

Figure 7. Comparison of base shears between the factory with and without LRBs under the seismic load of PGA = 0.32 g and T_s = 0.7 s and 0.14 s.

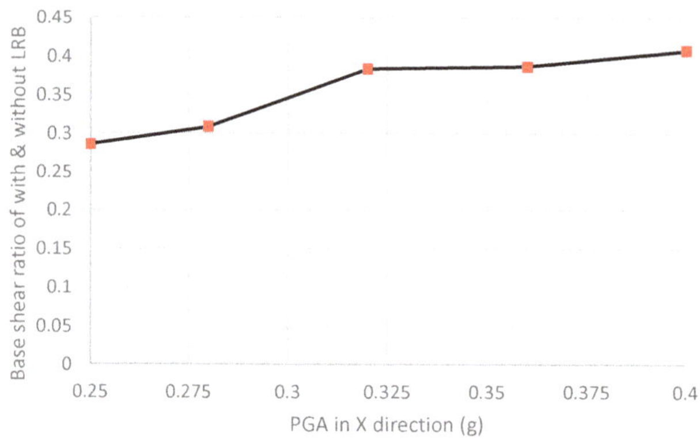

Figure 8. The base shear ratio changing with PGA under T_s of 0.6 s.

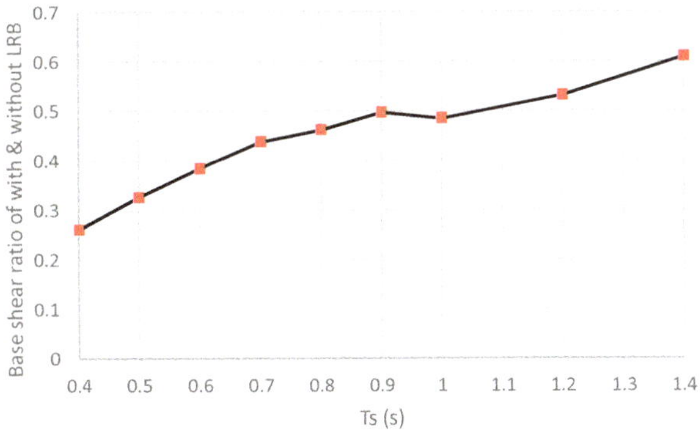

Figure 9. The base shear ratio changing with T_s under the PGA of 0.32 g.

4.2. Micro-Vibration Induced by Mobile Cranes

Floor micro-vibration induced by moving cranes inside the high-tech factory is the major environmental source which affects the production operation in high-tech factories. As shown in Figure 4, the rail and crane system on the second level was studied, where the crane moves back and forth on the 60-m rail system with a maximum crane speed of 3 m/s. Figure 10 shows the ambient vibration in X, Y, and Z directions at the 10 m location from the railway centerline of the moving crane, while the factory was arranged with and without LRBs. This figure shows that the vertical (Z) vibration induced by moving cranes is much larger than those in the in-plane (X and Y) directions. Moreover, the major vibrations that are above 40 dB, and between 15 to 40 Hz, in these three directions for the factory with and without LRBs are almost identical, in which these major vibrations between 15 to 40 Hz are the slab natural frequencies invoked by the vibration of the moving crane, more details can be referred to in [20]. The ambient vibrations at other frequencies are small but different from the factory with and without LRBs. One can still realize that the factory without LRBs has smaller ambient vibrations than that with LRBs, because the LRBs cause a big rigid body motion of the high-tech factory. Nevertheless, the moving crane does not change the major ambient vibrations between the factory with and without LRBs.

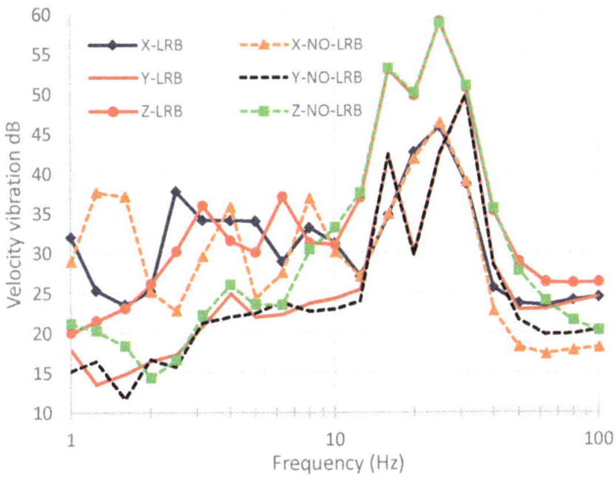

Figure 10. Velocity vibration dB at 10 m from the centerline of the crane railway on the first steel level for the factory with or without LRBs.

4.3. Micro-Vibration Simulation Under Wind Loads

In addition to vibration generated by moving cranes, wind-induced floor vibration in high-tech buildings is another major source of environmental loads affecting production operations. Therefore, we followed the reference [23] to study the wind induced vibration, and the LRB effect was investigated in this paper, where the analysis only included dead and wind loads and the seismic load was not used in this section. Since wind forces applied to the factory are space- and time-dependent, we used the wind speed simulation software TurbSim [27] to generate the space- and time-dependent wind speed field in the Y direction on the whole X-direction outer plane to compare the floor vibration of the factory with and without LRBs. Since the building is much longer in the X direction than that in the Y and Z directions, the Y-direction wind-induced vibration should be the largest, and we will thus only focus on this direction vibration. The normal turbulence model is used in the analysis with the average wind speed at the height of 30 m (V_{30m}) during 10 min, while V_{30m} is set to 5, 10, 15, 20, and 25 m/s for the five cases, and the turbulence standard deviation is set according to IEC-61400-1 in 2019 [28], as follows:

$$\sigma_1(m/s) = 0.16(0.75V_{30m} + 5.6) \tag{6}$$

In the setting of this program, an area of 500 m wide by 60 m high was arranged with 41 by 41 girders to find the turbulent wind speeds. The average wind speed in the vertical direction is according to the normal wind profile as below:

$$VZ = V_{30m}\left(\frac{Z}{30}\right)^{\alpha} \tag{7}$$

where Z (m) is the vertical height above the ground, and α equal to 0.14 is the power law exponent. Figure 11 shows the turbulent wind velocity at the height of 30 m on the building center, and it is noted that the wind velocity is time- and location-dependent. The wind pressure is determined as below:

$$P(X,Y,Z,t) = C_P\rho V(X,Y,Z,t)^2/2 \tag{8}$$

where $P(X,Y,Z,t)$ is the time- and space-dependent wind pressure, $V(X,Y,Z,t)$ is the time-and space-dependent wind speed from the TurbSim result, C_p (0.8) is the shape coefficient, and ρ (0.00128 t/m^3) is the air density. Finally, the time-history finite element analysis is performed to find the wind-induced vibrations on the three floors, which are shown in Figure 12 for the case of the average wind speed V_{30m} equal to 15 m/s. These figures indicate that the high-tech factory with LRBs will have much larger wind-induced vibration than that without LRBs, especially for the RC level that is located at the first level. The increased velocity vibration dBs for the RC level, the first steel level, and the second steel level are about 19, 6, and 4 dB, respectively. Therefore, this situation will bring great disadvantage to the use of LRB in high-tech factories. The reason for largely increasing the floor vibration induced by wind loads is that the initial stiffness of the LRB is considerably soft, so that the rigid body motion of the factory superstructure cannot be avoided due to the wind load. Even for a small wind load, which is still much larger than the load of moving cranes, the wind induced rigid body motion still causes problems for the factory with LRBs. We further analyzed the factory under different average wind speeds (V_{30m}) and then only selected the maximum dB from all the frequencies, as shown in Figure 13. This figure indicates a very similar conclusion as that of the average wind speed equal to 15 m/s not only for the steel levels but also for the RC level, while the average wind speed was set to a board range from 5 to 25 m/s. An interpolation scheme was used to find the requirement of micro vibration according to the guidelines for high-tech factories, and the result is shown in Table 3. The table can be used to estimate the wind-induced vibration for a high-tech factory approximately, although the result is dependent on the structure dimensions and member sizes. This table also indicates that using LRBs for the high-tech factory will highly increase the wind-induced vibration, especially for the vibration on the RC level. For the high-tech without LRBs, the RC level at the first floor can resistant vibration under a moderate wind field, but the steel levels above the RC level may not be qualified for such a wind field. To overcome this problem, the shade of adjacent buildings for the high-tech factory was proposed to resistant the wind induced vibration [23], where the height of the shading building should be more than 60% of the factory height. This shade method is still useful for the high-tech factory with LRBs.

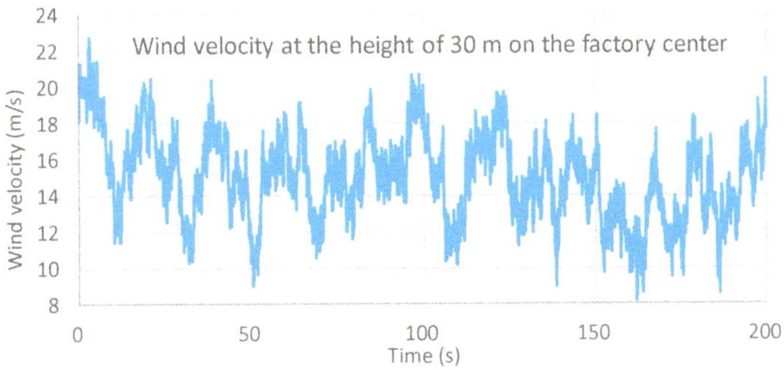

Figure 11. Turbulent wind velocity at the height of 30 m on the building center.

Table 3. The minimax average wind speed (m/s) during 10 min for the criteria of the micro-vibration for the studied high-tech factory.

Level	VC-D(48dB)	VC-C(54dB)	VC-B(60dB)	VC-A(66dB)	Type
1st Without LRB	16.8	23.4	>25	>25	RC
1st With LRB	5.9	8.8	12.6	17.6	RC
2nd Without LRB	7.3	10.3	14.7	20.9	Steel
2nd With LRB	4.4	7.3	10.4	14.7	Steel
3rd Without LRB	4.9	7.9	11.5	17	Steel
3rd With LRB	3.4	6.2	9.1	13	Steel

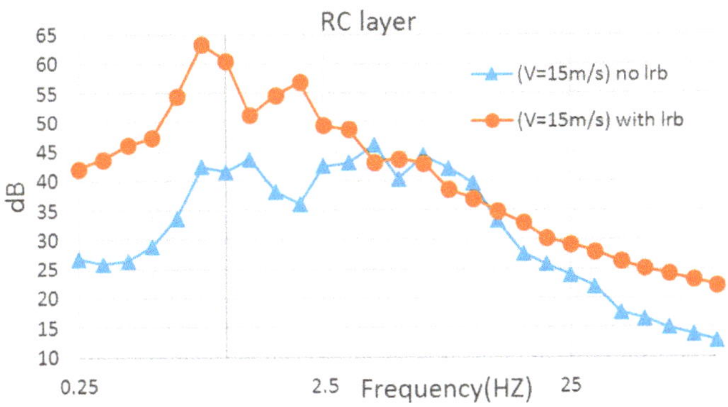

(**a**) At the first floor (RC level)

Figure 12. *Cont.*

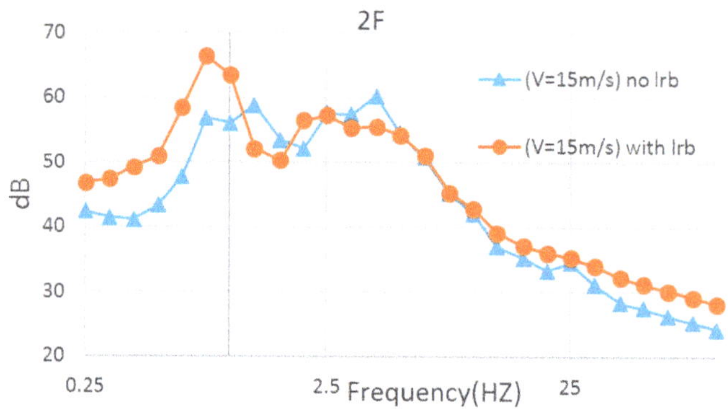

(b) At the second floor (the first steel level)

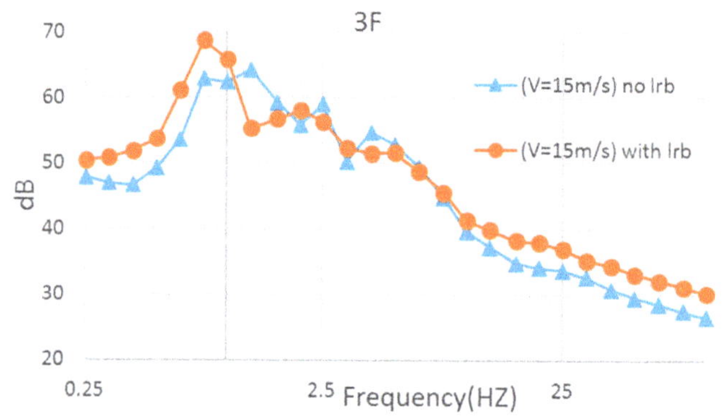

(c) At the third floor (the second steel level)

Figure 12. Velocity vibration dB at the first to third floors under the wind load with the average wind speed of 15 m/s for the factory with or without LRBs.

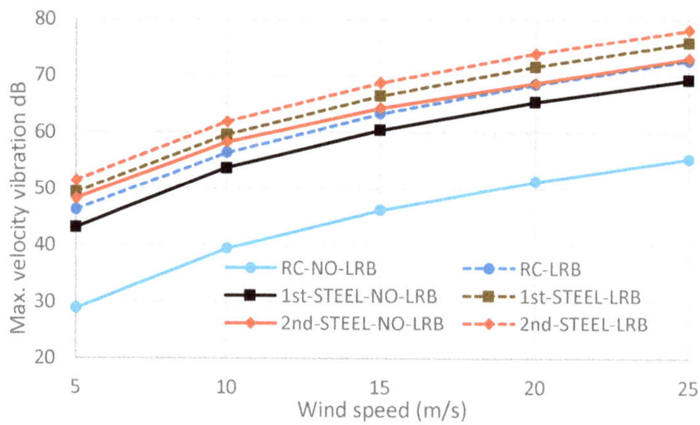

Figure 13. Maximum velocity vibration dB at the first to third floors under the wind load with the average wind speed from 5 to 25 m/s for the factory with or without LRBs.

5. Design of LRBs Concerning the Micro Vibration

The LRBs should possess large stiffness for frequent small or moderate earthquakes but small stiffness for extreme earthquakes. If the micro vibration is the major concern for the high-tech factory, the investigation of Section 4 indicates that the selection of LRBs should first consider the problem of the large micro vibration induced by the wind load. Thus, the LRB with a large initial stiffness (K_e) and a small ratio of the final stiffness over the initial stiffness (α) should be used, where the large K_e can resist wind loads and the small α can reduce seismic loads. However, this situation may cause difficulties in finding a suitable LRB, so we will first select LRBs with large K_e, where 5E5 kN/m (F_y = 300 kN and K_v = 8E7 kN/m) and 3E5 kN/m (F_y = 200 kN and K_v = 5E7 kN/m) are used at the bottom of the big and small columns, respectively. Then, α is set to 1%, 2.5%, 5%, and 7% for three cases. The artificial earthquake is set using the PGA of 0.32 g and T_s of 0.9 s (Figure 5), and the average speed of the turbulent wind load is set to 25 m/s. The finite element results are shown in Figures 14 and 15, where Figure 14 shows the velocity dB changing with frequencies for the wind load, in which the results are not dependent on α because the yield of LRBs is not obvious under the average wind speed of 25 m/s, and Figure 15 shows the base shear ratio ($R = Smax_{LRB}/ Smax_{NO\text{-}LRB}$) changing with α. The two figures indicate the following features:

(1) Figure 14 shows that the slab vibrations induced by the wind load are similar between the factories with and without LRBs, where the vibrations of the LRB factory are slightly large about 2 to 3 dB greater than those without LRBs. This improvement is significant compared to the result in Figure 11, because the large initial stiffness of the LRB resists the wind loads. Moreover, most of the LRBs are still not yielded, so the slab vibrations are independent of the LRB parameter α.

(2) Figure 15 shows that the α should be smaller at higher LRB initial stiffness to reduce the seismic load of high-tech factories. This situation may make it difficult to obtain a suitable LRB, for example, α in Figure 15 is less than 2%. Nevertheless, using the large initial stiffness and small α may reduce seismic responses but not increase the micro vibration for high-tech factories

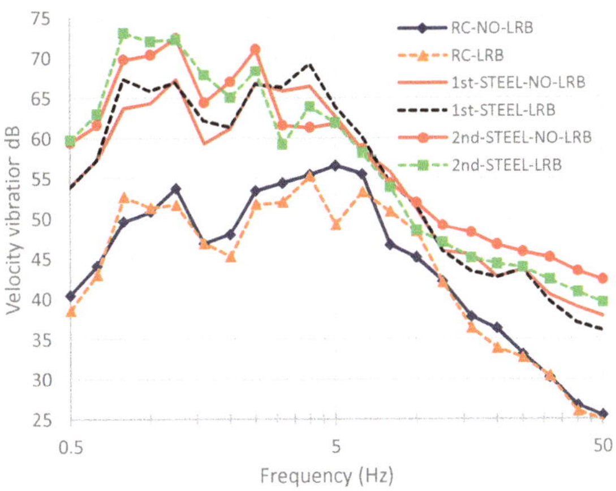

Figure 14. Velocity vibration dB changing with frequencies at the first to third floors under the wind load with the average speed of 25 m/s for the factory with or without LRBs (The dB values with LRBs due to α from 0.01 to 0.07 are almost identical.).

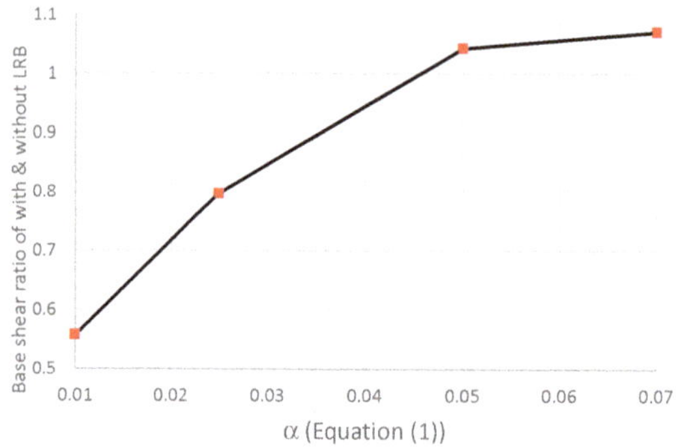

Figure 15. The base shear ratio changing with α (Equation (1)) under the PGA of 0.32 g and T_s of 0.9 s.

6. Conclusions

The important conclusions drawn from this work are the following:

(1) For the crane-induced vibration, the vertical vibration is much larger than that in the in-plane directions. Moreover, the major vibrations between 15 to 40 Hz, induced by the slab natural frequencies and the moving crane, are almost identical for the factory with and without LRBs, and the ambient vibrations at other frequencies are small. This is because the factory mass is much larger than that of the moving crane. Thus, the moving crane does not change the major ambient vibrations between the factory with and without LRBs.

(2) The high-tech factory with LRBs can decrease over 50% of the seismic base shear for earthquakes with $T_s \leq 1.0$, and for earthquakes with a long dominant period, such as $T_s = 1.4$ s, the LRB efficiency may decrease a little. However, the high-tech factory with LRBs may have much larger wind-induced vibration than that without LRBs, especially for the RC level that is the most critical for micro vibration. This is because the turbulent wind is fully loaded to the factory while the LRB initial stiffness is not large enough. Therefore, this situation will bring great disadvantage to the use of LRB in high-tech factories. To overcome this problem, the shade of adjacent buildings for the high-tech factory may be an alternative to resistant the wind-induced vibration.

(3) Because micro vibration is a major concern for high-tech factories, one should use the LRB with a large initial stiffness and a small ratio of the final stiffness over the initial stiffness (α). The large initial stiffness of the LRB can resist the wind loads, while the small α can reduce the seismic load of high-tech factories. This situation makes it difficult to obtain a suitable LRB but may reduce seismic responses while not increasing the micro vibration for high-tech factories.

Author Contributions: S.-H.J. was responsible for simulations and paper writing, and C.-C.Y. and W.-K.H. checked the data and helped analyze results. All authors have read and agreed to the published version of the manuscript.

Funding: This research received no external funding.

Conflicts of Interest: The authors declare no conflict of interest.

References

1. Turkington, D.H.; Carr, A.J.; Cooke, N.; Moss, P.J. Seismic design of bridges on lead-rubber bearings. *J. Struct. Eng.* **1989**, *115*, 3000–3016. [CrossRef]
2. Turkington, D.H.; Carr, A.J.; Cooke, N.; Moss, P.J. Design method for bridges on lead-rubber bearings. *J. Struct. Eng.* **1989**, *115*, 3017–3030. [CrossRef]
3. Fujita, T.; Sasaki, Y.; Fujimoto, S.; Tsuruya, C. Seismic isolation of industrial facilities using lead-rubber bearing. *JSME Int. J. Ser. 3 Vib. Control Eng. Eng. Ind.* **1990**, *33*, 427–434. [CrossRef]

4. Salic, R.B.; Garevski, M.A.; Milutinovic, Z.V. Response of lead-rubber bearing isolated structure. In Proceedings of the 14th World Conference on Earthquake Engineering (14WCEE), Beijing, China, 12–17 October 2008.
5. Kalpakidis, I.V.; Constantinou, M.C.; Whittaker, A.S. Modeling strength degradation in lead-rubber bearings under earthquake shaking. *Earthq. Eng. Struct. Dyn.* **2010**, *39*, 1533–1549. [CrossRef]
6. Kalpakidis, I.V.; Constantinou, M.C. Principles of scaling and similarity for testing of lead-rubber bearings. *Earthq. Eng. Struct. Dyn.* **2010**, *39*, 1551–1568. [CrossRef]
7. Islam, A.B.M.S.; Hussain, R.R.; Jumaat, M.Z.; Rahman, M.A. Nonlinear dynamically automated excursions for rubber-steel bearing isolation in multi-storey construction. *Autom. Constr.* **2013**, *30*, 265–275. [CrossRef]
8. Li, A.; Yang, C.; Xie, L.; Liu, L.; Zeng, D. Research on the Rational Yield Ratio of Isolation System and Its Application to the Design of Seismically Isolated Reinforced Concrete Frame-Core Tube Tall Buildings. *Appl. Sci.* **2017**, *7*, 1191. [CrossRef]
9. Warn, G.P.; Whittaker, A.S.; Constantinou, M.C. Vertical Stiffness of Elastomeric and Lead–Rubber Seismic Isolation Bearings. *J. Struct. Eng.* **2007**, *133*, 1227–1236. [CrossRef]
10. Weisman, J.; Warn, G.P. Stability of elastomeric and lead-rubber seismic isolation bearings. *J. Struct. Eng.* **2012**, *138*, 215–223. [CrossRef]
11. Al-Kutti, W.; Islam, A.B.M.S. Potential design of seismic vulnerable buildings incorporating lead rubber bearing. *Buildings* **2019**, *9*, 37. [CrossRef]
12. Nagarajaiah, S.; Reinhorn, A.M.; Constantinou, M.C. Nlinear dynamic analysis of 3-d-base-isolated structures. *J. Struct. Eng.* **1991**, *117*, 2035–2054. [CrossRef]
13. Huang, W.H.; Fenves, G.L.; Whittaker, A.S.; Mahin, S.A. Characterization of seismic isolation bearings for bridges from bi-directional testing. In Proceedings of the 12th World Conference on Earthquake Engineering, Auckland, New Zealand, 30 January–4 February 2000.
14. Abe, M.; Yoshida, J.; Fujino, Y. Multiaxial behaviors of laminated rubber bearings and their modeling. I: Experimental study. *J. Struct. Eng.* **2004**, *130*, 1119–1132. [CrossRef]
15. Falborski, T.; Jankowski, R. Experimental study on effectiveness of a prototype seismic isolation system made of polymeric bearings. *Appl. Sci.* **2017**, *7*, 808. [CrossRef]
16. Yuantien, C.C. Application of Biaxial Lead Rubber Bearings in High-Tech Factories. Master's Thesis, National Cheng-Kung University, Tainan, Taiwan, 2019. Available online: http://etds.lib.ncku.edu.tw/etdservice/view_metadata?etdun=U0026-1408201913481400 (accessed on 22 February 2020).
17. Benzoni, G.; Casarotti, C. Effects of vertical load, strain rate and cycling on the response of lead-rubber seismic isolators. *J. Earthq. Eng.* **2009**, *13*, 293–312. [CrossRef]
18. De Domenico, D.; Falsone, G.; Ricciardi, G. Improved response spectrum analysis of base isolated buildings: Asubstructure-based response spectrum method. *Eng. Struct.* **2018**, *162*, 198–212. [CrossRef]
19. Gordon, C.G. Generic criteria for vibration sensitive equipment 1991. In *Vibration Control in Microelectronics, Optics, and Metrology*; SPIE: Bellingham, WA, USA, 1992; Volume 1619, pp. 71–75.
20. Ju, S.H.; Kuo, H.H.; Yu, S.W.; Ni, S.H. Investigation of vibration induced by moving cranes in high-tech factories. *J. Low Freq. Noise Vib. Act. Control* **2019**. [CrossRef]
21. Ju, S.H. *Development a Nonlinear Finite Element Program with Rigid Link and Contact Element*; NSC-86-2213-E-006-063; Taiwan, 1997; Available online: http//myweb.ncku.edu.tw/~{}juju (accessed on 22 February 2020).
22. American Petroleum Institute (API). Planning, Design, Planning, Designing, and Constructing Fixed Offshore Platforms—Load and Resistance Factor Design. In *API RP 2A-LRFD*, 2nd ed.; American Petroleum Institute: Washington, DC, USA, 2019.
23. Ju, S.H.; Kuo, H.H. Experimental and numerical study of wind-induced vibration in high-tech factories. *J. Perform. Constr. Facil.* **2019**. [CrossRef]
24. MIT. *SIMQKE: A Program for Artificial Motion Generation: User's Manual and Documentation*; M.I.T. Department of Civil Engineering: Cambridge, MA, USA, 1976.
25. IBC. *International Building Code 2006*; International Code Council: Birmingham, AL, USA, 2006.
26. Idriss, I.M.; Sun, J.I. *User's Manual for SHAKE91: A Computer Program for Conducting Equivalent Linear Seismic Response Analyses of Horizontally Layered Soil Deposits*; Center for Geotechnical Modeling, Dept. of Civil and Environmental Engineering; University of California: Davis, CA, USA, 1993.

27. Jonkman, B.J.; Buhl, M. *Turbsim User's Guide v2.00.00*; Technical Report No. NRELEL-500-36970; National Renewable Energy Laboratory (NREL): Golden, CO, USA, 2004.
28. International Electrotechnical Commission. *IEC 61400-1: Wind Turbines—Part 1: Wind Energy Generation Systems–Part 1: Design Requirements*; International Electrotechnical Commission: Geneva, Switzerland, 2019.

applied sciences

MDPI

Article

Motion-Based Design of Passive Damping Systems to Reduce Wind-Induced Vibrations of Stay Cables under Uncertainty Conditions

Javier Naranjo-Pérez [1], Javier F. Jiménez-Alonso [2,*], Iván M. Díaz [2], Giuseppe Quaranta [3] and Andrés Sáez [1]

1 Department of Continuum Mechanics and Structural Analysis, Universidad de Sevilla, 41092 Seville, Spain; jnaranjo3@us.es (J.N.-P.); andres@us.es (A.S.)
2 Department of Continuum Mechanics and Theory of Structures, E.T.S. Ingenieros de Caminos, Canales y Puertos, Universidad Politécnica de Madrid, 28040 Madrid, Spain; ivan.munoz@upm.es
3 Department of Structural and Geotechnical Engineering, Sapienza University of Rome, 00184 Rome, Italy; giuseppe.quaranta@uniroma1.it
* Correspondence: jf.jimenez@upm.es; Tel.: +34-910-674-152

Received: 31 December 2019; Accepted: 28 February 2020; Published: 3 March 2020

Abstract: Stay cables exhibit both great slenderness and low damping, which make them sensitive to resonant phenomena induced by the dynamic character of external actions. Furthermore, for these same reasons, their modal properties may vary significantly while in service due to the modification of the operational and environmental conditions. In order to cope with these two limitations, passive damping devices are usually installed at these structural systems. Robust design methods are thus mandatory in order to ensure the adequate behavior of the stay cables without compromising the budget of the passive control systems. To this end, a motion-based design method under uncertainty conditions is proposed and further implemented in this paper. In particular, the proposal focuses on the robust design of different passive damping devices when they are employed to control the response of stay cables under wind-induced vibrations. The proposed method transforms the design problem into a constrained multi-objective optimization problem, where the objective function is defined in terms of the characteristic parameters of the passive damping device, together with an inequality constraint aimed at guaranteeing the serviceability limit state of the structure. The performance of the proposed method was validated via its application to a benchmark structure with vibratory problems: The longest stay cable of the Alamillo bridge (Seville, Spain) was adopted for this purpose. Three different passive damping devices are considered herein, namely: (i) viscous; (ii) elastomeric; and (iii) frictions dampers. The results obtained by the proposed approach are analyzed and further compared with those provided by a conventional method adopted in the Standards. This comparison illustrates how the newly proposed method allows reduction of the cost of the three types of passive damping devices considered in this study without compromising the performance of the structure.

Keywords: motion-based design; uncertainty conditions; constrained multi-objective optimization; reliability analysis; passive structural control; cable-stayed bridges

1. Introduction

One of the main elements that governs the dynamic behavior of cable-stayed bridges is their stay cables [1]. This structural system has both a high flexibility and a low damping, which makes it susceptible to suffer both from different vibratory problems [2] and exhibit significant changes in its modal properties induced by the modification of the operational and environmental conditions [3].

The vibratory problems observed in cables of cable-stayed bridges may be classified in terms of the structural elements excited during the vibration phenomenon into the following [2]: (i) local-global vibratory problems, in which the vibrations involve the excitation of both the cables and the deck of the structure [4]; and (ii) local vibratory problems, in which only the cables of the structure are excited laterally [5]. These vibratory problems may be caused by either of the following: (i) direct excitation sources, such as road traffic, wind [6] or earthquake action [7], or (ii) indirect excitation sources, such as linear internal resonances, parametric excitations or dynamic bifurcations.

In this paper, we focus on the case of wind-induced vibrations of stay cables, as this is the source problem of many vibratory issues reported in the literature [2]. As wind-induced vibrations can cause different structural problems on stay cables (like fatigue or comfort problems), two types of measures are normally adopted to mitigate the cable vibrations [8], consisting of either of the following: (i) modifying its natural frequency via the installation of a secondary net of cables [9]; or (ii) increasing its damping ratio via the installation of external control systems [2]. Such control systems for stay cables may be classified into three different groups [8]: (i) active [10]; (ii) semi-active [11]; and (iii) passive [12].

Active control systems for stay cables focus on controlling the dynamic response of the cable via the modification of its tensional state [13]. For this purpose, some kind of actuator, following the orders of a controller, acts on the cable in order to minimize the difference between the actual response of the cable (recorded by a sensor) and the allowable response value [14]. Although the theoretical research on the use of these devices has experienced a significant growth in recent years, their practical implementation in real cable-stayed bridges has been limited due to their high cost and the robustness problems associated with the power supply needed to guarantee their operation [2].

On the other hand, semi-active control systems focus on modifying the constitutive parameters of external damping devices deployed to control the response of the stay cable under external actions [15]. Among the different semi-active devices, magnetorheological dampers have been widely studied and implemented in real cable-stayed bridges [16]. Although semi-active damping devices outperform their passive damping counterparts [17] with a lower cost than active control systems, their efficiency is limited when they are employed under uncertainty conditions, since their performance highly depends on the control algorithm considered for the design [18].

Finally, passive control systems for stay cables focus on increasing the damping ratio of the cables via the installation of external devices, whose characteristic parameters are originally designed to mitigate the dynamic response of the structural system [19]. Due to the robustness of such passive damping devices [20], they have been installed successfully on numerous real cable-stayed bridges to reduce wind-induced vibrations [21]. Nevertheless, these devices present as main limitation, a lower flexibility to adapt the system response to the variability of both the external actions and the modification of the stay cable parameters induced by loading, when compared to the active and semi-active devices. In order to overcome this limitation, two strategies may be adopted as outlined: either (i) to install a hybrid control system [22]; or (ii) to design the passive damping device taking into account these uncertainty conditions via a robust design method [23].

Different design methods have been developed for this purpose. Among the different proposals, Kovacs was the first researcher to study the optimum design of viscous dampers for stay cables [24]. Subsequently, Pacheco et al. provided a universal curve which allows the representation of the modal damping of the first vibration mode of a taut cable in terms of the damping coefficient of the viscous damper [25]. The maximum of this curve corresponds to the optimum damping ratio of the taut cable when a viscous damper is installed on it. Later, Krenk et al. obtained an analytical expression for this curve [26]. Alternatively, other authors, such as Yoneda and Maeda, proposed an analytical model of the damped cable to determine the optimum parameters of the passive damper [27]. Although the design parameters obtained following any of these approaches are similar, so that they are currently employed for the practical design of passive damping devices, they fail to take into account a key aspect: the uncertainty associated with the variation of both the external actions and the modification of the modal properties of the stay cables [28].

In order to overcome this limitation, a motion-based design method [29] under uncertainty conditions is formulated, implemented and further validated in this paper. In fact, this proposal generalizes the formulation of a well-known design method, the so-called motion-based design method under deterministic conditions [30], to the abovementioned uncertainty conditions. The proposed motion-based design method under uncertainty conditions transforms the design problem into a constraint multi-objective optimization problem. Hence, the main objective of this problem is to find the optimum values of the characteristic parameters of the passive damping device which meet the design requirements for the structure. For this purpose, a multi-objective function is defined in terms of these parameters, together with an inequality constraint aimed at guaranteeing the compliance of the design requirements. Such design requirements are defined in terms of the vibration serviceability limit state of the structure. Since this serviceability limit state is defined under stochastic conditions, the failure probability of its compliance must be limited [31] and a reliability analysis must be performed [32,33]. For practical engineering applications [34], an equivalent reliability index is usually considered instead of the probability of failure. Thus, the formulation of the inequality constraint is realized in terms of the reliability index, which cannot exceed an allowable value [35]. For the computation of the reliability index, a sampling technique, the Monte Carlo method has been considered herein [36].

Finally, in order to validate the performance of the proposed method, it was applied to the robust design of three different passive damping devices (viscous, elastomeric, and friction dampers) where they are installed on the longest stay cable of the Alamillo bridge (Seville, Spain). To this end, only the effect of the rain–wind interaction phenomenon and the turbulent component of the wind action were considered. The results were compared with those obtained applying a conventional approach. This comparative study reveals that the proposed method allows the reduction of the cost of the passive damping devices while ensuring the structural reliability of the stay cable.

The manuscript is organized as follows: First, the motion-based design method under uncertainty conditions is described in detail. Next, a damper-cable interaction model under wind action, based on the finite element (FE) method, is presented. Subsequently, the performance of the proposed method is illustrated and further validated with a case-study (Alamillo bridge, Seville, Spain). In the final section, some concluding remarks are drawn to complete the paper.

2. Motion-Based Design of Structures under Uncertainty Conditions

2.1. Motion-Based Design of Structures under Deterministic Conditions

Structural optimization is a computational tool which can be used to assist engineering practitioners in the design of current structural systems [37]. Thus, this computational tool allows the optimum size, shape or topology of the structure to be found which meet the design requirements established by the designer/manufacturer/owner. Among the different structural optimization methods, the performance-based design method has been widely employed to design passive damping devices for civil engineering structures [23,30]. When the design requirements are defined in terms of the vibration serviceability limit state of the structure, the performance-based design method is denominated the motion-based design method [29]. This general design method was adapted herein for the design of passive damping devices when they are used to control the dynamic response of civil engineering structures. As assumption, all the variables, involved in this problem, are deterministic.

Thus, the motion-based design method under deterministic conditions transforms the design problem into a constrained multi-objective optimization problem. Therefore, the main objective of this problem is to find the optimum value of the characteristic parameters of the passive damping devices which guarantee an adequate serviceability structural behavior. For this purpose, a multi- objective function is minimized. The multi-objective function, $f(\theta)$, is defined in terms of the characteristic parameters, θ, of the considered passive damping devices. Additionally, the space domain is constrained including two restrictions in the optimization problem: (i) an inequality constraint, $g_{det}(\theta)$; and (ii) a search domain. $[\theta_{min}, \theta_{max}]$. As the relation between the objective function and the design variables is nonlinear, global optimization algorithms are normally considered to solve this constrained multi-objective optimization problem. [38]. Accordingly, the motion-based design problem under deterministic conditions can be formulated as follows:

$$\begin{array}{c}\text{Find } \theta \text{ to Minimize } f(\theta) \\ \text{Subjected to} \left\{ \begin{array}{c} g_{det}(\theta) \leq 0 \\ \theta_{\min} < \theta < \theta_{\max} \end{array} \right. \end{array} \tag{1}$$

where θ is the vector of the design variables; $f(\theta)$ is the multi-objective function to be minimized; θ_{min} and θ_{max} are the lower and upper bounds of the search domain; and $g_{det}(\theta)$ is a function which defines the inequality constraint.

Therefore, the key aspect of this optimization problem is the definition of the inequality constraint. In the case of slender civil engineering structures, whose design is conditioned by their dynamic response [29], the compliance of the vibration serviceability limit state can be considered for this purpose. According to the most advanced design guidelines [6,34], the vibration serviceability limit state of a structure is met if the movement of the structure, $d_s(\theta)$, which can be characterized by its displacement, velocity or acceleration, is lower than an allowable value, d_{lim}, defined in terms of the considered comfort requirements. Thus, the inequality constraint of the abovementioned optimization problem may be expressed as follows:

$$g_{det}(\theta) = \frac{d_s(\theta)}{d_{lim}} - 1 \leq 0 \tag{2}$$

Finally, as the result of this multi-objective optimization process, a set of possible solutions is obtained. This set of possible solutions is denominated the Pareto front. Accordingly, a subsequent decision-making problem must be solved, the selection of the best solution among the different elements of this Pareto front. Two possible alternatives are normally considered for this purpose [23]: (i) the selection of the best-balanced solution among all the elements of the Pareto front; and (ii) the consideration of additional requirements to solve this decision-making problem. The selection between both alternatives depends on the designer's own criterion and the particular conditions of the problem.

2.2. Motion-Based Design of Structures under Stochastic Conditions

In order to generalize the implementation of the motion-based design method to scenarios with stochastic conditions, it is necessary to consider during the design process the uncertainty associated with the variability of both the external actions and the modal properties of the structure. For this purpose, two types of methods are normally employed [33]: (i) probabilistic methods; and (ii) fuzzy logic methods. Between these two methods, a probabilistic approach was considered herein because engineering practitioners are more used to dealing with probability concepts than with fuzzy logic problems. Concretely, a structural reliability method [39] was adapted herein to deal with the aforementioned uncertainty. According to this method, the vibration serviceability limit state can be expressed as a probabilistic density function, $g_{unc}(\theta)$, which is defined in terms of the capacity of the structure, C_s, and the demand of the external actions, $D_a(\theta)$ (where both terms are random variables characterized by their probability density function). Thus, the vibration serviceability limit state can be defined as follows:

$$g_{unc}(\theta) = \left\{ \begin{array}{ll} C_s - D_a(\theta) & \text{if } g_{unc}(\theta) \text{ is assumed normally distributed} \\ \frac{C_s}{D_a(\theta)} & \text{if } g_{unc}(\theta) \text{ is assumed } \log-\text{normally distributed} \end{array} \right. \tag{3}$$

The above relation (Equation (3)) allows the computation of the probability of failure of the structural system, $p_f(\theta)$, to the vibration serviceability limit state. This probability of failure, $p_f(\theta)$, may be determined as follows:

$$p_f(\theta) = \begin{cases} \text{Prob}[g_{unc}(\theta) < 0] & \text{if } g_{unc}(\theta) \text{ is assumed normally distributed} \\ \text{Prob}[g_{unc}(\theta) < 1] & \text{f } g_{unc}(\theta) \text{is assumed log} - \text{normally distributed} \end{cases} \tag{4}$$

On the other hand, as it is shown in Figure 1, it is possible to characterize the probability of failure, $p_f(\theta)$, via an equivalent index, the so-called reliability index, $\beta_s(\theta)$.

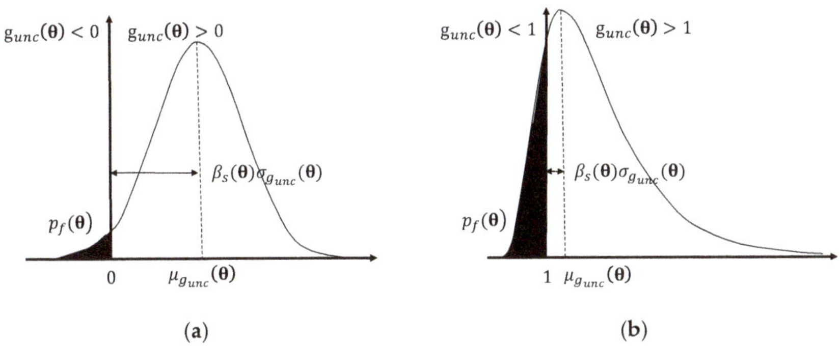

Figure 1. Probability density function of the vibration serviceability limit state, $g_{unc}(\theta)$: (**a**) $g_{unc}(\theta)$ follows a normal distribution; and (**b**) $g_{unc}(\theta)$ follows a log-normal distribution.

The relation between the probability of failure, $p_f(\theta)$, and the reliability index, $\beta_s(\theta)$, may be expressed as follows:

$$p_f(\theta) = \begin{cases} F_{g_{unc}}(0) = \Phi\left(-\frac{\mu_{g_{unc}}(\theta)}{\sigma_{g_{unc}}(\theta)}\right) = \Phi(-\beta_s(\theta)) & \text{normally distributed} \\ F_{g_{unc}}(1) = \Phi\left(\frac{\ln \mu_{C_s}/\mu_{D_a}(\theta)}{\sqrt{\sigma^2_{lnC_s} + \sigma^2_{lnD_a(\theta)}}}\right) = \Phi(-\beta_s(\theta)) & \text{log} - \text{normally distributed} \end{cases} \tag{5}$$

where $F_{g_{unc}}$ is the cumulative probability distribution function of $g_{unc}(\theta)$; $\mu_{g_{unc}}(\theta)$ and $\sigma_{g_{unc}}(\theta)$ are respectively the mean and standard deviation of $g_{unc}(\theta)$; Φ is the standard normal cumulative distribution function; μ_{C_s} and $\mu_{D_a}(\theta)$ are respectively the mean of the probabilistic distribution function of C_s and $D_a(\theta)$; and σ_{lnC_s} and $\sigma_{lnD_a(\theta)}$ are respectively the standard deviation of the log-normal distribution of C_s and $D_a(\theta)$.

In this manner, the use of the reliability index, $\beta_s(\theta)$, allows the computation of the vibration serviceability limit state under uncertainty conditions to be simplified. Hence, this design requirement is met if the reliability index, $\beta_s(\theta)$, is greater than the allowable reliability index, β_t, established by the designer/manufacturer/owner of the structure. In order to evaluate this inequality constraints, the reliability index, $\beta_s(\theta)$, is usually computed via sampling techniques and the recommended values of the allowable reliability index, β_t, can be found in literature [39]. In this study, Monte Carlo simulations [36] were considered in order to evaluate numerically the reliability index, $\beta_s(\theta)$, and the value proposed by the European guidelines [34] was considered for the allowable reliability index, β_t.

Finally, the motion-based design method under uncertainty conditions may be formulated as follows:

$$\begin{gathered} \text{Find } \theta \text{Minimize } f(\theta) \\ \text{Subjected to} \begin{cases} g_{unc}(\theta) = \frac{\beta_t}{\beta_s(\theta)} - 1 \le 0 \\ \theta_{min} < \theta < \theta_{max} \end{cases} \end{gathered} \tag{6}$$

According to this, one of the main virtues of the motion-based design method is highlighted. The method allows the deterministic and stochastic design problems to be dealt with using a similar formulation. Only the inequality constraint must be modified to adapt the formulation to the particular conditions of each problem. This virtue facilitates the implementation of this method for the robust design of passive damping devices when they are used to control the dynamic response of slender civil engineering structures.

3. Damper-Cable Interaction Model under Wind Action

The damper-cable interaction model, considered herein to evaluate the dynamic response of a stay cable damped by different passive control systems under wind action, is described in detail in this section. First, the interaction model based on the FE method is introduced. Later, the method employed to simulate the wind action is presented.

3.1. Modelling the Damper-Cable Interaction

The analysis of the dynamic behavior of stay cables has been studied extensively over the last four decades. Thus, analytical [40], numerical [30], and experimental studies [41] have been performed for this purpose. Among the different proposals, a numerical method, the FE method was considered herein to develop a damper-cable interaction model. This method presents three main advantages when it is implemented for this particular problem [30]: (i) its easy implementation for practical civil engineering applications; (ii) it allows a direct interaction of element with different constitutive laws (cable and dampers); and (iii) it simplifies the simulation of some effects such us the nonlinear behavior of the cable [40], the sag effect [42], and the influence of the external dampers on the modal properties of the cables (locking effect) [43].

The implementation of the FE method for this particular problem is based on the numerical integration of the weak formulation of the differential equilibrium equation of a vibrating stay cable in the lateral direction. Figure 2 shows an inclined cable of length, L [m], suspended between two supports at different level which presents a sag, d_c [m], with respect to the axis aligned with the two supports. The application of a small displacement causes the motion of a generic point from the self-weight configuration, P, to, P',where u_c and v_c represent the component of the movement of the cable respectively in the parallel and perpendicular direction to the axis traced between the two supports. The equation, which governs the vibration of a taut cable in the lateral direction under the assumptions of linear and flexible behavior, may be expressed as follows:

$$H\frac{\partial^2 v_c}{\partial x^2} - \frac{\partial^2}{\partial x^2}\left(EI\frac{\partial^2 v_c}{\partial x^2}\right) = m\frac{\partial^2 v_c}{\partial t^2} \tag{7}$$

where v_c is the lateral displacement of the cable [m]; H is the axial force of the cable [N]; EI is the bending stiffness of the cable [Nm^2] (where E is the Young's modulus [N/m^2] and I is the moment of inertia of the cross-section of the cable [m^4]); and m is the mass per unit length of the cable [kg/m]. According to the Equation (7), the vibration of the cable is governed by both its tensional state and its bending stiffness [44,45]. Additional phenomena can be simulated via the selection of the adequate finite-element. A nonlinear two-node element with six degrees of freedom per node has been considered herein to simulate the cable behavior. This element allows both the nonlinear geometrical and stress-stiffness behavior of the cable to be simulated adequately [46].

In order to take into account, the initial tensional and deformational state of a stay cable during either a modal or a transient analysis, a preliminary static nonlinear analysis must be performed. In this preliminary static analysis, the equilibrium form of the cable under its self-weight and a preliminary axial force is achieved. As a result of this analysis, both the stress and the shape of the cable are updated, which is a key aspect to simulate numerically its real behavior.

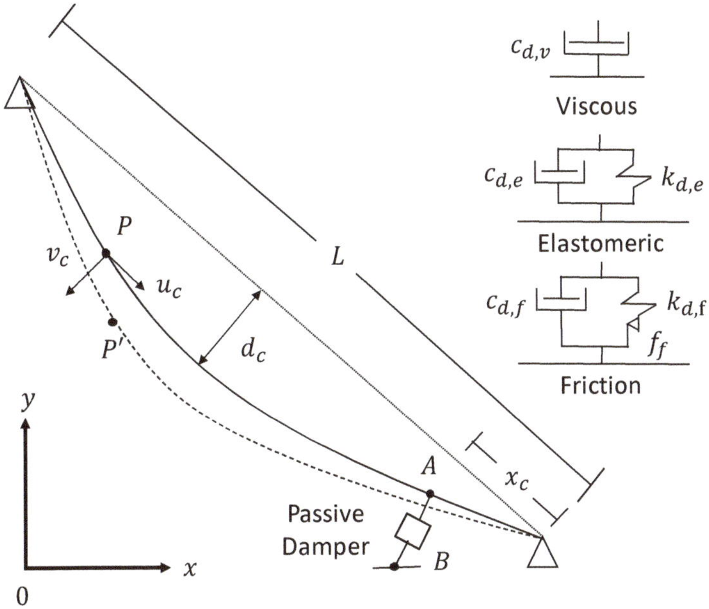

Figure 2. Damper-cable interaction model considered and mechanical model of each passive damper (viscous, elastomeric, and friction).

Subsequently, the modelling problem must focus on the simulation on the passive damping devices behavior. Three passive damping devices were considered herein (Figure 2). For these three passive damping devices, a linear constitutive law was assumed. The effect of these three passive damping devices on the cable may be simulated by an equivalent damping force. Each equivalent damping force is related to the energy that each damping device is able to dissipate, and it is opposed to the movement of the cable. Thus, each passive damping device has been modelled by a finite element whose behavior is equivalent to the corresponding damping force (Figure 2). This assumption has two advantages: (i) the relative movements between the damper and the cable, which govern the behavior of the damper, were obtained straight; and (ii) the effect of the dampers on the modal properties of the structure was taken into account directly.

First, the effect of a viscous damper is equivalent to a damping force which is proportional to a damping coefficient, $c_{d,v}$ [sN/m], and the relative velocity, $\dot{v}_r(t)$ [m/s], between the two extremes of the damper ($\dot{v}_r(t) = \dot{v}_{d,A}(t) - \dot{v}_{d,B}(t)$, where $\dot{v}_{d,A}(t)$ is the velocity of the extreme of the damper in contact with the cable and $\dot{v}_{d,B}(t)$ is the velocity of the extreme of the damper in contact with the deck, as it is illustrated in Figure 2. The viscous damping force of this damper, $F_{d,v}(t)$, may be expressed as [47]:

$$F_{d,v}(t) = c_{d,v}\dot{v}_r(t) \tag{8}$$

Second, the effect of the elastomeric damper may be simulated via the Kelvin–Voigt model. The equivalent viscoelastic damping force is characterized by two components: (i) a viscous damping component which is expressed in terms of a damping coefficient, $c_{d,e}$ [sN/m], and the relative velocity, $\dot{v}_r(t)$ [m/s]; and (ii) an elastic component which is expressed in terms of a stiffness coefficient, $k_{d,e}$ [N/m], and the relative displacement between the two extremes, $v_r(t)$ [m] ($v_r(t) = v_{d,A}(t) - v_{d,B}(t)$, where $v_{d,A}(t)$ is the displacement of the extreme of the damper in contact with the cable and $v_{d,B}(t)$ is the displacement of the extreme of the damper in contact with the deck, as it is illustrated in Figure 2. The viscoelastic damping force of this damper may be defined as [48,49]:

$$F_{d,e}(t) = c_{d,e}\dot{v}_r(t) + k_{d,e}v_r(t) \tag{9}$$

Finally, the effect of the friction damper may be mimicked via the extended Kelvin–Voigt model. The definition of the equivalent damping force involves three components: (i) a viscous damping component which is expressed in terms of a damping coefficient, $c_{d,f}$ [sN/m], and the relative velocity, $\dot{v}_r(t)$ [m/s]; (ii) an elastic component which is expressed in terms of a stiffness coefficient, $k_{d,f}$ [N/m], and the relative displacement, $v_r(t)$ [m], and (iii) a friction component defined in terms of a static friction force, f_f [N] (where, $f_f = \mu \cdot N$, being μ the friction coefficient [−] and N the normal force [N]) and a symbolic function, $\mathrm{sgn}(\dot{v}_r(t))$ (which returns −1, 0, and 1 in case $\dot{v}_r(t) < 0$, $\dot{v}_r(t) = 0$ and $\dot{v}_r(t) > 0$, respectively). The equivalent damping force of this damper may be expressed as [50]:

$$F_{d,f}(t) = c_{d,f}\dot{v}_r(t) + k_{d,f}v_r(t) + f_f \cdot \mathrm{sgn}(\dot{v}_r(t)) \tag{10}$$

These damping devices are usually located at a certain distance, x_c [m], of the lower anchorage of the stay cable (Figure 2) due to constructive limitations. Nevertheless, due to their mechanical characteristics, they can have influence on both the damping and the natural frequencies (locking effect) of the stay cable.

3.2. Modelling the Wind Action

Subsequently, the effect of the wind-induced forces was simulated numerically. The wind simulation was carried out under the assumption that the cable is a cylinder immersed in a turbulent flow [2]. Hence, the wind flow is composed of three components: (i) a mean wind velocity, $\mathcal{U}$ [m/s]; (ii) a fluctuating longitudinal velocity, $u(t)$ [m/s]; and (iii) a fluctuating transversal velocity, $v(t)$ [m/s].

The wind forces can be decomposed into a mean and a fluctuating component assuming the following hypothesis: (i) a quasi-steady behavior of the wind-induced forces; and (ii) small components of the turbulence with respect to the mean wind velocity, $\mathcal{U}$ [51]. The expression of these two components can be expressed as follows (assuming a linearized approximation [52]):

$$F_D(t) = F_D + f_{Du}(t) + f_{Dv}(t) \tag{11}$$

$$F_L(t) = F_L + f_{Lu}(t) + f_{Lv}(t) \tag{12}$$

where $F_D(t)$ is the drag force [N]; $F_L(t)$ is the lift force [N]; F_D is the mean wind drag force; F_L is the mean wind lift force; $f_{Du}(t)$ is the drag force induced by the longitudinal component of the wind; $f_{Lu}(t)$ is the lift force induced by the longitudinal component of wind; $f_{Dv}(t)$ is the drag force induced by the transversal component of wind; and $f_{Lv}(t)$ is the lift force induced by the transversal component of the wind. These magnitudes can be determined using the following relationships [2]:

$$F_D = 0.5\rho\mathcal{U}^2 DC_D \tag{13}$$

$$f_{Du}(t) = \rho\mathcal{U}u(t)DC_D \tag{14}$$

$$f_{Dv}(t) = 0.5\rho\mathcal{U}v(t)D\left(C'_D - C_L\right) \tag{15}$$

$$F_L = 0.5\rho\mathcal{U}^2 DC_L \tag{16}$$

$$f_{Lu}(t) = \rho\mathcal{U}u(t)DC_L \tag{17}$$

$$f_{Lv}(t) = 0.5\rho\mathcal{U}v(t)D\left(C_L - C'_D\right) \tag{18}$$

where ρ is the density of the air [kg/m^3]; D is the outer diameter of the cable [m]; C_D is the drag coefficient [−]; and C_L the lift coefficient [−]. The coefficients C'_D and C'_L are the derivative of C_D and C_L, respectively, with respect to the angle α neighboring β (Figure 3). As the section of the cable is assumed to be circular in this study, these derivatives are therefore null because of the symmetry, and hence these two coefficients can be neglected.

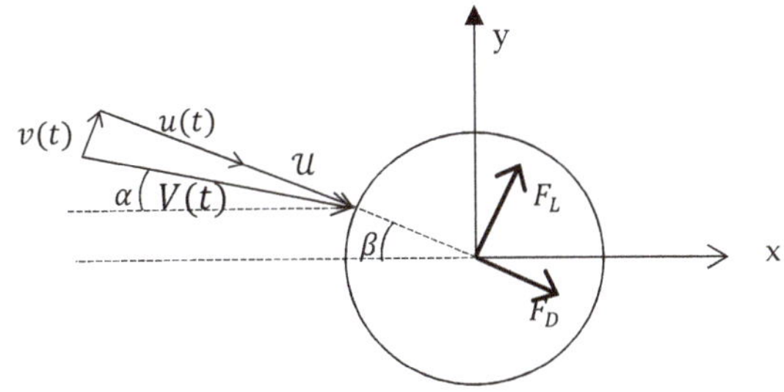

Figure 3. Reference coordinate system, drag force component, lift force component, and wind velocity components.

Finally, in order to determine the wind forces it is necessary to generate simulations of wind velocities. For this purpose, the wave superposition spectral-based method was considered [8]. This method allows the numerical determination of a series of wind velocities via the superposition of trigonometric functions. On the one hand, the amplitude of these functions is obtained in terms of a coherence function, which considers the spatial variability of the wind velocity, and the power spectral density function of the turbulent wind velocity. On the other hand, the phase of the trigonometric functions is generated randomly. The coherence function is defined using the relationship proposes by Davenport [53]. The power spectral density function proposed by the European guidelines [54] was considered herein.

4. Application Example

The proposed motion-based design method under uncertainty conditions was validated herein via the design of three passive damping devices when they are used to control the wind-induced vibrations of the longest cable of a real bridge. For this purpose, the Alamillo bridge (Seville, Spain) was considered (Figure 4). The length of the deck of this bridge is 200 m. Unlike most cable-stayed bridges, the Alamillo bridge has not back-stays. An inclination of its pylon of 32° with respect to the vertical axis compensates the lack of the back-stays [55]. A total of 26 stays (13 parallel pairs) with a longitudinal separation of 12 m guarantees an adequate connection between the deck and the pylon.

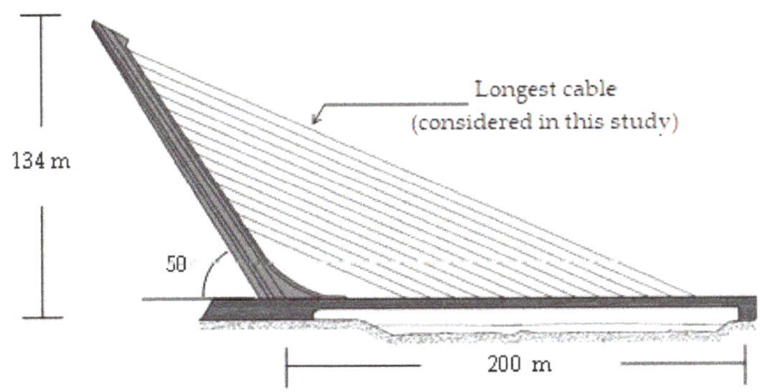

Figure 4. Illustrative scheme of the Alamillo bridge.

Previous research reported that the longest stay cable of this bridge, which has both a low damping and mass ratio, was prone to vibrate due to the wind action. Concretely, it was detected that the main sources of vibration of this cable were the rain–wind interaction phenomenon and the turbulent component of the wind action [56]. Therefore, this stay cable was considered as a benchmark to validate the performance of the proposed design method. For this purpose, three different passive damping devices (viscous, elastomeric and friction dampers) were designed according to the proposed method, and the results obtained were compared with the ones provided by a conventional method adopted by the Standards [6]. Additionally, the uncertainty associated with the variation of the modal properties of the cable due to the modifications of the operational and environmental conditions was taken into account in this design process. The development of this case-study was organized in the following steps: (i) a FE model of the cable was built and its numerical modal properties were obtained via a numerical modal analysis; (ii) a transient analysis was performed to evaluate the vibration serviceability limit state of the structure; (iii) as this limit state was not met, the three passive damping devices were designed according to both methods (the new proposal and the conventional one); and (iv) finally, the results obtained were compared and some conclusions were drawn to close the section.

4.1. FE Model and Numerical Modal Analysis

The FE model of the cable was built using the software Ansys [57]. The geometrical and mechanical properties of the cable under study were as follows: (i) its length, $L = 2.92 \times 10^2$ m; (ii) its outer diameter, $D = 0.20$ m; (iii) the effective area of its cross section, $A = 8.38 \times 10^{-3}$ m^2; (iv) the effective moment of inertia, $I = 5.58 \times 10^{-4}$ m^4; (v) its mass per unit length, $m = 60$ kg/m; (vi) an axial force, $H = 4.13 \times 10^6$ N; (vii) a Young's modulus, $E = 1.6 \times 10^{11}$ N/m^2; and (viii) the angle between the cable and the deck, $\gamma = 26°$. The cable was modelled by a mesh of 100 equal-length beam elements (BEAM188). In order to simulate numerically the sag effect, a nonlinear static analysis was previously performed. The objective of this preliminary analysis was to find both the initial tensional state and pre-deformed shape of the cable. The self-weight of the cable and its initial axial force were considered as loads for this preliminary nonlinear static analysis. Subsequently, the results of this analysis were used to update the geometry and tensional state of the cable. Later, the linear perturbation method was considered to perform the modal analysis [57]. Additionally, the stress stiffening effect was taken into account to perform this modal analysis.

As result of this numerical modal analysis, the first six natural frequencies were obtained. Table 1 shows the value of these first six natural frequencies (f_i being the natural frequencies of the ith vibration mode).

Table 1. Numerical natural frequencies of the cable.

Natural Frequency	f_1	f_2	f_3	f_4	f_5	f_6
Value [Hz]	0.452	0.905	1.351	1.802	2.254	2.706

4.2. Assessment of the Vibration Serviceability Limit State of the Cable under Uncertainty Conditions

As it was expected, according to the numerical natural frequencies obtained (Table 1), this cable was prone to vibrate under wind action due to both the turbulent component of the wind (the first two natural frequencies are lower than 1 Hz [58]) and the rain–wind interaction phenomenon (the six natural frequencies are lower than 3 Hz [6]). For this reason, the assessment of the vibration serviceability limit state of this stay cable was performed herein following the recommendations of the Federal Highway Administration (FHWA) guidelines [6].

On the one hand, in order to avoid the wind-induced vibrations associated with the rain–wind interaction phenomenon, it must be checked that the damping ratio of all the vibration modes, whose natural frequencies are lower than 3 Hz, are greater than a recommended value [6,59]. In order to determine this recommended value, the FHWA guidelines [6] establishes that the rain–wind interaction phenomenon can be neglected if the Scruton number, S_c, is greater than 10 for all the considered vibration modes. This condition may be expressed as follows:

$$S_{c,i} = \frac{m\xi_i}{\rho D^2} > 10 \tag{19}$$

where ξ_i is the damping ratio of the ith vibration mode.

Thus, this requirement is equivalent to guaranteeing a minimum damping ratio for each considered vibration mode. The minimum required damping ratio may be determined as follows:

$$\xi_i > \frac{10\rho D^2}{m} \tag{20}$$

As expected, due to the results of previous experimental tests, the damping ratio associated with the first six vibration modes of this cable did not meet this condition [56]. Hence, it was necessary to increase the value of these damping ratios. A passive damping device can be designed and installed on the cable for this purpose.

On the other hand, in order to analyze the effect of the turbulent component of wind action on the dynamic behavior of the cable, a transient analysis was performed. As a result of this transient analysis, the dynamic response of the cable under wind action can be obtained and the vibration serviceability limit state of the cable can be assessed. According to the FHWA guidelines [6], this limit state is met if the maximum displacement of the cable is lower than an allowable displacement which is defined in terms of the user tolerance. Table 2 shows the allowable displacement of the cable in terms of the design level required [6]. In this study, a recommended design level was established for the vibration serviceability limit state.

Table 2. User tolerance limits for the different design levels [6].

Design Level	Allowable Displacement [m] [1]
preferred	$0.5D$
recommended	$1.0D$
not to exceed	$2.0D$

[1] D is the outer diameter of the cable.

Additionally, as the dynamic response of the stay cable was sensitive to the variation of its modal properties associated with the change of the operational and environmental conditions during its overall life cycle, a reliability analysis about the compliance of the vibration serviceability limit state was performed. For this purpose, it was assumed that the axial force of the cable is a random variable normally distributed. According to the results provided by Stromquist-LeVoir et al., it could be also assumed that this random variable has a range of variation of ± 10% [60]. A sample of stay cables with different values of the axial force was generated. The vibration serviceability limit state was assessed on this sample. For this purpose, the vibration serviceability limit state must be reformulated in order to take into account the uncertainty conditions. According to this, this limit state is met if a reliability index, $\beta_s(\theta)$, is greater than an allowable reliability index, β_t.

In order to compute the reliability index, $\beta_s(\theta)$, the maximum displacement of the stay cable (obtained from the different transient analyses performed on the sample of stay cables), which constitutes the demand of the wind action, $D_a(\theta)$, and the allowable displacement of the stay cable (established by the FHWA guidelines [6]), which constitutes the capacity of the structure, C_ε, were determined. Additionally, as the wind action is defined according to a return period of 50 years, the corresponding value of the allowable reliability index is $\beta_t = 1.35$, according to the European guidelines [34].

As a numerical method in order to both determine the sample and compute the reliability index, $\beta_s(\theta)$, the Monte Carlo method was considered herein. A convergence analysis was performed to determine the size of the sample [61]. As a result of this convergence analysis, the size of the sample was established at 100.

Finally, in order to evaluate the demand of the wind action, $D_a(\theta)$, the wind forces must be determined. For this purpose, simulations of the wind velocities were generated. The simulation of these wind velocities was addressed employing the wave superposition spectral-based method [8]. Both the von Karma spectra and a coherence function, as they are defined by the European guidelines [54], were employed herein. The following design parameters were considered for the wind simulation [54]: (i) basic wind velocity, $v_{b,0} = 26$ m/s; (ii) a directional factor, $c_{dir} = 1$; (iii) a season factor, $c_{sea} = 1$; (iv) a orography factor, $c_{oro} = 1$; (v) an terrain type III category (which involves a terrain factor, $k_r = 0.216$; a roughness length, $z_o = 0.3$ m; and a minimum height, $z_{min} = 5$ m); (vi) a duration of each simulation of 300 s; and (vii) a time step of 5×10^{-3} s [62]. In this study, the wind velocities were generated at ten different heights of the cable (resulting from dividing the cable into ten equal-length segments), as Figure 5 depicts. This mesh density was considered for all the simulations conducted in the paper, in order to ensure that all the obtained results were consistent. Although preliminary analyses performed by the authors concluded that the meshing in Figure 5 was adequate for our aims (illustrating the performance of the proposed motion-based approach), the reader should be aware of the fact that the numerical simulation of the structural response under wind excitation depends on such mesh density, so that further analyses are recommended. A graphical user interface [63] was developed in the commercial software Matlab [64] to evaluate the wind action following the above guidelines.

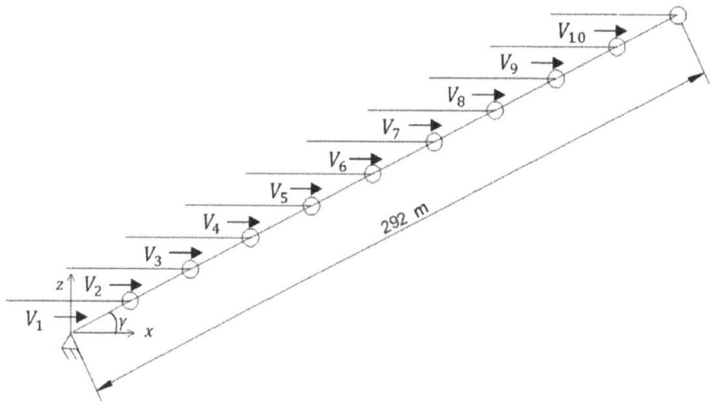

Figure 5. Representation of the ten different heights where the wind action is applied.

The application of Equations (11) and (12) allows the wind-induced forces in terms of the wind velocities to be computed. For this purpose, the following values for the characteristic parameters were adopted: (i) a density of the air, $\rho = 1.23$ kg/m^3; (ii) a drag coefficient, $C_D = 1.2$ [2]; and (iii) a lift coefficient, $C_L = 0.3$ [6].

Finally, a transient analysis (time history simulation) was performed for each element of the sample. The nonlinear geometrical behavior of the stay cables was considered for this analysis. A Newmark-beta method (an unconditionally stable method with parameters $\beta_m = 1/4$ and $\gamma_m = 1/2$) was considered to solve the transient analysis. Hence, the reliability index, $\beta_s(\theta)$, was computed from the results of this set of transient analysis. Subsequently, the vibration serviceability limit state of the stay cable under uncertainty conditions was assessed. Thus, the reliability index, $\beta_s(\theta)$, was lower than the allowable reliability index, β_t, so this limit state was not met.

In order to improve the dynamic behavior of this stay cables, different passive damping devices were installed at this stay cable. These passive damping devices were designed according to the proposed method. This design problem is described in next section.

4.3. Motion-Based Design of Passive Damping Devices under Uncertainty Conditions

Three different passive damping devices were considered for this study: (i) viscous damper; (ii) elastomeric damper; and (iii) friction damper. The FE method was employed to simulate the behavior of these damping devices. The software Ansys [57] was employed for this purpose. Figure 2 depicts the mechanical models, which simulate the behavior of each damper. For each passive damper, the following model was considered: (i) the viscous damper was modelled by a 1D element (COMBIN14) whose characteristic parameter was the damping coefficient, $c_{d,v}$ [sN/m]; the elastomeric damper was also modelled by a 1D element (COMBIN14) whose characteristic parameters were the damping coefficient, $c_{d,e}$ [sN/m], and the stiffness coefficient, $k_{d,e}$. [N/m]; and (iii) the frictioin damper was modelled by a 1D element (COMBIN40) whose characteristic parameters were the damping coefficient, $c_{d,f}$ [sN/m], the stiffness coefficient, $k_{d,f}$ [N/m], and the friction force, f_f [N].

Consequently, the different dampers were implemented in the numerical model and designed according to the motion-based design method under uncertainty conditions. The three dampers were installed at a length of $x_c = 0.03L$ according to the recommendations of Ref. [2]. The damper-cable interaction model is shown in Figure 2.

A search domain, $[\theta_{\min}, \theta_{\max}]$, for the characteristic parameters of the dampers was included in the optimization problem to ensure the physical meaning of the solutions obtained. The search domain was defined as follows: (i) the lower bound of the search domain, $\theta_{\min}$, was defined as $\theta_{\min} = \left[c_{min}, k_{min}, f_{f_{min}}\right]$ (where c_{min} is the minimum value of the damping coefficient; k_{min} is the minimum value of the stiffness coefficient, and $f_{f_{min}}$ is the minimum value of the friction force); and (ii) the upper bound of the search domain, $\theta_{\max}$, was defined as $\theta_{\max} = \left[c_{max}, k_{max}, f_{f_{max}}\right]$ (where c_{max} is the maximum value of the damping coefficient; k_{max} is the maximum value of the stiffness coefficient, and $f_{f_{max}}$ is the maximum value of the friction force).

The lower, c_{min}, and upper, c_{max}, bounds of the damping coefficient were determined considering both the requirement of the Scruton number [6] and the optimum damping coefficient of the Pacheco's universal curve [25]. According to this, the following bounds were established: (i) $c_{min} = 4.8 \times 10^4$ sN/m; and (ii) $c_{max} = 1.64 \times 10^5$ sN/m. This search range guarantees that any solution of this design problem avoids the occurrence of the rain–wind interaction phenomenon.

The search domain of the stiffness coefficient and the friction force were based on the results of previous research [2]. According to these results, the following bounds were established: (i) for the stiffness coefficient, $k_{min} = 5 \times 10^4$ N/m and $k_{max} = 5 \times 10^5$ N/m; and (ii) for the friction force, $f_{f_{min}} = 1 \times 10^4$ N and $f_{f_{max}} = 4 \times 10^4$ N.

In order to avoid falling into a local minimum, a global computational algorithm was considered for this optimization problem. Among the different computational algorithms, genetic algorithms were considered herein [65] for its simplicity and great efficiency to solve structural optimization problems.

Genetic algorithms are nature-inspired computational algorithms based on Darwin's natural selection theory. According to this, each possible value of the characteristic parameters of the damper is identified as a chromosome. Subsequently, each set of characteristic parameters is grouped into an individual (parameter vector). Later, the value of this parameter vector is improved via an iterative process where the value of the objective function is optimized. The optimization process can be summarized in the following steps: (i) an initial random population of parameter vectors is generated; (ii) the objective function is evaluated for all the individuals; (iii) a new population is created using three mechanisms (selection, crossover, and mutation); (iv) the objective function is evaluated for the individuals of the new population; (v) the steps (iii) and (iv) are repeated until some convergence criterion is met. The following parameters were considered for the considered genetic algorithms: (i) an initial population of 5 individuals; (ii) a crossover fraction of 0.4; (iii) a mutation fraction of 0.9; and (iv) a total number of iterations equal to 6.

As result of the optimization process, a Pareto front was obtained. Subsequently, a decision-making problem should be solved, the selection of the best solution among the different elements of the Pareto front. In order to address this problem, an additional condition was included. Among the different elements of the Pareto front, the element of the Pareto front with a lower value of the damping coefficient was selected as best solution. The commercial software Ansys [57] and Matlab [64] were used to solve this design problem. The results of the optimization problem are summarized in the next sub-sections.

4.3.1. Viscous Damper

First, the motion-based design of the viscous damper under uncertainty conditions was performed. The design problem of this viscous damper may be formulated as follows:

$$\begin{aligned} &\text{find } \theta = c_{d,v} \text{ to minimize } f(\theta) = c_{d,v} \\ &\text{subject to} \begin{cases} c_{\min} < c_{d,v} < c_{\max} \\ \beta_s(\theta) \geq \beta_t = 1.35. \end{cases} \end{aligned} \tag{21}$$

As result of the optimization process, the damping coefficient, $c_{d,v}$, was obtained. The optimum value obtained was $c_{d,v} = 1.06 \times 10^5$ sN/m. The reliability index for this solution was, $\beta_s(\theta) = 1.37$, which met the design requirements. Figure 6 shows the maximum displacement at the mid-span of the cable damped by the viscous damper for the different elements of the sample.

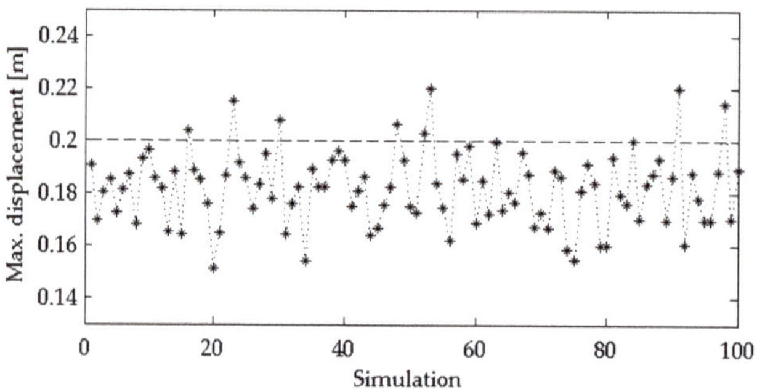

Figure 6. Maximum displacement at the mid-span of the stay cable damped by the viscous damper for the different elements of the sample.

4.3.2. Elastomeric Damper

Subsequently, the motion-based design of the elastomeric damper under uncertainty conditions may be addressed. The design problem of this elastomeric damper may be defined as follows:

$$\begin{aligned} &\text{find } \theta = \left[c_{d,e}, k_{d,e}\right] \text{ to minimize } \mathrm{f}(\theta) = [f_1, f_2] = \left[c_{d,e}, k_{d,e}\right] \\ &\text{subject to} \begin{cases} c_{\min} < c_{d,e} < c_{\max} \\ k_{\min} < k_{d,e} < k_{\max} \\ \beta_s(\theta) \geq \beta_t = 1.35\,. \end{cases} \end{aligned} \tag{22}$$

As result of the design process, the parameters of the elastomeric damper ($c_{d,e}$ and $k_{d,e}$) were obtained. The best solution among all the elements of the Pareto front was $c_{d,e} = 1.22 \times 10^5$ sN/m and $k_{d,e} = 1.30 \times 10^5$ N/m. The reliability index associated with this solution is, $\beta_s(\theta) = 1.49$, which met the design requirements. Figure 7 shows the maximum displacement at the mid-span of the cable damped by the elastomeric damper for the different elements of the sample.

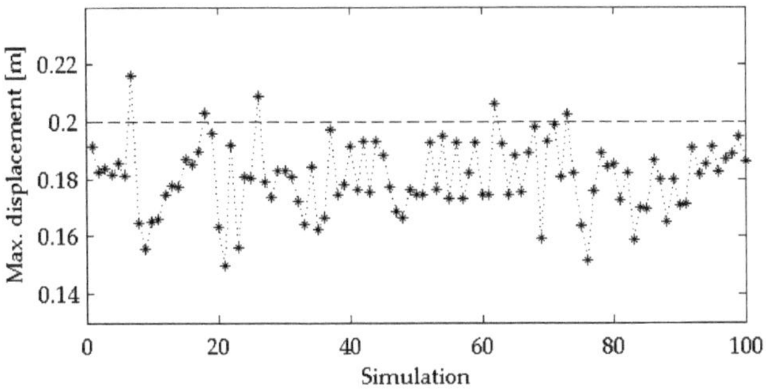

Figure 7. Maximum displacement at the mid-span of the stay cable with the elastomeric damper for the different elements of the sample.

4.3.3. Friction Damper

Finally, the motion-based design of the friction damper under uncertainty conditions was performed. The design problem of this friction damper may be formulated as follows:

$$\text{find } \boldsymbol{\theta} = \left[c_{d,e},\, k_{d,e},\, f_f\right] \text{ to minimize } \mathbf{f}(\boldsymbol{\theta}) = [f_1,\, f_2,\, f_3] = \left[c_{d,e},\, k_{d,e},\, f_f\right]$$
$$\text{subject to} \begin{cases} c_{\min} < c_{d,f} < c_{\max} \\ k_{\min} < k_{d,f} < k_{\max} \\ f_{f_{\min}} < k_f < f_{f_{\max}} \\ \beta_s(\boldsymbol{\theta}) \geq \beta_t = 1.35\,. \end{cases} \tag{23}$$

After the design process, the optimum value of the damping coefficient, stiffness coefficient, and friction force which characterize the friction damper were obtained. The optimum solution was $c_{d,f} = 1.24 \times 10^5$ sN/m, $k_{d,f} = 6.74 \times 10^4$ N/m and $f_f = 2.95 \times 10^4$ N. The reliability index associated with this solution was $\beta_s(\boldsymbol{\theta}) = 1.55$, which met the design requirements. Figure 8 shows the maximum displacement at the mid-span of the cable damped by the friction damper for the different elements of the sample.

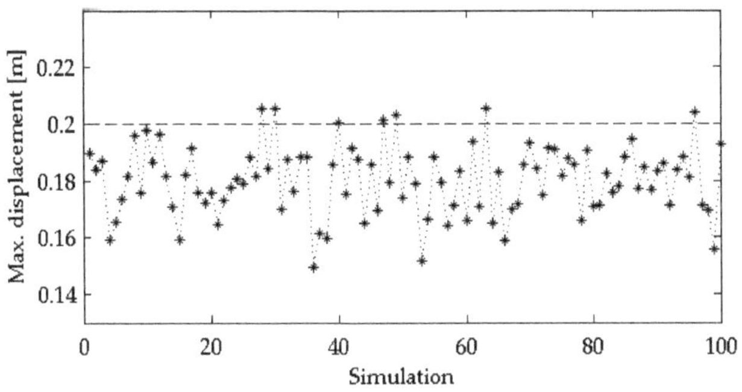

Figure 8. Maximum displacement at the mid-span of the stay cable damped by the friction damper for the different elements of the sample.

4.4. Discussion of Results

Finally, the performance of the proposed method was validated comparing the abovementioned results with the ones provided by a conventional one, the optimum damping coefficient of the Pacheco's universal curve [25]. This optimum value for a viscous damper can be determined using the following relationship:

$$c_{opt} = 0.10\frac{mL\omega_1}{\frac{x_c}{L}}, \tag{24}$$

where $\omega_1 = 2\pi f_1$ is the fundamental angular natural frequency of the stay cable [rad/s] and x_c is the distance between the anchorage of the cable and the point where the damper is implemented (Figure 2). As in the remaining cases, the viscous damper is located at the point, $x_c = 0.03L$, with respect to the lower anchorage. The optimum damping coefficient, according to this conventional method for the viscous damper was $c_{opt} = 1.64 \times 10^3$ sN/m.

Thus, two main conclusions may be obtained via the comparison of the abovementioned results: (i) the motion-based design method under uncertainty conditions allows reduction of the characteristic parameter of the viscous damper by about 35% with respect to the conventional method; and (ii) for this case-study, the viscous damper appears to be the best choice to control the dynamic response of the longest cable of the Alamillo bridge, as a minimum value of the damping coefficient was obtained for this passive damper. The proposed method allows a better adjustment to the design requirements of the problem, reducing, as consequence, the size and the cost of the passive damping devices. Hence, the performance of the motion-based design method, for this particular problem, has been validated.

5. Conclusions

Stay cables are prone to vibrate under wind-induced vibrations, so that passive damping devices are usually employed to control their response. Nevertheless, the performance of these damping devices is directly affected by the sensitivity of the stay cables to both the variability of the external actions and the modification of the constitutive modal properties of the cables induced by the changes of the operational and environmental conditions. Accordingly, it is necessary to establish design methods which overcome these limitations and can be easily implemented for practical engineering applications.

For this purpose, a motion-based design method under uncertainty conditions was proposed and implemented herein. In this approach, the design problem is transformed into a constrained multi-objective optimization problem. Thus, the different components of the multi-objective function are defined in terms of the characteristic parameters of the considered passive damping device; and an inequality constraint is additionally included to guarantee an acceptable probability of failure of the structural system. As design criterion to evaluate the probability of failure, the compliance of the vibration serviceability limit state (according to the FHWA guidelines) was considered. Therefore, the computation of the probability of failure was performed via a reliability index. In this manner, the compliance of the vibration serviceability limit state is met if the reliability index is greater than an allowable value (according to the European guidelines). A sampling technique, the Monte Carlo method, was considered to determine numerically this index.

The performance of the method was validated numerically via its implementation for the design of three different passive damping devices (viscous, elastomeric, and friction dampers) when they are used to control the wind-induced vibrations of the longest stay cable of the Alamillo bridge (Seville). To this end, the effects of the rain-wind interaction phenomenon and the turbulent component of the wind action were considered as excitation sources. Additionally, and for comparison purposes, the passive damping devices were also designed according to a conventional method. As result of this study, a clear reduction of the values of the characteristic parameters of the dampers was obtained when the motion-based design method was applied, when compared to the results of the conventional method. Thus, the proposed method allows improvement of the design of passive damping devices for stay cables under wind-induced vibrations considering uncertainty conditions. This improvement is reflected in a reduction of both the size and the budget of the devices, which facilitates its installation. Nevertheless, despite the good performance of the proposed approach, further studies are recommended to validate experimentally the long-term behavior of passive damping devices designed according to this proposal.

Author Contributions: Conceptualization, J.N.-P. and J.F.J.-A.; Methodology, J.F.J.-A.; Programming, J.N.-P. and G.Q.; Software, J.N.-P. and G.Q.: Validation, J.F.J.-A. and I.M.D.; Writing—Original Draft Preparation, J.N.- P. and J.F.J.-A.; Writing—Review & Editing, J.F.J.-A.; Supervision, I.M.D. and A.S.; Funding Acquisition, I.M.D. All authors have read and agreed to the published version of the manuscript.

Funding: This work was partially funded by the Ministry of Science, Innovation and Universities (Government of Spain) under the Research Project SEED-SD (RTI2018-099639-B-I00). Additionally, the co-author, Javier Naranjo-Pérez, was supported by the research contract, USE-17047-G, provided by the Universidad de Sevilla.

Conflicts of Interest: The authors declare no conflict of interest.

References

1. Abdel-Ghaffar, A.M.; Khalifa, M.A. Importance of Cable Vibration in Dynamics of Cable-Stayed Bridges. *J. Eng. Mech.* **1991**, *117*, 2571–2589. [CrossRef]
2. Caetano, E. *Cable Vibrations in Cable-Stayed Bridges*; IABSE: Zürich, Switzerland, 2007; Volume 9.
3. Lepidi, M.; Gattulli, V. Static and dynamic response of elastic suspended cables with thermal effects. *Int. J. Solids Struct.* **2012**, *49*, 1103–1116. [CrossRef]
4. Caetano, E.; Cunha, A.; Gattulli, V.; Lepidi, M. Cable–deck dynamic interactions at the International Guadiana Bridge: On-site measurements and finite element modelling. *Struct. Control Health Monit.* **2008**, *15*, 237–264. [CrossRef]
5. Domaneschi, M.; Martinelli, L. Extending the benchmark cable-stayed bridge for transverse response under seismic loading. *J. Bridge Eng.* **2013**, *19*, 04013003. [CrossRef]
6. *Wind-Induced Vibration of Stay Cables*; FHWA-HRT-05-083; Federal Highway Administration: New York, NY, USA, 2007.
7. Dyke, S.J.; Caicedo, J.M.; Turan, G.; Bergman, L.A.; Hague, S. Phase I benchmark control problem for seismic response of cable-stayed bridges. *J. Struct. Eng.* **2003**, *129*, 857–872. [CrossRef]
8. Cremona, C. *Comportement Au Vent Des Ponts*; Presses De L'ecole Natinale Des Ponts Et Chaussées; Association Francaise De Génie Civil: Paris, France, 2002.
9. Caracoglia, L.; Zuo, D. Effectiveness of cable networks of various configurations in suppressing stay-cable vibration. *Eng. Struct.* **2009**, *31*, 2851–2864. [CrossRef]
10. Bossens, F.; Preumont, A. Active tendon control of cable-stayed bridges: A large-scale demonstration. *Earthq. Eng. Struct. Dyn.* **2001**, *30*, 961–979. [CrossRef]
11. Johnson, E.A.; Baker, G.A.; Spencer, B.F., Jr.; Fujino, Y. Semiactive Damping of Stay Cables. *J. Eng. Mech.* **2007**, *133*, 1–11. [CrossRef]
12. Ali, H.-E.M.; Abdel-Ghaffar, A.M. Seismic Passive Control of Cable-Stayed Bridges. *Shock Vib.* **1995**, *2*, 918721. [CrossRef]
13. Rodellar, J.; Mañosa, V.; Monroy, C. An active tendon control scheme for cable-stayed bridges with model uncertainties and seismic excitation. *J. Struct. Control* **2002**, *9*, 75–94. [CrossRef]
14. Huang, P.; Wang, X.; Wen, Q.; Wang, W.; Sun, H. Active Control of Stay Cable Vibration Using a Giant Magnetostrictive Actuator. *J. Aerosp. Eng.* **2018**, *31*, 04018074. [CrossRef]
15. Zhou, H.J.; Sun, L.M. Damping of stay cable with passive-on magnetorheological dampers: A full-scale test. *Int. J. Civ. Eng.* **2013**, *11*, 154–159.
16. Chen, Z.H.; Lam, K.H.; Ni, Y.Q. Enhanced damping for bridge cables using a self-sensing MR damper. *Smart Mater. Struct.* **2016**, *25*, 085019. [CrossRef]
17. YeganehFallah, A.; Attari, N.K.A. Robust control of seismically excited cable stayed bridges with MR dampers. *Smart Mater. Struct.* **2017**, *26*, 035056. [CrossRef]
18. Zhao, Y.-L.; Xu, Z.-D.; Wang, C. Wind vibration control of stay cables using magnetorheological dampers under optimal equivalent control algorithm. *J. Sound Vib.* **2019**, *443*, 732–747. [CrossRef]
19. Xu, Y.L.; Zhou, H.J. Damping cable vibration for a cable-stayed bridge using adjustable fluid dampers. *J. Sound Vib.* **2007**, *306*, 349–360. [CrossRef]
20. Shi, X.; Zhu, S. Magnetic negative stiffness dampers. *Smart Mater. Struct.* **2015**, *24*, 072002. [CrossRef]
21. Zhou, P.; Fang, Q. Match of Negative Stiffness and Viscous Damping in A Passive Damper for Cable Vibration Control. Available online: https://www.hindawi.com/journals/sv/2019/3208321/ (accessed on 30 September 2019).

22. Caracoglia, L.; Jones Nicholas, P. Passive hybrid technique for the vibration mitigation of systems of interconnected stays. *J. Sound Vib.* **2007**, *307*, 849–864. [CrossRef]
23. Jiménez-Alonso, J.F.; Sáez, A. Robust optimum design of tuned mass dampers to mitigate pedestrian-induced vibrations using multi-objective genetic algorithms. *Struct. Eng. Int.* **2017**, *27*, 492–501. [CrossRef]
24. Kovacs, I. Zur frage der seilschwingungen und der seildämpfung. *Bautechnik* **1982**, *59*, 325–332.
25. Pacheco, B.M.; Fujino, Y.; Sulekh, A. Estimation curve for modal damping in stay cables with viscous damper. *J. Struct. Eng.* **1993**, *119*, 1961–1979. [CrossRef]
26. Krenk, S. Vibrations of a taut cable with an external damper. *J. Appl. Mech.* **2000**, *67*, 772–776. [CrossRef]
27. Yoneda, M.; Maeda, K. A study on practical estimation method for structural damping of stay cables with dampers. *Doboku Gakkai Ronbunshu* **1989**, *1989*, 455–458. [CrossRef]
28. Ontiveros-Pérez, S.P.; Miguel, L.F.F.; Miguel, L.F.F. Robust Simultaneous Optimization of Friction Damper for the Passive Vibration Control in a Colombian Building. *Procedia Eng.* **2017**, *199*, 1743–1748. [CrossRef]
29. Connor, J.J. *Introduction to Structural Motion Control, MIT-Prentice Hall Series on Civil, Environmental and Systems Engineering*; Prentice Hall: Bergen, NJ, USA, 2003.
30. Naranjo-Pérez, J.; Jiménez-Manfredi, J.; Jiménez-Alonso, J.; Sáez, A. Motion-based design of passive damping devices to mitigate wind-induced vibrations in stay cables. *Vibration* **2018**, *1*, 269–289. [CrossRef]
31. Hao, P.; Wang, B.; Li, G.; Meng, Z.; Wang, L. Hybrid Framework for Reliability-Based Design Optimization of Imperfect Stiffened Shells. *AIAA J.* **2015**, *53*, 2878–2889. [CrossRef]
32. Wang, L.; Wang, X.; Li, Y.; Lin, G.; Qiu, Z. Structural time-dependent reliability assessment of the vibration active control system with unknown-but-bounded uncertainties. *Struct. Control Health Monit.* **2017**, *24*, e1965. [CrossRef]
33. Rathi, A.K.; Chakraborty, A. Reliability-based performance optimization of TMD for vibration control of structures with uncertainty in parameters and excitation. *Struct. Control Health Monit.* **2017**, *24*, e1857. [CrossRef]
34. *EN 1990 Eurocode 0: Basis of Structural Design*; European Committee for Standardization: Brussels, Belgium, 2005.
35. Wang, L.; Liu, J.; Yang, Y. A nonprobabilistic time-variant reliability-based optimization approach to the reliable active controller design of structural vibration considering convex uncertainties. *Struct. Control Health Monit.* **2018**, *25*, e2269. [CrossRef]
36. Binder, K.; Heermann, D. *Monte Carlo Simulation in Statistical Physics: An Introduction; Graduate Texts in Physics*, 5th ed.; Springer-Verlag: Berlin/Heidelberg, Germany, 2010; ISBN 978-3-642-03162-5.
37. Arora, J.S. *Optimization of Structural and Mechanical Systems*; World Scientific Publishing Co. Pte. Ltd.: 5 Toh Tuck Link, Singapore, 2007.
38. Nocedal, J.; Wright, S.J. *Numerical Optimization*; Springer Verlag: New York, NY, USA, 1999.
39. Holickỳ, M. *Reliability Analysis for Structural Design*; African Sun Media: Stellenbosch, South Africa, 2009.
40. Warnitchai, P.; Fujino, Y.; Susumpow, T. A non-linear dynamic model for cables and its application to a cable structure system. *J. Sound Vib.* **1995**, *187*, 695–712. [CrossRef]
41. Caetano, E.; Cunha, A.; Taylor, C.A. Investigation of dynamic cable–deck interaction in a physical model of a cable-stayed bridge. Part I: Modal analysis. *Earthq. Eng. Struct. Dyn.* **2000**, *28*, 481–498. [CrossRef]
42. Johnson, E.A.; Christenson, R.E.; Spencer, B.F., Jr. Semiactive damping of cables with sag. *Comput. Aided Civ. Infrastruct. Eng.* **2003**, *18*, 132–146. [CrossRef]
43. Zhou, Q.; Nielsen, S.R.; Qu, W.L. Semiactive control of three-dimensional vibrations of an inclined sag cable with magnetorheological dampers. *J. Sound Vib.* **2006**, *296*, 1–22. [CrossRef]
44. Mehrabi, A.B.; Tabatabai, H. Unified finite difference formulation for free vibration of cables. *J. Struct. Eng.* **1998**, *124*, 1313–1322. [CrossRef]
45. Hoang, N.; Fujino, Y. Analytical study on bending effects in a stay cable with a damper. *J. Eng. Mech.* **2007**, *133*, 1241–1246. [CrossRef]
46. Yu, Z.; Xu, Y.L. Mitigation of three-dimensional vibration of inclined sag cable using discrete oil dampers—I. Formulation. *J. Sound Vib.* **1998**, *214*, 659–673. [CrossRef]
47. Connor, J.; Laflamme, S. Optimal Passive Damping Distribution. In *Structural Motion Engineering*; Springer: Berlin/Heidelberg, Germany, 2014; pp. 141–197.
48. Xu, Z.-D.; Shen, Y.-P.; Zhao, H.-T. A synthetic optimization analysis method on structures with viscoelastic dampers. *Soil Dyn. Earthq. Eng.* **2003**, *23*, 683–689. [CrossRef]

49. Zhou, H.; Sun, L.; Xing, F. Free vibration of taut cable with a damper and a spring. *Struct. Control Health Monit.* **2014**, *21*, 996–1014. [CrossRef]
50. Seong, J.-Y.; Min, K.-W. An analytical approach for design of a structure equipped with friction dampers. *Procedia Eng.* **2011**, *14*, 1245–1251. [CrossRef]
51. Hong, S. Time Domain Buffeting Analysis of Large-Span Cable-Stayed Bridge. Master's Thesis, FEUP. University of Porto, Porto, Portugal, 2009.
52. Solari, G. Gust-excited vibrations. In *Wind-Excited Vibrations of Structures*; Springer: Berlin/Heidelberg, Germany, 1994; pp. 195–291.
53. Davenport, A.G. The Dependence of Wind Loads on Meteorological Parameter. *Wind Eff. Build. Struct.* **1968**, *1*, 19–82.
54. *EN 1991-1-4 Eurocode 1. Actions on Structures—Part 1–4: General Actions—Wind Actions*; European Committee for Standardization: Brussels, Belgium, 2005.
55. Casas, J.R.; Aparicio, A.C. Monitoring of the Alamillo cable-stayed bridge during construction. *Exp. Mech.* **1998**, *38*, 24–28. [CrossRef]
56. Casas, J.R.; Aparicio, A.C. Rain–wind-induced cable vibrations in the Alamillo cable-stayed bridge (Sevilla, Spain). Assessment and remedial action. *Struct. Infrastruct. Eng.* **2010**, *6*, 549–556. [CrossRef]
57. Ansys Mechanical 19.0. Ansys Inc. 2019. Available online: http://www.ansys.com/ (accessed on 1 October 2019).
58. Simiu, E.; Scanlan, R.H. *Wind Effects on Structures: Fundamentals and Applications to Design*, 3rd ed.; John Willey & Sons, Inc.: New York, NY, USA, 1996.
59. *EN 1993-1-11 Eurocode 3: Design of Steel Structures-Part 1–11: Design of Structures with Tension Components*; European Committee for Standardization: Brussels, Belgium, 2006.
60. Stromquist-LeVoir, G.; McMullen, K.F.; Zaghi, A.E.; Christenson, R. Determining Time Variation of Cable Tension Forces in Suspended Bridges Using Time-Frequency Analysis. *Adv. Civ. Eng.* **2018**, *2018*. [CrossRef]
61. Naranjo-Pérez, J.; Jiménez-Alonso, J.F.; Díaz, I.M.; Sáez, A. Motion-based design of viscous dampers for cable-stayed bridges under uncertainty conditions. In Proceedings of the 5th International Conference on Mechanical Models in Structural Engineering, CMMoST 2019, Alicante, Spain, 23–25 Ocober 2019.
62. Park, S.; Bosch, H.R. *Mitigation of Wind-Induced Vibration of Stay Cables: Numerical Simulations and Evaluations*; Federal Highway Administration: Hampton, VA, USA, 2014.
63. Jurado, D. Simulación Estocástica De Cargas Para Análisis Dinámico De Estructuras En Ingeniería Civil. Master's Thesis, Universidad de Sevilla, Seville, Spain, 2017.
64. Matlab Inc. Matlab R2019b. Available online: http://www.mathworks.com/ (accessed on 15 October 2019).
65. Koh, C.G.; Perry, M.J. *Structural Identification and Damage Detection Using Genetic Algorithms: Structures and Infrastructures Book Series*; CRC Press: Boca Raton, FL, USA, 2017; Volume 6.

applied sciences

MDPI

Article

Shared Tuned Mass Dampers for Mitigation of Seismic Pounding

Rajesh Rupakhety, Said Elias * and Simon Olafsson

Earthquake Engineering Research Centre, Faculty of Civil and Environmental Engineering, School of Engineering and Natural Sciences, University of Iceland, Austurvegur 2a, 800 Selfoss, Iceland; rajesh@hi.is (R.R.); simon@hi.is (S.O.)

* Correspondence: said@hi.is

Received: 17 January 2020; Accepted: 6 March 2020; Published: 11 March 2020

Abstract: This study explores the effectiveness of shared tuned mass damper (STMD) in reducing seismic pounding of adjacent buildings. The dynamics of STMDs is explored through numerical simulations of buildings idealized as single and multiple degree of freedom oscillators. An optimization method proposed in the literature is revisited. It is shown that the optimization results in two different solutions. The first one corresponds to the device being tuned to one of the buildings it is attached to. The second solution corresponds to a very stiff system where the TMD mass hardly moves. This solution, which has been described as an STMD in the literature, is shown to be impractical due to its high stiffness and use of a heavy stationary mass that plays no role in response mitigation but adds unnecessary load to the structure. Furthermore, it is shown that the second solution is equivalent to a viscous coupling of the two buildings. As for the properly tuned solution, i.e., the first solution, sharing the device with an adjacent building was found to provide no added benefits compared to when it is placed on one of the buildings. Based on results from a large set of real earthquake ground motions, it is shown that sharing a TMD mass with an adjacent building, in contrary to what is reported in the literature, is not an effective strategy.

Keywords: adjacent buildings; seismic pounding; tuned mass damper

---

1. Introduction

Seismic pounding refers to the collision of adjacent structures during earthquakes. When closely spaced structures vibrate out of phase, they might pound against each other. Pounding can cause local damage at the contact region. In addition, the impact generates short-duration acceleration pulse that can adversely affect the pounding buildings and their contents (see, for example, Abdel Raheem [1], and references therein). Such effects have been documented during past earthquakes. Miari et al. [2] and Abdel Raheem et al. [3] provide very good reviews of seismic pounding and refer to many studies describing pounding observed during past earthquakes. Various methods of modeling pounding and computation of structural response have been discussed in the literature (see, for example, Dimitrakopoulos, [4]).

Pounding can be avoided if adjacent buildings are adequately separated, but there is large uncertainty in just how much separation is adequate. Architectural and financial factors often dictate separation distance between buildings in metropolitan areas where land is expensive and scarce. Mitigation of seismic pounding is, therefore, an important structural engineering problem. One approach is to reduce the consequences of pounding, while the other is to reduce probabilities of pounding. Collision shear walls have been found to be effective in reducing the consequences of pounding (Anagnostopoulos and Karamaneas, [5]). Elastic gap devices (see, for example, Dicleli, [6]), have been shown to be effective in mitigating damage caused by pounding.

Several solutions for reducing probabilities of pounding between adjacent structures have been proposed in the literature. These solutions range from connecting adjacent buildings with springs and viscous/viscoelastic dashpots (Jankowski and Mahmoud, [7]; Richardson et al., [8,9]; Patel and Jangid, [10,11]; and Tubaldi, [12]). Friction dampers and viscous fluid dampers have also been proposed for mitigation of pounding risk (see, for example, Pratesi et al., [13] and Sorace and Terenzi, [14]).

Active, semi-active, and fuzzy control using magneto-rheological dampers have also been shown to be effective for seismic pounding mitigation (see, for example, Abdeddaim et al., [15]; Uz and Hadi, [16]). Pounding risk may be reduced by controlling the vibrations of adjacent structures. Tuned mass dampers (TMDs) for vibration control have been extensively studied in the literature (see, for example, Cao and Li, [17] or the review by Elias and Matsagar, [18]).

The risk of pounding in base-isolated buildings is higher due to large displacements concentrated at the isolation level (Agagnostopoulos and Spiliopoulos, [19]). Controlling displacement demands on base-isolated structures has been investigated using different strategies in the literature. These strategies include the use of tuned liquid column dampers [20], and tuned mass dampers located at different locations of the host buildings [20–23]. Coupling of base isolation system with tuned mass damper inerter systems is reported in De Domenico and Ricciardi [24], Hashimoto et al. [25] and De Domenico et al. [26]. Other strategies for displacement control of base-isolated buildings include the use of gap dampers [27].

While several studies on active and semi-active dampers for pounding mitigation are available in the literature, very few have tested passive TMDs for this purpose. The only work we are aware of in this regard is that of Abdullah et al. [28]. The idea was extended to semi-active TMD by Kim [29]. Abdullah et al. [28] presented the concept of a shared TMD to reduce vibrations and seismic pounding between adjacent buildings. Their optimal solution is to connect the TMD mass with a spring to one of the buildings and with a dashpot to the other building. If effective, this device could be advantageous due to its simplicity and low cost compared to active control schemes. One of the main appeals of this scheme is that the mass of the shared TMD is only half the mass of two TMDs installed on individual buildings.

Abdullah et al. [28] reported that the device is very effective in controlling vibrations and pounding between adjacent buildings. Upon closer inspection, we find that the solution presented by Abdullah et al. [28] is neither a tuned mass damper, nor optimal in mitigating pounding, and needs revisiting. The main objective of this paper is to investigate the dynamics of adjacent buildings connected by a tuned mass damper. We use single degree of freedom (SDOF) and multiple degree of freedom (MDOF) representations of buildings and several earthquake ground motions for numerical simulations to investigate the parameters of TMDs for optimal control of structural displacements and pounding.

2. Conceptual and Mathematical Model of Shared TMDs (STMD)

For conceptual convenience, let us consider two single degree of freedom (SDOF) systems connected by a shared TMD as shown in Figure 1. The SDOFs are simplified representations of buildings A and B, and their masses, stiffnesses, and damping coefficients are denoted as m_A, m_B; k_A, k_B; and c_A, c_B, respectively. The mass, stiffness and damping coefficient of the tuned mass damper are denoted by m, k, and c, respectively. An alternate scheme is to connect the TMD mass to building A with a dashpot and to building B with a spring.

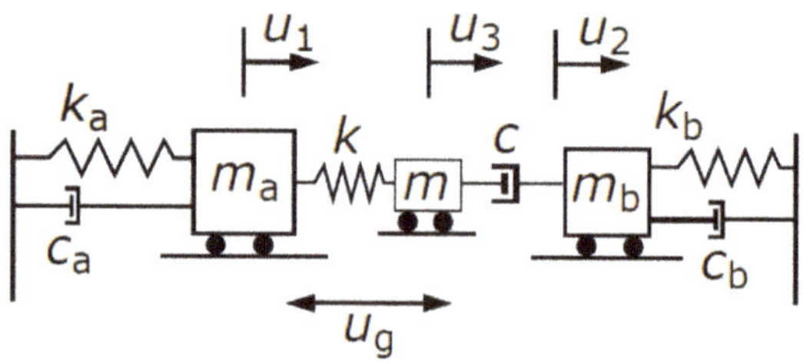

Figure 1. Schematic representation of a tuned mass damper (TMD) shared by two buildings modeled as single degree of freedom (SDOF) systems.

The system mass, stiffness, and damping matrices are given below.

$$\mathbf{K} = \begin{bmatrix} k_A + k & 0 & -k \\ 0 & k_B & 0 \\ -k & 0 & k \end{bmatrix} \tag{1}$$

$$\mathbf{M} = \begin{bmatrix} m_A & 0 & 0 \\ 0 & m_B & 0 \\ 0 & 0 & m \end{bmatrix} \tag{2}$$

$$\mathbf{C} = \begin{bmatrix} c_A & 0 & 0 \\ 0 & c_B + c & -c \\ 0 & -c & c \end{bmatrix} \tag{3}$$

and the state matrix is defined as

$$\mathbf{S} = \begin{bmatrix} 0 & \mathbf{I} \\ -\mathbf{M}^{-1}\mathbf{K} & -\mathbf{M}^{-1}\mathbf{C} \end{bmatrix} \tag{4}$$

where $\mathbf{I}$ is an identity matrix. The performance (or cost) function, as defined by Abdullah et al. (2001) is the trace of a matrix $\mathbf{L}$, which satisfies the following Lyapunov equation.

$$\mathbf{LS} + \mathbf{S}^T\mathbf{L} + \mathbf{Q} = 0, \tag{5}$$

where $\mathbf{Q}$ is a weighting matrix that assigns different importance to the displacements (and/or velocities) at the different degrees of freedom of the system. The design problem is then to estimate k and c that minimize the trace of $\mathbf{L}$, denoted hereafter by J. The weighting matrix is taken as $\mathbf{Q}_{11} = 1$ and all other elements as 0.

3. Numerical Study

We present several numerical studies, starting with SDOF systems subjected to some example ground motions. MDOF systems and a large set of ground motions are considered next. The section starts with an example of an optimization problem to gain insight into the properties of STMDs.

3.1. An Optimization Example

An example of the optimization problem is presented here. The properties of buildings A and B are fixed for numerical simulations. Their frequencies are taken as 6.04 and 7.58 rad/s, and the damping ratio is 5% for both. The frequencies are the same as those of the two buildings studied by Abdullah et al. [28]. The mass of the two buildings are equal, and the mass of the TMD is expressed as a fraction of the mass of building A, i.e., $m = \rho_A m_A$. The frequency of the TMD is denoted by $\omega = \sqrt{\frac{k}{m}}$, and the frequency ratio is defined as $f = \frac{\omega}{\omega_A}$. The damping ratio of the TMD, expressed in terms of the mass and frequency of the TMD, is denoted by ζ. The variables of interest are therefore f and ζ.

Figure 2 shows the variation of the cost function with the frequency ratio and damping ratio of the TMD. The cost function has two minima, which correspond to two solutions. There is a local minimum near a frequency ratio of 1. This minimum is more prominent for lower damping ratios. The global minima occur at much larger frequency ratios, ranging from about 10 to about 90 for damping ratios from 0.4 to 0.05. It is interesting to note that the amplitudes of the global minima are not very sensitive to damping ratios of the TMD.

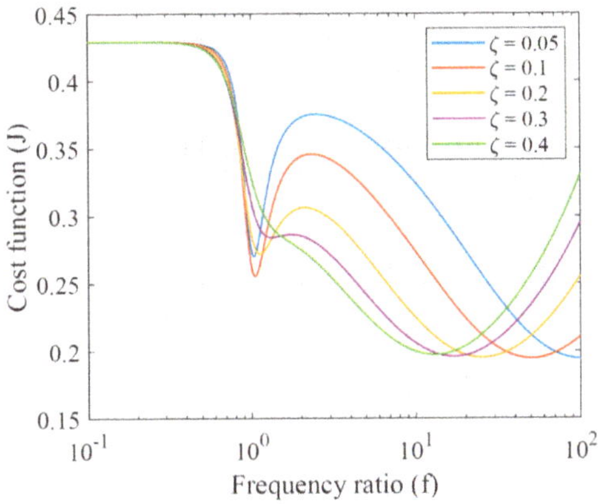

Figure 2. Variation of the cost function with frequency ratio and damping ratio of the TMD. Themass of the TMD is taken as 3% of the mass of building A in this example.

It is then interesting to investigate whether the solution that minimizes J corresponds to minimum displacements of the two buildings and minimizes the separation distance required to prevent pounding. This distance is the largest negative relative displacement between the two buildings and is hereafter referred to as the pounding distance. To investigate this, we use the north-south component of the El Centro ground motion from the 1940 Imperial Valley Earthquake and simulate the response of the two buildings and STMD with varying frequency and damping ratios.

The results are shown in Figure 3. The maximum displacement of the two buildings are shown on the top panel. The maximum stroke of the TMD is shown in the bottom left panel. Here, stroke is defined as the relative displacement of the TMD mass from building A. Pounding distance is shown in the bottom right panel. For comparison, the response of uncontrolled buildings is shown with black dashed lines. Another scenario considered is where an optimally tuned (based on Sadek, [30]) TMD is placed on building A whereas building B is uncontrolled. In this case, the frequency tuning ratio is 0.96 and the damping ratio is 0.22. The results corresponding to this are shown with dashed blue lines. The results show that the local minima, similar to those observed in Figure 2, occurs in the displacement response of building A when the TMD is tuned to the frequency of the building. However, the combination of f and ζ that minimizes J does not necessarily minimize the displacement of building A. For example, for f approximately equal to 1, a damping ratio of 10% is required to minimize J, whereas 5% damping ratio minimizes the peak displacement of building A. This can be explained by the larger stroke of the device when the damping ratio is low, as is clearly seen in the bottom right panel of Figure 3. The optimal solution near $f \approx 1$ is not effective in controlling the peak response of building B.

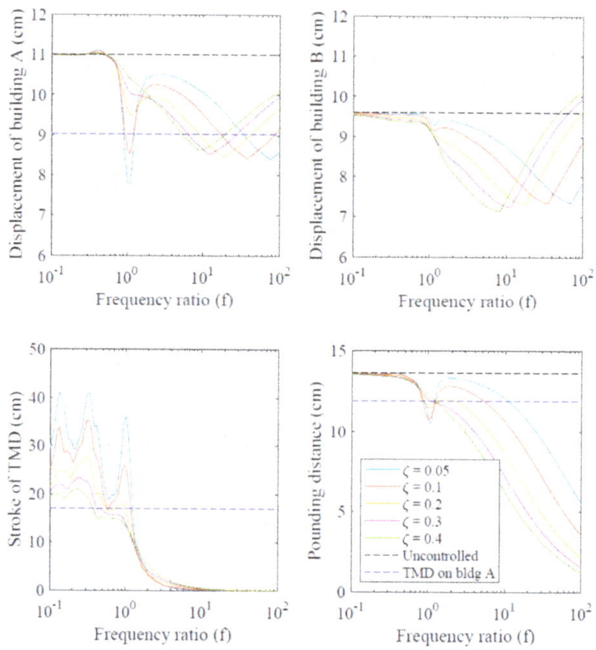

Figure 3. Peak displacement of building A, displacement of building B, stroke of the TMD, and minimum separation.

It is also interesting to note that the locally optimal solution ($f \approx 1$) of the shared TMD is more effective in controlling peak response of building A, and with a much smaller damping ratio than when the TMD is placed only on building A. This indicates that sharing a TMD with an adjacent building can increase its effectiveness without adversely affecting the response of the adjacent building. This, however, comes at a cost of increased stroke of the device, which is due to low damping values of the locally optimal STMD, as is evident from the bottom left panel of Figure 3. The stroke of the locally optimal STMD is much larger than the stroke of the TMD placed on building A.

As the frequency ratio is increased, the displacement response of both the buildings show a decreasing trend followed by an increasing trend. This trend is also visible in the cost function shown in Figure 2. If we consider the minima (based on the cost function) at a higher frequency ratio as the global minima, the optimal frequency ratio decreases with increasing damping, and the optimal solutions are more effective than TMD placed on building A alone. However, these optimal solutions correspond to very stiff springs in the TMD. For a damping ratio as large as 40%, the optimal frequency ratio is about 10. With a mass ratio of 3%, the stiffness of the TMD is about 3 times the stiffness of building A. From a practical point of view, it might not feasible to construct a TMD which is so much stiffer than the building it rests on. Such a solution, although not realistic, will be hereafter referred to as a globally optimal solution referring to the cost function.

At the locally optimal solution, the shared TMD reduces the minimum separation distance to some extent. As the frequency ratio and damping ratio are increased, the pounding distance decreases further. With a very stiff TMD and sufficiently high damping ratio, it appears the pounding distance can be reduced to a large extent. While this seems like a plausible mitigation strategy against pounding, it is deficient in both conceptual and practical terms. In practical terms, having a TMD 3 times stiffer than the building itself is not attractive. The conceptual deficiency is that such a stiff device cannot be considered as a TMD. This can be better understood by inspecting the results shown in the bottom left panel of Figure 3. Irrespective of the damping ratio, as the frequency ratio increases above 1, the stroke of the TMD reduces rapidly, and for a frequency ratio of 10, the device does not move relative to building A. Therefore, the device is not acting like a TMD and any reduction in response is due to the coupling of the two buildings with a viscous damping device. It is also interesting to note that unlike the cost function which increases after the global minima, the pounding distance decreases with increasing frequency ratio.

3.2. Equivalence of STMD and Viscous Coupling of the Two Buildings

Based on the discussion above, it is apparent that a very stiff TMD is equivalent, in terms of structural response reduction, to a viscous dashpot coupling the two buildings. To illustrate this point better, we modify the system shown in Figure 1 by lumping the TMD mass on building A, and connecting the two buildings by the same viscous dashpot as that used in the STMD. The resulting system has only two degrees of freedom with masses ($m_a + m$) and m_b connected by a dashpot c. The mass ($m_a + m$) is connected to the ground by spring and dashpot k_a and c_a, while m_b is connected to the ground by spring and dashpot k_b and c_b. The peak response of this modified system is divided by the peak response of the system with STMD. The ratio of peak displacement of building A is presented, for the El Centro ground motion, as a function of frequency ratio and damping ratio in Figure 4. When the shared TMD is near resonance with building A, it is more effective than viscous coupling alone. However, as the frequency ratio is increased, the two systems approach each other, and for a frequency ratio greater than about 20, there is no difference between the two systems. This helps us conclude that STMD optimized on the global minima of the cost function is not a tuned mass damper, as the mass is neither tuned to the structure, nor does it play any part in reducing structural response. In this sense, it is illogical to use additional TMD mass and a very stiff spring element, when the same effect can be obtained by just connecting the two buildings with a viscous dashpot.

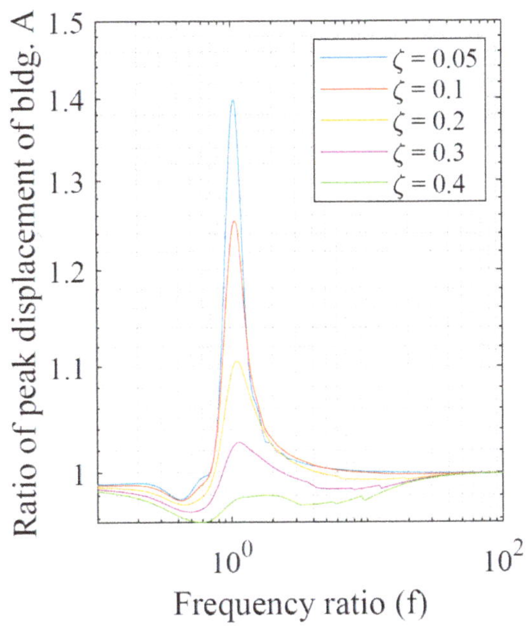

Figure 4. The ratio of maximum displacement of building A when the two buildings are connected by a viscous dashpot and an STMD.

3.3. Examination with MDOF Systems

Here we examine the results and conclusions obtained for two SDOF buildings with MDOF models of two buildings. The selected buildings are the same as those used by Abdullah et al. [28]. Building A has a mass of 3.1752 × 10^5 kg, and stiffness of 3.404 × 10^8 N/m on each floor. Building B has a mass of 3.6287 × 10^5 kg and stiffness of 6.127 × 10^8 on each floor. Both buildings have 8 floors with a damping coefficient of 100 Ns/m on each floor. Fundamental frequencies of buildings A and B are 6.04, and 7.58 rad/s, respectively. The mass of the TMD is 54,430 kg. Abdullah et al. [28] present two cases as optimal solutions. Case 1 corresponds to attaching the TMD mass to building A by a dashpot and building B by a spring. Case 2 corresponds to a spring connection with building A and dashpot connection to building B. Case 2 was found to be more effective and is considered here for further analysis. Abdullah et al. [28] arrived at this solution by specifying weights equal to 1 for the displacements in the top four floors of building A and 0 for all other displacements and velocities in defining the **Q** matrix

The cost function for this example is presented in Figure 5. The frequency ratio, for this system, is defined as the frequency of the STMD divided by the fundamental frequency of building A. Like the cost function for SDOF buildings shown in Figure 2, the cost function of these MDOF buildings sharing a TMD displays two minima for each damping ratio. The first minima, which is larger than the second one, occurs near the resonance of STMD mass with the first mode of building A. The second minimum occurs at a much larger frequency ratio. The optimal stiffness presented by Abdullah et al. (2001) is 1.786 × 10^8 N/m, which is about 50% of the stiffness of each of the floors of building A. Based on this stiffness, the frequency of STMD is 57.28 rad/s, which corresponds to a frequency ratio of 9.48. The optimal damping coefficient reported by Abdullah et al. [28] is 1.809E6, which, for this frequency ratio corresponds to a damping ratio of 0.29 for the STMD. It is then clear that the optimal solution reported in Abdullah et al. [28] corresponds to the global minima of the cost function (see purple line in Figure 4).

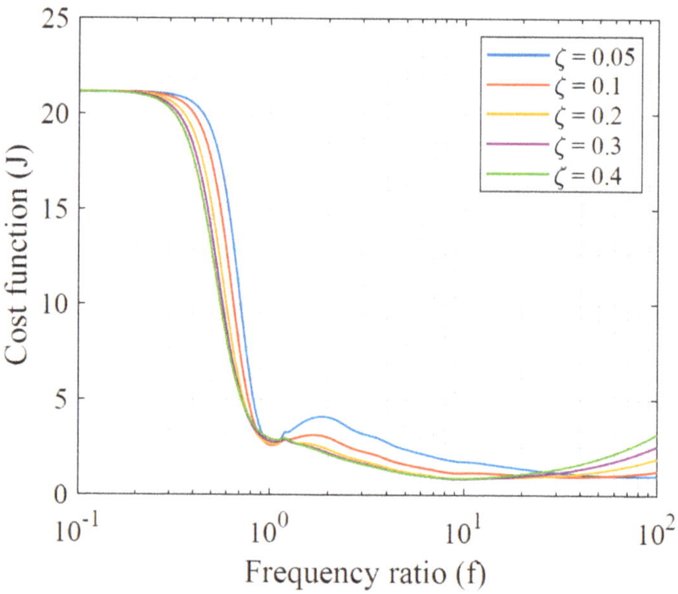

Figure 5. The cost function for the two eight-storey buildings presented in Abdullah et al. [28] as a function of frequency and damping ratios.

Variation of the peak displacements of buildings A and B with frequency ratio and damping ratio is shown in Figure 6. The solution considered optimal by Abdullah et al. [28] is indicated in the figure with red dots. The response of building A, to which the STMD mass is connected by a spring, is optimally controlled when the frequency ratio is close to 1, and the damping ratio is low. This solution, however, is not effective in controlling the response of building B. The optimal solution, based on the global minima of J, as denoted by red dots, is effective in controlling pounding as well as reducing the response of both the buildings. However, the solution uses a very stiff spring, and the TMD mass does not move relative to building A.

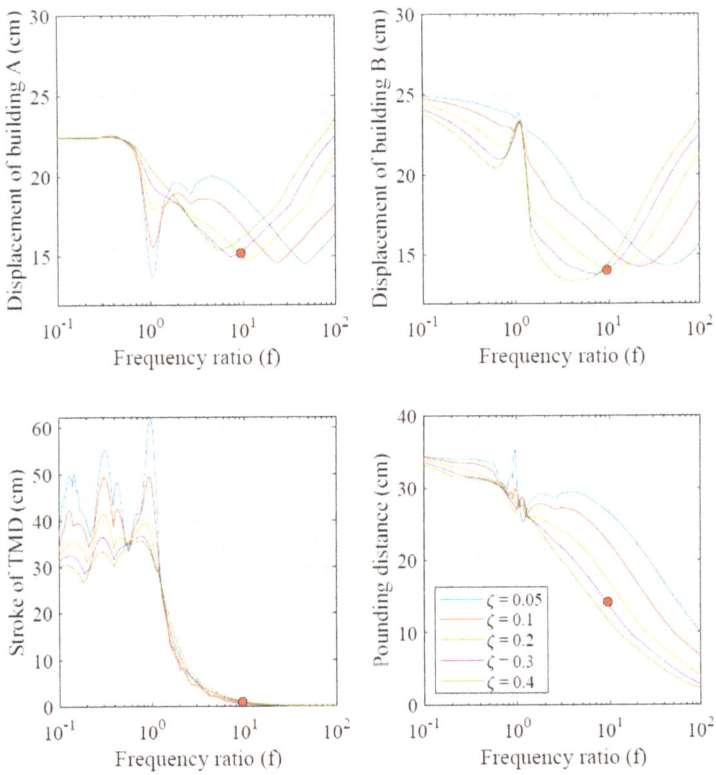

Figure 6. Peak roof displacements, device stroke, and the minimum separation distance between two buildings connected by an STMD. The optimal solution presented in Abdullah [28] is shown with red dots. The results correspond to El Centro ground motion.

This can be confirmed from the low stroke of the device as shown in Figure 6. The STMD mass, in the solution presented by Abdullah et al. [28] is essentially motionless and does not function as intended. The solution, therefore, is not a tuned mass damper as its frequency is much larger than the frequency of the buildings it is attached to. In fact, it can be verified that the same response reduction can be achieved by coupling the two buildings with a dashpot. If we consider the optimal solution as the one near resonance of STMD with building A, the STMD is not very effective in controlling pounding between the buildings.

3.4. Effectiveness of STMDs

Based on the results and arguments presented above, the stiff solution cannot be considered as a shared TMD. However, we can define a shared TMD in such a way that it is tuned to the frequency of building A, for example. This solution would correspond to the frequency ratio in the vicinity of 1. An optimal STMD in this context would correspond to the one that minimizes the cost function J in the vicinity of resonance with building A. The example presented above shows that this STMD marginally reduces the pounding distance for the El Centro ground motion. To test if this observation can be generalized to different ground motions and quantify the effectiveness in reducing pounding distance, we use several ground motion records for dynamic analysis of the system. The ground motion records are taken from the European Strong Motion Database [31]. The database contains 462 ground motion records from 110 earthquakes recorded in Europe and the Middle East. The ground motions are obtained from earthquakes with a moment magnitude of 4.2 to 7.6. The range of epicentral distance is 1 to 558 km, and that of peak ground acceleration (PGA) is 0.9% to 91% of acceleration due to gravity. The strong motion duration, defined as the interval covering 5–95% of Arias Intensity is in the range of 1.2 s to 139 s. Response spectra and duration of these ground motions have been presented in Rupakhety and Sigbjörnsson [32,33]. More details about this database can be found in Ambraseys et al. [34].

Optimal parameters of the STMD are estimated by minimizing the cost function using a genetic algorithm. Population size is taken as 50 and the initial population follows uniform distribution in the specified bounds. Elite count is set to 5% of the population. Crossover fraction is set to 0.8 and migration fraction is set to 0.2. The selection criterion is stochastic uniform and mutation are Gaussian. The 'ga' function in Matlab™ with default options was used to implement the algorithm. Different scenarios are simulated: (i) both the buildings are uncontrolled; (ii) TMD is used in building A only, the parameters of the TMD are optimized based on Sadek's (1997) equations; (iii) the STMD is connected with a spring to building A, and a dashpot to building B. These cases are denoted hereafter as UC (uncontrolled), AC (building A controlled), and SC (controlled with STMD), respectively. In all cases, the TMD mass is taken as 1% of the mass of building A.

Peak roof displacement of building A and pounding distance corresponding to the 462 earthquake ground motions are presented in Figure 7. The response of the uncontrolled and controlled structures are plotted on the abscissa and ordinate, respectively. The diagonal lines in the figure correspond to the equal response of uncontrolled and controlled structures. The results show that the TMDs result in reduction of peak roof displacement of building A in almost all the ground motions. When the displacement of the uncontrolled structure is small, the response reduction obtained by the TMD is not very significant. However, as the ground motions get more demanding, the effectiveness of the TMDs is higher. For the most severe case (that producing the highest response of the uncontrolled structure), the TMD reduces the response by about 50%. However, it is interesting to note that when the TMD is shared with building B, its effectiveness in controlling the response of building A remains almost unchanged for all the ground motions considered in this study. This shows that sharing the TMD with building B does not reduce its effectiveness. The shared TMD is not effective in controlling the response of building B. Neither does it provide any advantage in reducing the pounding distance between the two buildings. The optimal STMD, therefore, does not provide any clear advantage over the TMD installed on building A alone.

The mean (over 462 ground motions) ratios between the pounding distance of the AC controlled and SC controlled structures to that of the uncontrolled structure are 0.79 and 0.76, respectively. Controlling building A alone is, on average, about 3% less effective than using an STMD. The stiffnesses of AC and SC schemes are 8.96×10^5 N/m and 9.68×10^5 N/m, respectively. Their damping coefficients are 4.73×10^4 kg/s and 2.27×10^4 kg/s. This shows that the SC scheme results in similar performance like the AC scheme, but with a much smaller dashpot. This might be, in theory, an advantage of sharing the device. However, there might be practical difficulties in properly connecting the dashpot between the two buildings, and suitability of the scheme needs to be judged on a case by case basis.

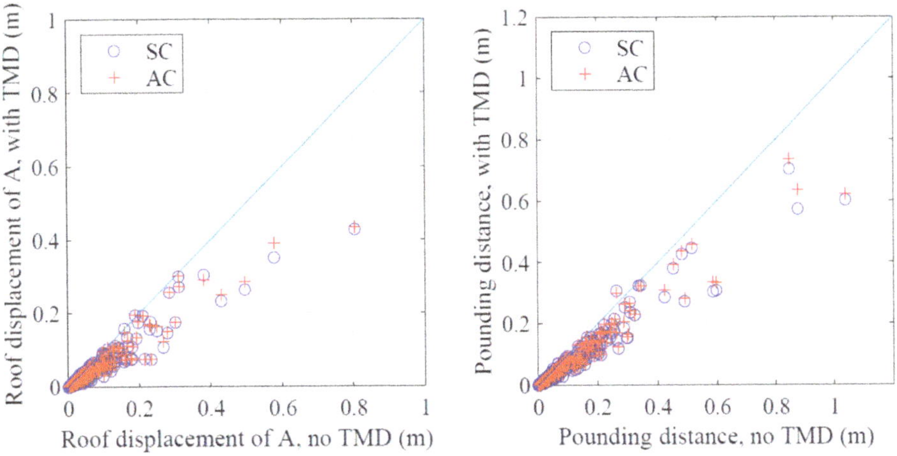

Figure 7. Roof displacement of building A (**left**) and minimum separation distance to avoid pounding (**right**) when subjected to 462 earthquake ground motions. The abscissa and ordinate correspond to uncontrolled and controlled structures, respectively.

To illustrate the effectiveness of AC and SC schemes in reducing the minimum separation distance between the two buildings, reduction ratio, which is defined as minimum separation distance of the controlled building divided by that of the uncontrolled building is shown in Figure 8. The smaller this ratio, the more effective is the control scheme. Effectiveness is investigated against different ground motion and structural response parameters. Only those ground motions with PGA larger than 0.2 g are used in the figure. Here, g represents acceleration due to gravity. Ground motions weaker than this are not considered as they were found to cause a low risk of pounding. Figure 8c shows two different clusters of data, one below ~25 s, and the other to its right. The cluster to the right has very few data points, and any apparent trend in this cluster is not reliable. When duration is less than 25 s, the Pearson correlation coefficient between reduction ratio and duration is −0.44 with a p-value of 0.0011, which shows that the correlation is statistically significant, and not by random chance. Figure 8b shows that the effectiveness of the TMDs is independent of PGA. There seems to be a weak negative correlation between the pounding distance of the uncontrolled structure and the effectiveness of TMD schemes. The Pearson correlation coefficient was found to be −0.17 with a large p-value of 0.21, which means that any apparent trend in the data is statistically insignificant. The same observation can be made in Figure 8d, which shows that the effectiveness of TMD schemes is independent of spectral displacement at the fundamental period of the uncontrolled structure. In all these cases, there is no significant difference between the SC and AC schemes, which means that sharing the TMD does not seem to provide additional benefit in preventing pounding.

We next compare the effectiveness of the SC scheme to that of the AC scheme for different mass ratios of the TMD. Mass of the TMD is taken as 1%, 2%, and 3% of the total mass of building A. The optimal solutions for these mass ratios were found with frequency ratios of 1.02, 1.04, and 1.04 and damping ratios of 7%, 11%, and 13%. The effectiveness of the devices is quantified by the mean ratios of pounding distance of SC and AC schemes to that of uncontrolled structure, denoted by r_{SC} and r_{AC} respectively. A value of this ratio less than 1 means the control systems, on average, reduce pounding distance. These ratios are shown in Figure 9. As the mass ratio increases, the performance of SC is better than that of AC. However, even for a mass ratio of 3%, which corresponds to 24% of floor mass, the SC scheme is only about 6% better than the AC scheme. We repeated the analysis with different frequencies of the two buildings and arrived at similar results.

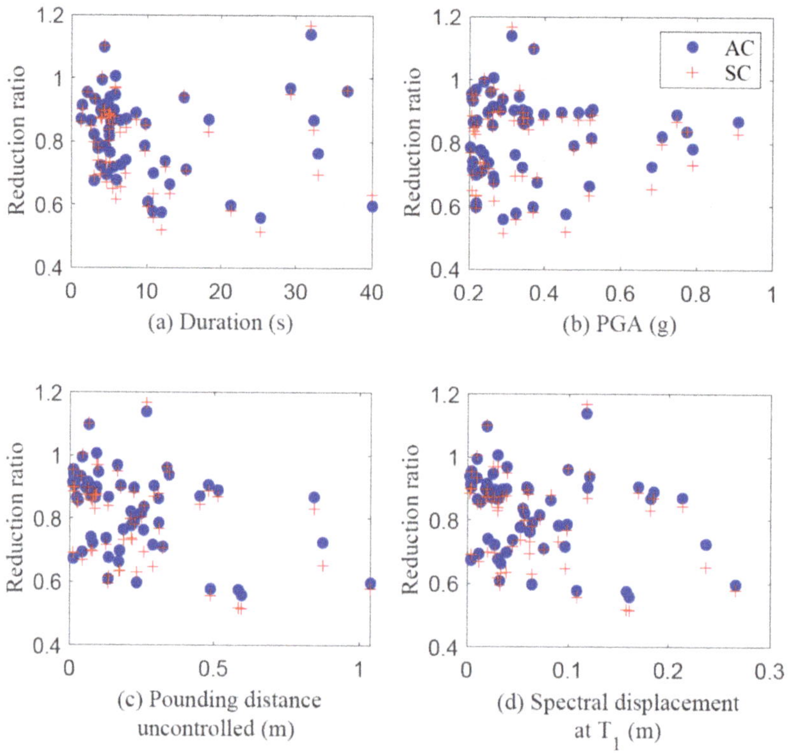

Figure 8. Effectiveness of tuned mass dampers in reducing the minimum separation distance to prevent pounding. The variation of effectiveness with different ground motion parameters is presented for those ground motion whose PGA is larger than 20% of acceleration due to gravity.

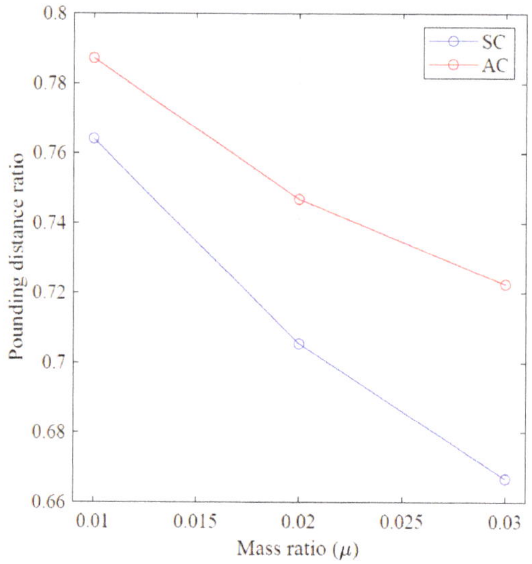

Figure 9. Ratio of pounding distance of the two buildings with controlled with STMD (SC) and building A controlled (AC) schemes to that of uncontrolled buildings.

4. Conclusions

This study revisits the idea of shared TMD for mitigation of seismic pounding between adjacent buildings. Dynamics of an optimal solution reported to be effective in Abdullah et al. [28] is investigated in detail to arrive at the following conclusions.

1. The solution proposed in the literature does not act like a tuned mass damper. The mass of the device is not tuned to the structure it is connected to; it is very stiff and as a result, the device does not move. Any reduction in response resulting from such devices is due to the viscous coupling of the two buildings rather than the tuned vibration of the STMD mass. Our results show that a similar level of effectiveness can be achieved without the TMD mass, simply by coupling the two buildings by a viscous dashpot. This eliminates the need for extra costs required to support the additional mass of the TMD.
2. The cost function for optimization of STMDs displays two minima, one near resonance of the device with one of the buildings, and the other at a frequency much larger than that of the buildings. The solution near the resonance is a tuned mass damper, while the other one is equivalent to the viscous coupling of two buildings.
3. For a large set of 462 ground motions, an optimal TMD placed on one of the buildings alone was found to be almost as effective as an STMD tuned to the building both in controlling roof displacement of the building and pounding distance with an adjacent building. The trouble of sharing the TMD with another building is therefore not worthwhile.

Author Contributions: All authors discussed and agreed upon the idea, and made scientific contributions. All authors have read and agreed to the published version of the manuscript.

Funding: This research was funded by University of Iceland Research Fund.

Acknowledgments: We acknowledge constructive and insightful comments and discussions from two anonymous reviewers and the editors, which helped improve the paper. The authors acknowledge support from the University of Iceland Research Fund.

Conflicts of Interest: The authors declare no conflict of interest.

References

1. Abdel Raheem, S.E. Mitigation measures for earthquake induced pounding effects on seismic performance of adjacent buildings. *Bull. Earthq. Eng.* **2014**, *12*, 1705–1724. [CrossRef]
2. Miari, M.; Choong, K.K.; Jankowski, R. Seismic pounding between adjacent buildings: Identification of parameters, soil interaction issues and mitigation measures. *Soil Dyn. Earthq. Eng.* **2019**, *121*, 135–150. [CrossRef]
3. Abdel Raheem, S.E.; Fooly, M.Y.; Shafy, A.G.; Taha, A.M.; Abbas, Y.A.; Latif, M.M. Numerical simulation of potential seismic pounding among adjacent buildings in series. *Bull. Earthq. Eng.* **2019**, *17*, 439–471. [CrossRef]
4. Dimitrakopoulos, E.; Makris, N.; Kappos, A.J. Dimensional analysis of the earthquake-induced pounding between adjacent structures. *Earthq. Eng. Struct. Dyn.* **2009**, *38*, 867–886. [CrossRef]
5. Anagnostopoulos, S.A.; Karamaneas, C.E. Use of collision shear walls to minimize seismic separation and to protect adjacent buildings from collapse due to earthquake-induced pounding. *Earthq. Eng. Struct. Dyn.* **2008**, *37*, 1371–1388. [CrossRef]
6. Dicleli, M. Performance of seismic-isolated bridges with and without elastic-gap devices in near-fault zones. *Earthq. Eng. Struct. Dyn.* **2008**, *37*, 935–954. [CrossRef]
7. Jankowski, R.; Mahmoud, S. Linking of adjacent three-storey buildings for mitigation of structural pounding during earthquakes. *Bull. Earthq. Eng.* **2016**, *14*, 3075–3097. [CrossRef]
8. Richardson, A.; Walsh, K.K.; Abdullah, M.M. Closed-form equations for coupling linear structures using stiffness and damping elements. *Struct. Control Health Monit.* **2013**, *20*, 259–281. [CrossRef]
9. Richardson, A.; Walsh, K.K.; Abdullah, M.M. Closed-form design equations for controlling vibrations in connected structures. *J. Earthq. Eng.* **2013**, *17*, 699–719. [CrossRef]
10. Patel, C.C.; Jangid, R.S. Seismic response of dynamically similar adjacent structures connected with viscous dampers. *IES J. Part A Civ. Struct. Eng.* **2010**, *3*, 1–3. [CrossRef]

11. Patel, C.C.; Jangid, R.S. Dynamic response of identical adjacent structures connected by viscous damper. *Struct. Control Health Monit.* **2014**, *21*, 205–224. [CrossRef]
12. Tubaldi, E.; Barbato, M.; Ghazizadeh, S. A probabilistic performance-based risk assessment approach for seismic pounding with efficient application to linear systems. *Struct. Saf.* **2012**, *36*, 14–22. [CrossRef]
13. Pratesi, F.; Sorace, S.; Terenzi, G. Analysis and mitigation of seismic pounding of a slender R/C bell tower. *Eng. Struct.* **2014**, *71*, 23–34. [CrossRef]
14. Sorace, S.; Terenzi, G. Damped interconnection-based mitigation of seismic pounding between adjacent R/C buildings. *Int. J. Eng. Technol.* **2013**, *5*, 406. [CrossRef]
15. Abdeddaim, M.; Ounis, A.; Djedoui, N.; Shrimali, M. Reduction of pounding between buildings using fuzzy controller. *Asian J. Civ. Eng.* **2016**, *17*, 985–1005.
16. Uz, M.E.; Hadi, M.N. Optimal design of semi active control for adjacent buildings connected by MR damper based on integrated fuzzy logic and multi-objective genetic algorithm. *Eng. Struct.* **2014**, *69*, 135–148. [CrossRef]
17. Cao, L.; Li, C. Tuned tandem mass dampers-Inerters with broadband high effectiveness for structures under white noise base excitations. *Struct. Control Health Monit.* **2019**, *26*, e2319. [CrossRef]
18. Elias, S.; Matsagar, V. Research developments in vibration control of structures using passive tuned mass dampers. *Annu. Rev. Control* **2017**, *44*, 129–156. [CrossRef]
19. Anagnostopoulos, S.A.; Spiliopoulos, K.V. An investigation of earthquake induced pounding between adjacent buildings. *Earthq. Eng. Struct. Dyn.* **1992**, *21*, 289–302. [CrossRef]
20. De Domenico, D.; Ricciardi, G. Earthquake-resilient design of base isolated buildings with TMD at basement: Application to a case study. *Soil Dyn. Earthq. Eng.* **2018**, *113*, 503–521. [CrossRef]
21. Stanikzai, M.H.; Elias, S.; Matsagar, V.A.; Jain, A.K. Seismic response control of base-isolated buildings using tuned mass damper. *Aust. J. Struct. Eng.* **2019**, 1–2. [CrossRef]
22. Stanikzai, M.H.; Elias, S.; Matsagar, V.A.; Jain, A.K. Seismic Response Control of Base-Isolated Buildings Using Multiple Tuned Mass Dampers. *Struct. Des. Tall Spec. Build* **2019**, *28*, e1576. [CrossRef]
23. Stanikzai, M.H.; Elias, S.; Rupakhety, R. Seismic Response Mitigation of Base-Isolated Buildings. *Appl. Sci.* **2020**, *10*, 1230. [CrossRef]
24. De Domenico, D.; Ricciardi, G. An enhanced base isolation system equipped with optimal tuned mass damper inerter (TMDI). *Earthq. Eng. Struct. Dyn.* **2018**, *47*, 1169–1192. [CrossRef]
25. Hashimoto, T.; Fujita, K.; Tsuji, M.; Takewaki, I. Innovative base-isolated building with large mass-ratio TMD at basement for greater earthquake resilience. *Future Cities Environ.* **2015**, *1*, 9. [CrossRef]
26. De Domenico, D.; Ricciardi, G.; Zhang, R. Optimal design and seismic performance of tuned fluid inerter applied to structures with friction pendulum isolators. *Soil Dyn. Earthq. Eng.* **2020**, *132*, 106099. [CrossRef]
27. De Domenico, D.; Gandelli, E.; Quaglini, V. Effective base isolation combining low-friction curved surface sliders and hysteretic gap dampers. *Soil Dyn. Earthq. Eng.* **2020**, *130*, 105989. [CrossRef]
28. Abdullah, M.M.; Hanif, J.H.; Richardson, A.; Sobanjo, J. Use of a shared tuned mass damper (STMD) to reduce vibration and pounding in adjacent structures. *Earthq. Eng. Struct. Dyn.* **2001**, *30*, 1185–1201. [CrossRef]
29. Kim, H.S. Seismic response control of adjacent buildings coupled by semi-active shared TMD. *Int. J. Steel Struct.* **2016**, *16*, 647–656. [CrossRef]
30. Sadek, F.; Mohraz, B.; Taylor, A.W.; Chung, R.M. A method of estimating the parameters of tuned mass dampers for seismic applications. *Earthq. Eng. Struct. Dyn.* **1997**, *26*, 617–635. [CrossRef]
31. Ambraseys, N.N.; Smit, P.; Douglas, J.; Margaris, B.; Sigbjörnsson, R.; Olafsson, S.; Suhadolc, P.; Costa, G. Internet site for European strong-motion data. *Boll. Di Geofis. Teor. Ed Appl.* **2004**, *45*, 113–129.
32. Rupakhety, R.; Sigbjörnsson, R. Rotation-invariant measures of earthquake response spectra. *Bull. Earthq. Eng.* **2013**, *11*, 1885–1893. [CrossRef]
33. Rupakhety, R.; Sigbjörnsson, R. Rotation-invariant mean duration of strong ground motion. *Bull. Earthq. Eng.* **2014**, *12*, 573–584. [CrossRef]
34. Ambraseys, N.N.; Douglas, J.; Sigbjörnsson, R.; Berge-Thierry, C.; Suhadolc, P.; Costa, G.; Smit, P.M. Dissemination of European strong-motion data, Volume 2. In Proceedings of the 13th World Conference on Earthquake Engineering, Vancouver, CO, Canada, 1–6 August 2004.

applied sciences

MDPI

Article

Distributed Passive Actuation Schemes for Seismic Protection of Multibuilding Systems

Francisco Palacios-Quiñonero [1,*], Josep Rubió-Massegú [1], Josep M. Rossell [1] and Hamid Reza Karimi [2]

1 Department of Mathematics, Universitat Politècnica de Catalunya, EPSEM, Av. Bases de Manresa 61-73, 08242 Manresa, Spain; josep.rubio@upc.edu (J.R.-M.); josep.maria.rossell@upc.edu (J.M.R.)
2 Department of Mechanical Engineering, Politecnico di Milano, via La Masa 1, 20156 Milan, Italy; hamidreza.karimi@polimi.it
* Correspondence: francisco.palacios@upc.edu; Tel.: +34-938-777-302

Received: 25 February 2020; Accepted: 26 March 2020; Published: 31 March 2020

Abstract: In this paper, we investigate the design of distributed damping systems (DDSs) for the overall seismic protection of multiple adjacent buildings. The considered DDSs contain interstory dampers implemented inside the buildings and also interbuilding damping links. The design objectives include mitigating the buildings seismic response by reducing the interstory-drift and story-acceleration peak-values and producing small interbuilding approachings to decrease the risk of interbuilding collisions. Designing high-performance DDS configurations requires determining convenient damper positions and computing proper values for the damper parameters. That allocation-tuning optimization problem can pose serious computational difficulties for large-scale multibuilding systems. The design methodology proposed in this work—(i) is based on an effective matrix formulation of the damped multibuilding system; (ii) follows an H_∞ approach to define an objective function with fast-evaluation characteristics; (iii) exploits the computational advantages of the current state-of-the-art genetic algorithm solvers, including the usage of hybrid discrete-continuous optimization and parallel computing; and (iv) allows setting actuation schemes of particular interest such as full-linked configurations or nonactuated buildings. To illustrate the main features of the presented methodology, we consider a system of five adjacent multistory buildings and design three full-linked DDS configurations with a different number of actuated buildings. The obtained results confirm the flexibility and effectiveness of the proposed design approach and demonstrate the high-performance characteristics of the devised DDS configurations.

Keywords: energy-dissipation systems; distributed damping systems; optimal placement; multibuilding systems; seismic protection; hybrid genetic algorithm; parallel computing; pounding protection

1. Introduction

Over the last few years, an increasing research effort has been invested in the analysis, design and implementation of distributed damping systems (DDSs) for seismic protection of buildings and civil structures [1,2]. That kind of passive energy-dissipation systems is formed by a set of damping elements installed at suitable locations of the structure. DDSs are simple, reliable and robust and, when properly designed, are able to produce a remarkable reduction of the overall seismic vibrational response [3–5]. Broadly speaking, three main issues related to the damping elements have to be addressed in DDS design—(i) *technical setup determination*, (ii) *allocation* and (iii) *tuning*. Determination of the technical setup is a preliminary step, which is strongly conditioned by the particular problem under consideration and involves selecting a particular kind of damping devices and, possibly, setting other technical characteristics such as the type of damper mounting, maximum damping capacity or maximum stroke [6,7]. It can also include some overall characteristics as the total number of allowed dampers or the overall damping capacity. The allocation problem requires determining a suitable set of structural positions to implement the dampers, and the tuning problem consists in computing adequate values for the damper parameters. For a given technical setup, the allocation and tuning issues are clearly interlinked and both must be simultaneously solved in order to obtain a DDS design with high-performance characteristics. The combined allocation-tuning problem can be formulated as a constrained optimization problem, with a set of decision variables that describes the different allocation schemes and parameter values of the damping devices, an objective function that allows evaluating the suitability of the corresponding DDSs, and a system of constraints that incorporate relevant features of the considered technical setup. To solve allocation-tuning optimization problems (ATOPs), a wide variety of computational strategies have been proposed, which include specialized optimization methods [8–12] and adaptations of general-purpose optimization algorithms [13–15]. Specialized optimization methods are specifically designed for particular kinds of ATOPs and, sometimes, can produce remarkably effective results for the considered problem. However, it should be highlighted that solving ATOPs for large structural systems can be a hard computational task, due to a number of factors such as high dimensionality, the combination of discrete and continuous decision variables, the presence of complex structural constraints, and the computational cost associated to the evaluation of the objective function. In this context, taking advantage of state-of-the-art general-purpose optimization solvers can be an element of critical relevance.

The general goal of the paper is to design high-performance DDSs for seismic protection of multibuilding systems (MBSs) formed by a row of m adjacent buildings as the one schematically displayed in Figure 1. There certainly exists a large number of works on DDS design for single buildings in the literature; some recent discussions on the topic can be found in References [5–7,15,16]. For the particular case of $m = 2$ adjacent buildings, the number of references is remarkably smaller but yet significant. Works in this line include different kinds of interbuilding damping devices [17–19] and actuation schemes [20–22]. In contrast, to our best knowledge, the more general case of DDS design for MBSs with $m \geq 3$ adjacent buildings remains practically unexplored.

To keep the complexity of the considered problem within a reasonable level, in this work we assume that all the buildings have identical dynamic characteristics [20,21]. That choice allows simplifying the notations in the overall mathematical model and can help to clarify the effects produced by the DDS, which otherwise could be confounded by the action of the distinct building responses. Moreover, as rows of adjacent identical buildings is a quite common arrangement in residential areas, the selected MBS configuration can be considered as a case of potential practical interest [23]. Also for simplicity, the buildings are modeled as linear planar frames and the damping devices are assumed to be fluid viscous dampers (FVDs), which have proved to be effective energy-dissipation elements in structural vibration control and bring the modeling advantage of admitting a reasonably linear representation [7]. Attending to the structural placement, the DDS can contain two different kinds of dampers—(i) *interstory dampers*, which are implemented between consecutive stories of the same building and produce a resistant force proportional to the corresponding interstory velocity and (ii) *interbuilding dampers*, which are implemented as linking elements between stories located at the same level in adjacent buildings and produce a resistant force proportional to the relative velocity of the linked stories. The DDS can also include two types of building damping configurations of particular interest—(i) *nonactuated buildings*, which do not contain any interstory damper and (ii) *linked buildings*, which are linked to all their adjacent buildings by means of interbuilding dampers. For instance, in the MBS presented in Figure 1, buildings $\mathcal{B}^{(2)}$ and $\mathcal{B}^{(3)}$ are nonactuated; buildings $\mathcal{B}^{(1)}$, $\mathcal{B}^{(2)}$ and $\mathcal{B}^{(3)}$ are linked; $\mathcal{B}^{(4)}$ is only partially linked and $\mathcal{B}^{(5)}$ is unlinked.

It should be observed that the DDS implementation can be carried out without internal modifications of nonactuated buildings, which can be an important factor in retrofitting. The relevance of linked configurations lies in the fact that they can help to mitigate the vibrational response of nonactuated buildings and can also provide an effective protection against interbuilding impacts (pounding) [22,23]. To incorporate those aspects in the DDS design and reducing the number of optimization variables, we introduce the *schemes of allowed damper positions*, which specify the interstory and interbuilding locations where the dampers can be implemented. Finally, we complete the technical setup determination by setting the maximum damping capacity of the dampers, the overall maximum damping capacity of the DDS and the total number of allowed damping elements.

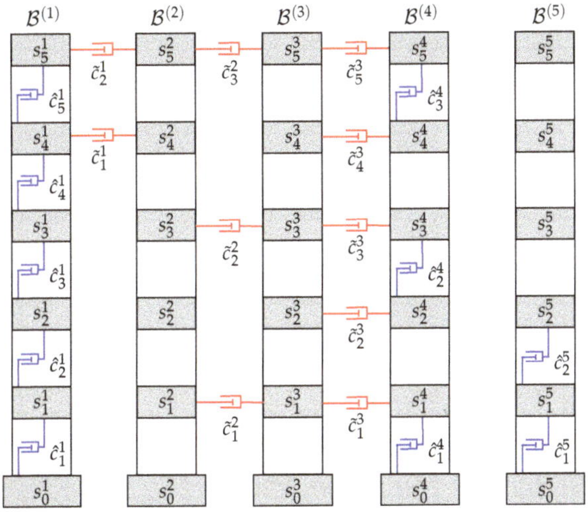

Figure 1. System of $m = 5$ adjacent buildings equipped with a distributed set of interstory and interbuilding dampers.

In order to define the ATOP, the generic goal of improving the overall seismic response of the MBS is formulated as a triplet of particular design objectives: (i) reducing the magnitude of the buildings interstory drifts, (ii) reducing the magnitude of the buildings floor accelerations and (iii) reducing the interbuilding approachings. Objectives (i) and (ii) are applicable to single-building designs, and are respectively associated to the protection of the buildings structural and nonstructural elements [16,24]. Objective (iii) is specific to MBS designs and is associated to avoiding interbuilding impacts, which can produce severe damage to both structural and nonstructural elements. In order to obtain a computationally effective procedure, we select the overall vector of interstory drifts as controlled output and follow a single-objective H_∞ approach [25,26]. That choice sets the reduction of interstory drifts as the primary objective and, at the same time, is able to produce positive results in mitigating the story accelerations [22]. As for design objective (iii), reduction of interbuilding approachings can be attained by enforcing a full-linked configuration in the optimization constraints [27,28]. Computationally, the selected H_∞ objective-function avoids conducting numerical simulations of seismic time-responses and admits a fast evaluation using the `hinfnorm` function of the Matlab Robust Control Toolbox [29,30]. The ATOP solutions are obtained with the genetic algorithm (GA) solver provided by the Matlab Global Optimization Toolbox, which allows using hybrid sets of discrete and continuous optimization variables, permits defining a sufficiently wide variety of optimization constraints and facilitates an easy implementation of parallel computing [31]. To demonstrate the flexibility of the proposed methodology, three different DDSs are designed for the seismic protection of a MBS formed by $m = 5$ adjacent five-story buildings. After that, a proper set of numerical simulations are conducted using the full-scale 180-component of the El Centro 1940 seismic record as ground acceleration disturbance. The obtained results corroborate the effectiveness of the proposed design procedure and confirm its computational efficiency for large-scale problems.

The content of the rest of the paper is as follows—in Section 2, a general mathematical model for plain and damped MBSs is presented. The main elements of the optimization procedure are discussed in Section 3. In Section 4, three different DDSs for a five-building system are designed. In Section 5, the corresponding seismic time-responses are computed and compared. Finally, in Section 6, some brief conclusions and future research lines are provided.

Remark 1. *To reduce the problem complexity and facilitate an effective computational solution, a number of model simplifications have been introduced in the paper, which should be carefully considered for a proper understanding of the scope and applicability of the proposed design methodology. Specifically, buildings are considered as linear multistory planar frames with stories of the same height. In the multibuilding systems, all the buildings are identical, with the same mass, stiffness and damping matrices, and there are no vertical differences between stories placed at the same level in different buildings. The damping elements are ideal linear dampers and the damping constants can take any real value in a prescribed interval. The interbuilding separations are assumed to be large enough to avoid interbuilding impacts. Additionally, the effect of some relevant elements such as the soil-structure interaction and the seismic wave propagation have been neglected.*

Remark 2. *The considered problem requires a complex system of notations. To obtain a more organized and clear presentation, throughout this paper symbols related to interstory elements will be usually marked with* hats *and those corresponding to interbuilding elements will be signaled with* tildes. *When convenient,* overlines *will be used to distinguish some elements related to the overall MBS. Thus, for example, $\hat{n}_j$ will denote the number of interstory dampers in building $\mathcal{B}^{(j)}$, the number of interbuilding dampers between buildings $\mathcal{B}^{(j)}$ and $\mathcal{B}^{(j+1)}$ will be indicated by $\tilde{n}_j$, and the overall number of degrees of freedom in the MBS will be represented by $\overline{n}$.*

2. Mathematical Model

In this section we present mathematical models for the dynamical response of MBSs equipped with a distributed set of interstory and interbuilding FVDs. The proposed models are fully formulated in matrix form and include state-space representations to facilitate an efficient computational implementation, which is a factor of critical relevance in the practical application of the design procedure.

2.1. Plain Building Model

Let us consider a MBS system formed by a row of m adjacent n-story buildings with identical dynamic characteristics. In the plain configuration, we assume that the buildings damping is only due to the effect of the structural damping, and the dynamical response of building $\mathcal{B}^{(j)}$ can be modeled in the following form:

$$\mathbf{M}_b\ddot{\mathbf{q}}^{(j)}(t) + \mathbf{C}_b\dot{\mathbf{q}}^{(j)}(t) + \mathbf{K}_b\mathbf{q}^{(j)}(t) = -\mathbf{M}_b[\mathbf{1}]_{n\times 1}\ddot{w}(t), \tag{1}$$

where $\mathbf{q}^{(j)}(t) = [q_1^j(t), \ldots, q_n^j(t)]^T$ is the vector of story displacements with respect to the ground (see Figure 2); $\mathbf{M}_b$, $\mathbf{C}_b$ and $\mathbf{K}_b$ are the building mass, damping and stiffness matrices, respectively, which are common to all the buildings; $[\mathbf{1}]_{n\times 1}$ is a column vector of size n with all its entries equal to one; and $\ddot{w}(t)$ is the acceleration of the seismic ground disturbance. The building mass matrix has the diagonal form

$$\mathbf{M}_b = \text{diag}[m_1, \ldots, m_n] = \begin{bmatrix} m_1 & & \\ & \ddots & \\ & & m_n \end{bmatrix}, \tag{2}$$

where m_i is the mass of the ith story. The stiffness matrix has the following tridiagonal structure:

$$\mathbf{K}_b = \begin{bmatrix} k_1+k_2 & -k_2 & & & \\ -k_2 & k_2+k_3 & -k_3 & & \\ & \cdots & \cdots & \cdots & \\ & & -k_{n-1} & k_{n-1}+k_n & -k_n \\ & & & -k_n & k_n \end{bmatrix}, \tag{3}$$

where k_i is the stiffness coefficient of the ith story. The stiffness matrix can be computed in the form

$$\mathbf{K}_b = \mathbf{P}\,\text{diag}[k_1, \ldots, k_n]\,\mathbf{P}^T, \tag{4}$$

where $\text{diag}[k_1, \ldots, k_n]$ is the diagonal matrix defined by the story stiffness coefficients and $\mathbf{P} \in \mathbb{R}^{n \times n}$ is the upper band-diagonal matrix

$$\mathbf{P} = \begin{bmatrix} 1 & -1 & & & \\ & 1 & -1 & & \\ & & \cdots & \cdots & \\ & & & 1 & -1 \\ & & & & 1 \end{bmatrix}, \tag{5}$$

with the following elements:

$$\begin{cases} [P]_{i,i} = 1, & i = 1, \ldots, n, \\ [P]_{i,i+1} = -1, & i = 1, \ldots, n-1, \\ [P]_{i,j} = 0, & \text{otherwise.} \end{cases} \tag{6}$$

When the story damping coefficients c_i, $i = 1, \ldots, n$ are known, the structure of the building damping matrix $\mathbf{C}_b$ is similar to the structure of the stiffness matrix in Equation (3) and can be computed in the following form:

$$\mathbf{C}_b = \mathbf{P}\,\text{diag}[c_1, \ldots, c_n]\,\mathbf{P}^T. \tag{7}$$

Frequently, however, the damping coefficients c_i cannot be properly determined and an approximate damping matrix $\mathbf{C}_b$ is computed by setting a suitable damping ratio on some of the building vibration modes [32]. Specifically, for the five-building model used in the numerical examples discussed in Sections 4 and 5, the matrix $\mathbf{C}_b$ has been computed as a Rayleigh damping matrix by setting a 2% of relative damping in the first and fifth modes (see Equation (70)).

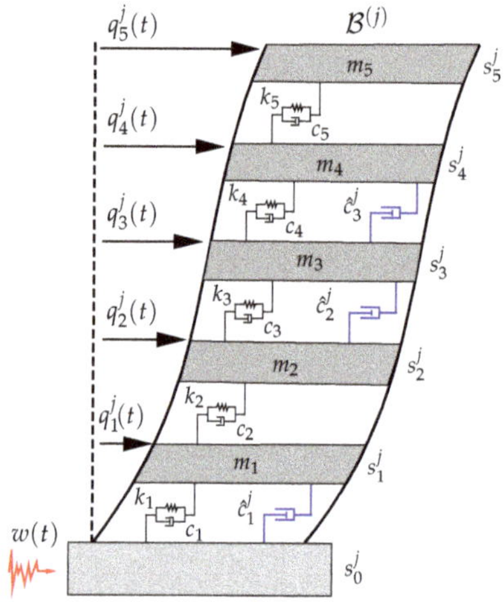

Figure 2. Schematic mechanical model of a five-story building equipped with a distributed system of $\hat{n}_j = 3$ supplemental interstory dampers implemented at the story levels $\hat{p}_1^j = 1$, $\hat{p}_2^j = 3$ and $\hat{p}_3^j = 4$.

2.2. Interstory and Interbuilding Damping Models

The dynamical response of building $\mathcal{B}^{(j)}$ can be improved by introducing a system of $0 < \hat{n}_j \leq n$ additional interstory dampers $\widehat{\mathcal{C}}^{(j)} = [\hat{p}^{(j)}, \hat{c}^{(j)}]$, where $\hat{c}^{(j)} = [\hat{c}_1^j, \ldots, \hat{c}_{\hat{n}_j}^j]$ is the list of damping coefficients and $\hat{p}^{(j)} = [\hat{p}_1^j, \ldots, \hat{p}_{\hat{n}_j}^j]$ is a list that contains the story levels at which the dampers are implemented. The dynamical response of building $\mathcal{B}^{(j)}$ equipped with the additional interstory damping system $\widehat{\mathcal{C}}^{(j)}$ can be modeled in the form

$$\mathbf{M}_b\ddot{\mathbf{q}}^{(j)}(t) + \left[\mathbf{C}_b + \widehat{\mathbf{C}}^{(j)}\right]\dot{\mathbf{q}}^{(j)}(t) + \mathbf{K}_b\mathbf{q}^{(j)}(t) = -\mathbf{M}_b[\mathbf{1}]_{n\times 1}\ddot{w}(t), \tag{8}$$

where the damping matrix $\widehat{\mathbf{C}}^{(j)}$ can be computed as

$$\widehat{\mathbf{C}}^{(j)} = \widehat{\mathbf{P}}^{(j)}\,\mathrm{diag}[\hat{c}_1^j, \ldots, \hat{c}_{\hat{n}_j}^j]\,\{\widehat{\mathbf{P}}^{(j)}\}^T \tag{9}$$

by considering the diagonal matrix $\mathrm{diag}[\hat{c}_1^j, \ldots, \hat{c}_{\hat{n}_j}^j]$ and the location matrix $\widehat{\mathbf{P}}^{(j)} \in \mathbb{R}^{n\times\hat{n}_j}$ formed by the columns of the matrix $\mathbf{P}$ indicated in the placement list $\hat{p}^{(j)}$. Thus, for example, the system of additional interstory dampers in Figure 2 contains $\hat{n}_j = 3$ dampers located at the story positions $\hat{p}^{(j)} = [1,3,4]$. The corresponding placement and diagonal matrices are, respectively,

$$\widehat{\mathbf{P}}^{(j)} = \begin{bmatrix} 1 & 0 & 0 \\ 0 & -1 & 0 \\ 0 & 1 & -1 \\ 0 & 0 & 1 \\ 0 & 0 & 0 \end{bmatrix}, \quad \mathrm{diag}[\hat{c}_1^j, \hat{c}_2^j, \hat{c}_3^j] = \begin{bmatrix} \hat{c}_1^j & 0 & 0 \\ 0 & \hat{c}_2^j & 0 \\ 0 & 0 & \hat{c}_3^j \end{bmatrix} \tag{10}$$

and the matrix of additional interstory damping has the following form:

$$\widehat{\mathbf{C}}^{(j)} = \begin{bmatrix} \hat{c}_1^j & 0 & 0 & 0 & 0 \\ 0 & \hat{c}_2^j & -\hat{c}_2^j & 0 & 0 \\ 0 & -\hat{c}_2^j & \hat{c}_2^j + \hat{c}_3^j & -\hat{c}_3^j & 0 \\ 0 & 0 & -\hat{c}_3^j & \hat{c}_3^j & 0 \\ 0 & 0 & 0 & 0 & 0 \end{bmatrix}. \tag{11}$$

Improving the dynamical response of nonactuated buildings and reducing the pounding risk between adjacent buildings can be achieved by introducing proper systems of additional interbuilding dampers $\widetilde{\mathcal{C}}^{(j)} = [\tilde{p}^{(j)}, \tilde{c}^{(j)}]$, where $\tilde{c}^{(j)} = [\tilde{c}_1^j, \ldots, \tilde{c}_{\tilde{n}_j}^j]$ is the list of damping coefficients, $\tilde{p}^{(j)} = [\tilde{p}_1^j, \ldots, \tilde{p}_{\tilde{n}_j}^j]$ is a list that contains the story levels between buildings $\mathcal{B}^{(j)}$ and $\mathcal{B}^{(j+1)}$ at which the interbuilding dampers are implemented, and $\tilde{n}_j$ is the number of dampers in $\widetilde{\mathcal{C}}^{(j)}$. To compute the dynamical response of the buildings subjected to the action of interbuilding damping systems we define the interbuilding damping matrix $\widetilde{\mathbf{C}}^{(j)}$ associated to the damping system $\widetilde{\mathcal{C}}^{(j)}$ as

$$\widetilde{\mathbf{C}}^{(j)} = \widetilde{\mathbf{P}}^{(j)} \operatorname{diag}[\tilde{c}_1^j, \ldots, \tilde{c}_{\tilde{n}_j}^j] \, \{\widetilde{\mathbf{P}}^{(j)}\}^T, \tag{12}$$

where $\operatorname{diag}[\tilde{c}_1^j, \ldots, \tilde{c}_{\tilde{n}_j}^j]$ is the diagonal matrix defined by the list of damping coefficients $\tilde{c}^{(j)}$ and the location matrix $\widetilde{\mathbf{P}}^{(j)} \in \mathbb{R}^{n\times\tilde{n}_j}$ contains the columns of the identity matrix $\mathbf{I}_n$ indicated in the list of damper positions $\tilde{p}^{(j)}$. Thus, for example, the system of additional interbuilding dampers $\widetilde{\mathcal{C}}^{(2)}$ implemented between buildings $\mathcal{B}^{(2)}$ and $\mathcal{B}^{(3)}$ in Figure 1 contains $\tilde{n}_2 = 3$ dampers with coefficients $\tilde{c}^{(2)} = [\tilde{c}_1^2, \tilde{c}_2^2, \tilde{c}_3^2]$, which are located at the story levels $\tilde{p}^{(2)} = [1, 3, 5]$. The corresponding placement and diagonal matrices are, respectively,

$$\widetilde{\mathbf{P}}^{(2)} = \begin{bmatrix} 1 & 0 & 0 \\ 0 & 0 & 0 \\ 0 & 1 & 0 \\ 0 & 0 & 0 \\ 0 & 0 & 1 \end{bmatrix}, \quad \operatorname{diag}[\tilde{c}_1^2, \tilde{c}_2^2, \tilde{c}_3^2] = \begin{bmatrix} \tilde{c}_1^2 & 0 & 0 \\ 0 & \tilde{c}_2^2 & 0 \\ 0 & 0 & \tilde{c}_3^2 \end{bmatrix} \tag{13}$$

and the matrix of additional interbuilding damping is

$$\widetilde{\mathbf{C}}^{(2)} = \begin{bmatrix} \tilde{c}_1^2 & 0 & 0 & 0 & 0 \\ 0 & 0 & 0 & 0 & 0 \\ 0 & 0 & \tilde{c}_2^2 & 0 & 0 \\ 0 & 0 & 0 & 0 & 0 \\ 0 & 0 & 0 & 0 & \tilde{c}_3^2 \end{bmatrix}. \tag{14}$$

For an m-building system equipped with the set of interstory and interbuilding damping systems $\widehat{\mathcal{C}}^{(1)}, \ldots, \widehat{\mathcal{C}}^{(m)}, \widetilde{\mathcal{C}}^{(1)}, \ldots, \widetilde{\mathcal{C}}^{(m-1)}$, the dynamical response of building $\mathcal{B}^{(j)}$ can be described by the model

$$\mathbf{M}_b\ddot{\mathbf{q}}^{(j)}(t) + \mathbf{f}_d^{(j)}(t) + \mathbf{K}_b\mathbf{q}^{(j)}(t) = -\mathbf{M}_b[\mathbf{1}]_{n\times 1}\ddot{w}(t), \tag{15}$$

where the term $\mathbf{f}_d^{(j)}(t)$ denotes the total damping force acting on $\mathcal{B}^{(j)}$. Using the damping matrix $\widetilde{\mathbf{C}}^{(j)}$ defined in Equation (12), the total damping force $\mathbf{f}_d^{(1)}(t)$ acting on the initial building $\mathcal{B}^{(1)}$ can be written in the following form.

$$\mathbf{f}_d^{(1)}(t) = \left[\mathbf{C}_b + \widehat{\mathbf{C}}^{(1)} + \widetilde{\mathbf{C}}^{(1)}\right] \dot{\mathbf{q}}^{(1)}(t) - \widetilde{\mathbf{C}}^{(1)}\dot{\mathbf{q}}^{(2)}(t); \tag{16}$$

for an interior building $\mathcal{B}^{(j)}, j = 2, \ldots, m-1$, we have

$$\mathbf{f}_d^{(j)}(t) = -\widetilde{\mathbf{C}}^{(j-1)}\dot{\mathbf{q}}^{(j-1)}(t) + \left[\mathbf{C}_b + \widehat{\mathbf{C}}^{(j)} + \widetilde{\mathbf{C}}^{(j-1)} + \widetilde{\mathbf{C}}^{(j)}\right]\dot{\mathbf{q}}^{(j)}(t) - \widetilde{\mathbf{C}}^{(j)}\dot{\mathbf{q}}^{(j+1)}(t); \tag{17}$$

and for the final building $\mathcal{B}^{(m)}$, we obtain

$$\mathbf{f}_d^{(m)}(t) = -\widetilde{\mathbf{C}}^{(m-1)}\dot{\mathbf{q}}^{(m-1)}(t) + \left[\mathbf{C}_b + \widehat{\mathbf{C}}^{(m)} + \widetilde{\mathbf{C}}^{(m-1)}\right]\dot{\mathbf{q}}^{(m)}(t). \tag{18}$$

Remark 3. *For notational convenience, we will assume that a nonactuated building $\mathcal{B}^{(j)}$ is equipped with an empty system of interstory dampers $\widehat{\mathcal{C}}^{(j)} = [\hat{p}^{(j)}, \hat{c}^{(j)}]$, where $\hat{p}^{(j)}$ and $\hat{c}^{(j)}$ are empty lists and the number of dampers is $\hat{n}_j = 0$. In that case, we agree that $\widehat{\mathbf{C}}^{(j)} = [\mathbf{0}]_{n\times n}$. Analogously, when there are no dampers implemented between buildings $\mathcal{B}^{(j)}$ and $\mathcal{B}^{(j+1)}$ we will consider an empty system of interbuilding dampers $\widetilde{\mathcal{C}}^{(j)} = [\tilde{p}^{(j)}, \tilde{c}^{(j)}]$, where $\tilde{p}^{(j)}$ and $\tilde{c}^{(j)}$ are empty lists. Also in that case, we have $\tilde{n}_j = 0$ and set $\widetilde{\mathbf{C}}^{(j)} = [\mathbf{0}]_{n\times n}$.*

2.3. Overall Multibuilding Model

By considering the overall vector of displacements

$$\mathbf{q}(t) = \begin{bmatrix} \mathbf{q}^{(1)}(t) \\ \vdots \\ \mathbf{q}^{(m)}(t) \end{bmatrix}, \tag{19}$$

the dynamical response of the overall multibuilding system can be described by the model

$$\mathbf{M}\,\ddot{\mathbf{q}}(t) + \left(\mathbf{C} + \widehat{\mathbf{C}} + \widetilde{\mathbf{C}}\right)\dot{\mathbf{q}}(t) + \mathbf{K}\,\mathbf{q}(t) = -\mathbf{M}\,[\mathbf{1}]_{\overline{n}\times 1}\ddot{w}(t), \tag{20}$$

where $\overline{n} = m \cdot n$ is the total number of degrees of freedom; $\mathbf{M} \in \mathbb{R}^{\overline{n}\times\overline{n}}$ and $\mathbf{K} \in \mathbb{R}^{\overline{n}\times\overline{n}}$ are the overall mass and stiffness matrices, respectively, which have the following block-diagonal structure:

$$\mathbf{M} = \text{diag}\left[\mathbf{M}_b, \overset{(m)}{\ldots}, \mathbf{M}_b\right] = \begin{bmatrix} \mathbf{M}_b & & \\ & \ddots & \\ & & \mathbf{M}_b \end{bmatrix}, \; \mathbf{K} = \text{diag}\left[\mathbf{K}_b, \overset{(m)}{\ldots}, \mathbf{K}_b\right] = \begin{bmatrix} \mathbf{K}_b & & \\ & \ddots & \\ & & \mathbf{K}_b \end{bmatrix}; \tag{21}$$

$\mathbf{C} \in \mathbb{R}^{\overline{n}\times\overline{n}}$ and $\widehat{\mathbf{C}} \in \mathbb{R}^{\overline{n}\times\overline{n}}$ are the overall matrices of structural and interstory damping, respectively, which also have a block-diagonal structure

$$\mathbf{C} = \text{diag}\left[\mathbf{C}_b, \overset{(m)}{\ldots}, \mathbf{C}_b\right] = \begin{bmatrix} \mathbf{C}_b & & \\ & \ddots & \\ & & \mathbf{C}_b \end{bmatrix}, \; \widehat{\mathbf{C}} = \text{diag}\left[\widehat{\mathbf{C}}^{(1)}, \ldots, \widehat{\mathbf{C}}^{(m)}\right] = \begin{bmatrix} \widehat{\mathbf{C}}^{(1)} & & \\ & \ddots & \\ & & \widehat{\mathbf{C}}^{(m)} \end{bmatrix}; \tag{22}$$

and $\widetilde{\mathbf{C}} \in \mathbb{R}^{\overline{n}\times\overline{n}}$ is a matrix that models the overall interbuilding damping and has the following block-tridiagonal structure:

$$\widetilde{\mathbf{C}} = \begin{bmatrix} \widetilde{\mathbf{C}}^{(1)} & -\widetilde{\mathbf{C}}^{(1)} & & & \\ -\widetilde{\mathbf{C}}^{(1)} & \widetilde{\mathbf{C}}^{(1)} + \widetilde{\mathbf{C}}^{(2)} & -\widetilde{\mathbf{C}}^{(2)} & & \\ & \cdots & \cdots & \cdots & \\ & & -\widetilde{\mathbf{C}}^{(m-2)} & \widetilde{\mathbf{C}}^{(m-2)} + \widetilde{\mathbf{C}}^{(m-1)} & -\widetilde{\mathbf{C}}^{(m-1)} \\ & & & -\widetilde{\mathbf{C}}^{(m-1)} & \widetilde{\mathbf{C}}^{(m-1)} \end{bmatrix}. \quad (23)$$

The matrix $\widetilde{\mathbf{C}}$ can be computed as

$$\widetilde{\mathbf{C}} = \mathbf{Q}\, \mathrm{diag}\left[\widetilde{\mathbf{C}}^{(1)}, \ldots, \widetilde{\mathbf{C}}^{(m-1)}\right] \mathbf{Q}^T, \quad (24)$$

where $\mathbf{Q} \in \mathbb{R}^{\bar{n} \times (m-1)n}$ is the block-matrix

$$\mathbf{Q} = \begin{bmatrix} \mathbf{I}_n & & & \\ -\mathbf{I}_n & \mathbf{I}_n & & \\ & \cdots & \cdots & \\ & & -\mathbf{I}_n & \mathbf{I}_n \\ & & & -\mathbf{I}_n \end{bmatrix}, \quad (25)$$

defined by blocks $[\mathbf{Q}]_{i,j}$, $i = 1, \ldots, m$, $j = 1, \ldots, m-1$ with the following form:

$$\begin{cases} [\mathbf{Q}]_{i,i} = \mathbf{I}_n, & i = 1, \ldots, m-1, \\ [\mathbf{Q}]_{i+1,i} = -\mathbf{I}_n, & i = 1, \ldots, m-1, \\ [\mathbf{Q}]_{i,j} = [\mathbf{0}]_{n \times n}, & \text{otherwise}, \end{cases} \quad (26)$$

where $[\mathbf{0}]_{n \times n}$ denotes the null matrix of dimension $n \times n$. For example, the five-building system in Figure 1 includes three non-empty interbuilding damping systems $\widetilde{\mathcal{C}}^{(1)}$, $\widetilde{\mathcal{C}}^{(2)}$ and $\widetilde{\mathcal{C}}^{(3)}$ with $\tilde{n}_1 = 2$, $\tilde{n}_2 = 3$, and $\tilde{n}_3 = 5$ dampers, respectively. It also includes one empty damping system $\widetilde{\mathcal{C}}^{(4)}$, which, according to Remark 3, has the damping matrix $\widetilde{\mathbf{C}}^{(4)} = [\mathbf{0}]_{5 \times 5}$. In this case, we obtain the position and coefficient block-matrices

$$\mathbf{Q} = \begin{bmatrix} \mathbf{I}_5 & [\mathbf{0}]_{5\times5} & [\mathbf{0}]_{5\times5} & [\mathbf{0}]_{5\times5} \\ -\mathbf{I}_5 & \mathbf{I}_5 & [\mathbf{0}]_{5\times5} & [\mathbf{0}]_{5\times5} \\ [\mathbf{0}]_{5\times5} & -\mathbf{I}_5 & \mathbf{I}_5 & [\mathbf{0}]_{5\times5} \\ [\mathbf{0}]_{5\times5} & [\mathbf{0}]_{5\times5} & -\mathbf{I}_5 & \mathbf{I}_5 \\ [\mathbf{0}]_{5\times5} & [\mathbf{0}]_{5\times5} & [\mathbf{0}]_{5\times5} & -\mathbf{I}_5 \end{bmatrix}, \quad \mathrm{diag}\left[\widetilde{\mathbf{C}}^{(1)}, \widetilde{\mathbf{C}}^{(2)}, \widetilde{\mathbf{C}}^{(3)}, [\mathbf{0}]_{5\times5}\right] \quad (27)$$

and the overall interbuilding damping matrix has the following block structure:

$$\widetilde{\mathbf{C}} = \begin{bmatrix} \widetilde{\mathbf{C}}^{(1)} & -\widetilde{\mathbf{C}}^{(1)} & [\mathbf{0}]_{5\times5} & [\mathbf{0}]_{5\times5} & [\mathbf{0}]_{5\times5} \\ -\widetilde{\mathbf{C}}^{(1)} & \widetilde{\mathbf{C}}^{(1)} + \widetilde{\mathbf{C}}^{(2)} & -\widetilde{\mathbf{C}}^{(2)} & [\mathbf{0}]_{5\times5} & [\mathbf{0}]_{5\times5} \\ [\mathbf{0}]_{5\times5} & -\widetilde{\mathbf{C}}^{(2)} & \widetilde{\mathbf{C}}^{(2)} + \widetilde{\mathbf{C}}^{(3)} & -\widetilde{\mathbf{C}}^{(3)} & [\mathbf{0}]_{5\times5} \\ [\mathbf{0}]_{5\times5} & [\mathbf{0}]_{5\times5} & -\widetilde{\mathbf{C}}^{(3)} & \widetilde{\mathbf{C}}^{(3)} & [\mathbf{0}]_{5\times5} \\ [\mathbf{0}]_{5\times5} & [\mathbf{0}]_{5\times5} & [\mathbf{0}]_{5\times5} & [\mathbf{0}]_{5\times5} & [\mathbf{0}]_{5\times5} \end{bmatrix}. \quad (28)$$

2.4. State-Space Model and Output Variables

By considering the state vector

$$\mathbf{x}(t) = \begin{bmatrix} \mathbf{q}(t) \\ \dot{\mathbf{q}}(t) \end{bmatrix}, \tag{29}$$

the overall dynamical response of the multibuilding system can be described by a first-order state-space model

$$\dot{\mathbf{x}}(t) = (\mathbf{A}_b + \mathbf{A}_d)\mathbf{x}(t) + \mathbf{B}\,\ddot{w}(t), \tag{30}$$

where the matrices

$$\mathbf{A}_b = \begin{bmatrix} [\mathbf{0}]_{\overline{n}\times\overline{n}} & \mathbf{I}_{\overline{n}} \\ -\mathbf{M}^{-1}\mathbf{K} & -\mathbf{M}^{-1}\mathbf{C} \end{bmatrix}, \quad \mathbf{B} = \begin{bmatrix} [\mathbf{0}]_{\overline{n}\times 1} \\ -[\mathbf{1}]_{\overline{n}\times 1} \end{bmatrix} \tag{31}$$

are constant matrices that model the dynamical response of the nonactuated MBS and

$$\mathbf{A}_d = \begin{bmatrix} [\mathbf{0}]_{\overline{n}\times\overline{n}} & [\mathbf{0}]_{\overline{n}\times\overline{n}} \\ [\mathbf{0}]_{\overline{n}\times\overline{n}} & -\mathbf{M}^{-1}(\widehat{\mathbf{C}} + \widetilde{\mathbf{C}}) \end{bmatrix} \tag{32}$$

is a matrix that reflects the action of the overall system of added dampers.

The vector of interstory drifts $\hat{\mathbf{r}}^{(j)}(t) = [\hat{r}_1^j(t), \ldots, \hat{r}_n^j(t)]^T$ corresponding to building $\mathcal{B}^{(j)}$ has the following components:

$$\begin{cases} \hat{r}_1^j(t) = q_1^j(t), \\ \hat{r}_i^j(t) = q_i^j(t) - q_{i-1}^j(t), \quad i = 2, \ldots, n. \end{cases} \tag{33}$$

Using the matrix $\mathbf{P}$ in Equation (5), the vector $\hat{\mathbf{r}}^{(j)}(t)$ can be computed as

$$\hat{\mathbf{r}}^{(j)}(t) = \mathbf{P}^T \mathbf{q}^{(j)}(t), \tag{34}$$

and the overall vector of interstory drifts corresponding to the multibuilding system

$$\hat{\mathbf{r}}(t) = \begin{bmatrix} \hat{\mathbf{r}}^{(1)}(t) \\ \vdots \\ \hat{\mathbf{r}}^{(m)}(t) \end{bmatrix} \tag{35}$$

can be obtained in the form $\hat{\mathbf{r}}(t) = \mathbf{C}_{\hat{r}}\,\mathbf{x}(t)$ with the output-matrix

$$\mathbf{C}_{\hat{r}} = \begin{bmatrix} \mathrm{diag}[\mathbf{P}, \overset{(m)}{\ldots}, \mathbf{P}]^T & [\mathbf{0}]_{\overline{n}\times\overline{n}} \end{bmatrix}. \tag{36}$$

The vector of total accelerations of the jth building

$$\mathbf{a}^{(j)}(t) = [a_1^j(t), \ldots, a_n^j(t)]^T = \ddot{\mathbf{q}}^{(j)}(t) + [\mathbf{1}]_{n\times 1}\ddot{w}(t) \tag{37}$$

contains the story accelerations with respect to an inertial reference frame. Using the state vector $\mathbf{x}(t)$, the overall vector of total accelerations

$$\mathbf{a}(t) = \begin{bmatrix} \mathbf{a}^{(1)}(t) \\ \vdots \\ \mathbf{a}^{(m)}(t) \end{bmatrix} \tag{38}$$

can be computed in the form

$$\mathbf{a}(t) = -\begin{bmatrix} \mathbf{M}^{-1}\mathbf{K} & \mathbf{M}^{-1}(\mathbf{C}+\widehat{\mathbf{C}}+\widetilde{\mathbf{C}}) \end{bmatrix} \mathbf{x}(t). \tag{39}$$

The vector of interbuilding approachings between buildings $\mathcal{B}^{(j)}$ and $\mathcal{B}^{(j+1)}$

$$\tilde{\mathbf{r}}^{(j)}(t) = [\tilde{r}_1^j(t), \ldots, \tilde{r}_n^j(t)]^T = -\{\mathbf{q}^{(j+1)}(t) - \mathbf{q}^{(j)}(t)\} \tag{40}$$

describes the variation of the interbuilding separation at the different story levels. As schematically illustrated in Figure 3, for a given interbuilding gap Δ_j, the value $\Delta_j - \tilde{r}_i^j(t)$ indicates the separation between stories s_i^j and s_i^{j+1}. Hence, positive values of the interbuilding approaching $\tilde{r}_i^j(t)$ will produce a reduction of the interbuilding separation, and a value $\tilde{r}_i^j(t) > \Delta_j$ will indicate an interbuilding impact at the ith story level. To avoid the computational complexity associated to building impacts, we will assume that the interbuilding gaps $\Delta_j, j = 1, \ldots, m-1$ are large enough to avoid pounding. In that case, the approaching peak-value

$$\tilde{r}_{\text{max}}^{(j)} = \max_{1 \le i \le n} \left(\max_{t \ge 0} \tilde{r}_i^j(t) \right) \tag{41}$$

obtained in numerical simulations of the seismic response can be taken as a lower bound of safe interbuilding separation between buildings $\mathcal{B}^{(j)}$ and $\mathcal{B}^{(j+1)}$ for the considered seismic excitation. Using the matrix $\mathbf{Q}$ in Equation (25), the overall vector of interbuilding approachings

$$\tilde{\mathbf{r}}(t) = \begin{bmatrix} \tilde{\mathbf{r}}^{(1)}(t) \\ \vdots \\ \tilde{\mathbf{r}}^{(m-1)}(t) \end{bmatrix} \tag{42}$$

can be computed from the state vector in the form

$$\tilde{\mathbf{r}}(t) = \begin{bmatrix} \mathbf{Q}^T & [\mathbf{0}]_{(m-1)n \times \overline{n}} \end{bmatrix} \mathbf{x}(t). \tag{43}$$

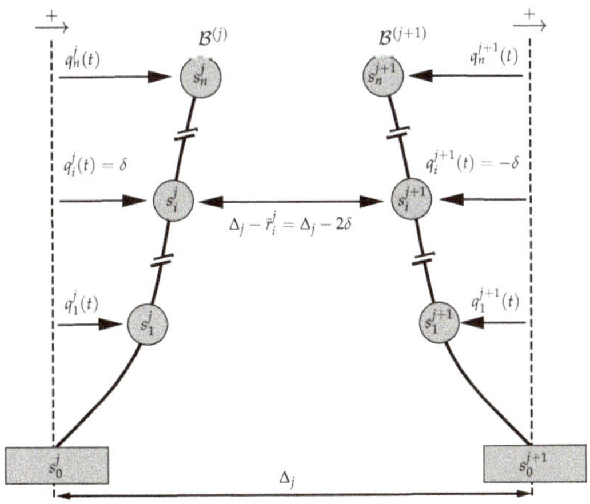

Figure 3. Interbuilding separation of the stories s_i^j and s_i^{j+1} corresponding to the interbuilding approaching $\tilde{r}_i^j(t) = -\{q_i^{j+1}(t) - q_i^j(t)\} = 2\delta$ for adjacent buildings $\mathcal{B}^{(j)}$ and $\mathcal{B}^{(j+1)}$ with an interbuilding gap Δ_j.

3. Optimization Procedure

Due to the complexity of the considered problem, the structure of the optimization variables and constraints is a critical issue for computational efficiency. The selected genetic algorithm (GA) optimization solver permits simultaneously work with discrete and continuous variables. In the proposed optimization procedure, discrete binary variables are used to indicate the damper allocations while damping capacities are described by continuous variables. Also, two different kinds of constraints are employed: (i) *preliminary constraints*, which include the scheme of allowed damper positions $\overline{\sigma}$ and the total number of damping elements n_d and (ii) *solving constraints*, which are regular optimization constraints imposed on the optimization variables. The preliminary constraints are established prior to the solving phase and determine the number and type of decision variables. The solving constraints are associated to the optimization solver and are used in the solving phase to enforce the binary character of allocation variables and controlling the feasibility of solutions with respect to the selected technical setup. In this section, we provide a detailed description of the schemes of allowed damper positions and the construction of the multibuilding model corresponding to a particular vector of optimization variables. Next, we discuss the main features of the objective function and the solving constraints.

3.1. Allowed Damper Positions, Dampers Allocation and Damping Coefficients

In order to determine the structure of the interstory damping systems, we introduce the scheme of allowed damper positions $\hat{\sigma} = [\hat{\sigma}^{(1)}, \ldots, \hat{\sigma}^{(m)}]$, where $\hat{\sigma}^{(j)} = [\hat{\sigma}_1^j, \ldots, \hat{\sigma}_{\hat{\eta}_j}^j]$ is a list that indicates the story levels in building $\mathcal{B}^{(j)}$ at which additional interstory dampers can be implemented and $\hat{\eta}_j$ is the number of such allowed positions. When no additional interstory dampers are allowed in building $\mathcal{B}^{(j)}$, we agree that $\hat{\sigma}^{(j)}$ is an empty list with $\hat{\eta}_j = 0$ elements. Analogously, to specify the structure of the interbuilding damping system, we introduce the scheme of allowed positions $\tilde{\sigma} = [\tilde{\sigma}^{(1)}, \ldots, \tilde{\sigma}^{(m-1)}]$, where $\tilde{\sigma}^{(j)} = [\tilde{\sigma}_1^j, \ldots, \tilde{\sigma}_{\tilde{\eta}_j}^j]$ is a list that indicates the story levels between buildings $\mathcal{B}^{(j)}$ and $\mathcal{B}^{(j+1)}$ at which additional interbuilding dampers can be implemented and $\tilde{\eta}_j$ is the number of those allowed positions. Also in this case, the value $\tilde{\eta}_j = 0$ indicates that $\tilde{\sigma}^{(j)}$ is an empty list and that no interbuilding dampers can be implemented between $\mathcal{B}^{(j)}$ and $\mathcal{B}^{(j+1)}$. The overall scheme of allowed damper positions

$$\overline{\sigma} = [\hat{\sigma}^{(1)}, \ldots, \hat{\sigma}^{(m)}, \tilde{\sigma}^{(1)}, \ldots, \tilde{\sigma}^{(m-1)}] \tag{44}$$

has $\overline{\eta} = \hat{\eta} + \tilde{\eta}$ elements with $\hat{\eta} = \sum_{j=1}^{m} \hat{\eta}_j$ and $\tilde{\eta} = \sum_{j=1}^{m-1} \tilde{\eta}_j$. To illustrate the introduced definitions, let us consider the scheme of allowed damper positions for the three-building system displayed in Figure 4, which permits implementing interstory dampers at the story levels 1–3 in buildings $\mathcal{B}^{(1)}$ and $\mathcal{B}^{(3)}$ (blue dashed squares) and interbuilding dampers at the interbuilding levels 4 and 5 (red dotted squares). In this case, we have

$$\begin{array}{lllll} \hat{\sigma}^{(1)} = [1,2,3] & \hat{\sigma}^{(2)} = [\,] & \hat{\sigma}^{(3)} = [1,2,3] & \tilde{\sigma}^{(1)} = [4,5] & \tilde{\sigma}^{(2)} = [4,5] \\ \hat{\eta}_1 = 3 & \hat{\eta}_2 = 0 & \hat{\eta}_3 = 3 & \tilde{\eta}_1 = 2 & \tilde{\eta}_2 = 2 \end{array} \tag{45}$$

with a total number of allowed damper positions $\overline{\eta} = 10$.

To define the damper placement positions of an admissible interstory damping system for building $\mathcal{B}^{(j)}$, we introduce the allocation list $\hat{\alpha}^{(j)} = [\hat{\alpha}_1^j, \ldots, \hat{\alpha}_{\hat{\eta}_j}^j]$, where $\hat{\alpha}_i^j$ is a binary variable that takes the values

$$\hat{\alpha}_i^j = \begin{cases} 1 & \text{if an interstory damper is implemented at position } \hat{\sigma}_i^j \text{ in } \mathcal{B}^{(j)}, \\ 0 & \text{otherwise.} \end{cases} \tag{46}$$

The number of interstory dampers implemented in $\mathcal{B}^{(j)}$ is $\hat{n}_j = \sum_{i=1}^{\hat{\eta}_j} \hat{\alpha}_i^j$ and $\hat{n} = \sum_{j=1}^{m} \hat{n}_j$ is the total number of interstory dampers. Analogously, the damper placement positions for an admissible interbuilding damping system between buildings $\mathcal{B}^{(j)}$ and $\mathcal{B}^{(j+1)}$ are defined by the allocation list of binary variables $\tilde{\alpha}^{(j)} = [\tilde{\alpha}_1^j, \ldots, \tilde{\alpha}_{\tilde{\eta}_j}^j]$, where $\tilde{\alpha}_i^j = 1$ indicates that an interbuilding damper is implemented at level $\tilde{\sigma}_i^j$. The number of dampers implemented between $\mathcal{B}^{(j)}$ and $\mathcal{B}^{(j+1)}$ is $\tilde{n}_j = \sum_{i=1}^{\tilde{\eta}_j} \tilde{\alpha}_i^j$ and $\tilde{n} = \sum_{j=1}^{m-1} \tilde{n}_j$ is the total number of interbuilding dampers. For an admissible damping system with n_d dampers, the actual damper placement positions can be described by an overall allocation list

$$\overline{\alpha} = [\hat{\alpha}^{(1)}, \ldots, \hat{\alpha}^{(m)}, \tilde{\alpha}^{(1)}, \ldots, \tilde{\alpha}^{(m-1)}] \tag{47}$$

with the constraint $\hat{n} + \tilde{n} = n_d$. For the scheme of allowed damper positions in Equation (45), the damping system displayed in Figure 4 has the following lists of allocated dampers:

$$\begin{array}{lllll} \hat{\alpha}^{(1)} = [1,1,0] & \hat{\alpha}^{(2)} = [\,] & \hat{\alpha}^{(3)} = [1,0,1] & \tilde{\alpha}^{(1)} = [0,1] & \tilde{\alpha}^{(2)} = [1,1] \\ \hat{n}_1 = 2 & \hat{n}_2 = 0 & \hat{n}_3 = 2 & \tilde{n}_1 = 1 & \tilde{n}_2 = 2. \end{array} \tag{48}$$

The total number of allocated interstory and interbuilding dampers are $\hat{n} = 4$ and $\tilde{n} = 3$, respectively. The overall list of allocated dampers

$$\overline{\alpha} = [\overbrace{1,1,0}^{\hat{\alpha}^{(1)}}, \overbrace{1,0,1}^{\hat{\alpha}^{(3)}}, \overbrace{0,1}^{\tilde{\alpha}^{(1)}}, \overbrace{1,1}^{\tilde{\alpha}^{(2)}}] \tag{49}$$

contains $\overline{\eta} = 10$ elements and indicates that the overall number of allocated dampers is $n_d = 7$.

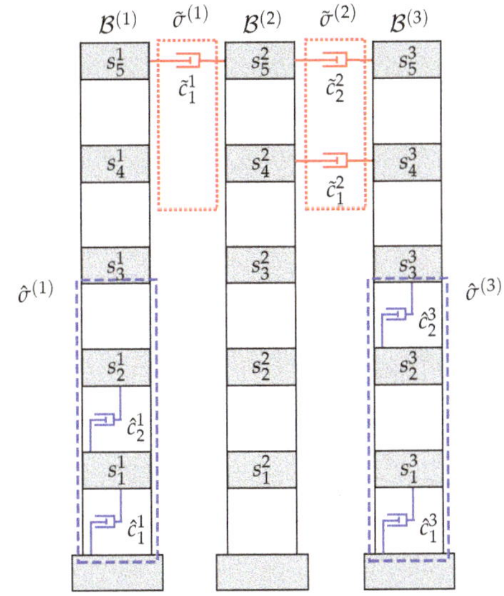

Figure 4. Schemes of allowed damper positions for a three-building system. Interstory scheme $\hat{\sigma} = [\hat{\sigma}^{(1)}, \hat{\sigma}^{(2)}, \hat{\sigma}^{(3)}]$ with $\hat{\eta}_1 = 3$, $\hat{\eta}_2 = 0$ and $\hat{\eta}_3 = 3$ (blue dashed rectangles). Interbuilding scheme $\tilde{\sigma} = [\tilde{\sigma}^{(1)}, \tilde{\sigma}^{(2)}]$ with $\tilde{\eta}_1 = 2$ and $\tilde{\eta}_2 = 2$ (red dotted rectangles).

3.2. Optimization Variables and Associated Multibuilding Model

The optimization variables are organized in a list $v = [\overline{\alpha}, d]$, where $\overline{\alpha} = [\overline{\alpha}_1, \ldots, \overline{\alpha}_{\overline{\eta}}]$ is the list of dampers allocation and $d = [d_1, \ldots, d_{n_d}]$ is the list of damping coefficients. To carry out the optimization process, for a given list of optimization variables $v = [\overline{\alpha}, d]$, we have to compute the damping schemes $\widehat{\mathcal{C}}^{(1)}, \ldots, \widehat{\mathcal{C}}^{(m)}$, $\widetilde{\mathcal{C}}^{(1)}, \ldots, \widetilde{\mathcal{C}}^{(m-1)}$ and the associated damping matrices $\widehat{\mathbf{C}}^{(1)}, \ldots, \widehat{\mathbf{C}}^{(m)}$, $\widetilde{\mathbf{C}}^{(1)}, \ldots, \widetilde{\mathbf{C}}^{(m-1)}$. To simplify that task, we will denote by $\pi(L)$ the list of positions of the nonzero elements in the list $L = [l_1, \ldots, l_p]$. We will also use the shorthands $sum(L) = \sum_{i=1}^{p} l_i$; $L(i_1 : i_2) = [l_{i_1}, \ldots, l_{i_2}]$ for $i_1 \leq i_2$; and $L(\pi) = [l_{\pi_1}, \ldots, l_{\pi_k}]$ for a list of positions $\pi = [\pi_1, \ldots, \pi_k]$.

To obtain the matrices of interstory damping $\widehat{\mathbf{C}}^{(j)}$, $j = 1, \ldots, m$, we consider the lists of allowed interstory damper positions $\hat{\sigma}^{(j)} = [\hat{\sigma}_1^j, \ldots, \hat{\sigma}_{\hat{\eta}_j}^j]$, $j = 1, \ldots, m$. For j values with $\hat{\eta}_j = 0$, there can not be any interstory damper implemented in building $\mathcal{B}^{(j)}$ and we directly set $\hat{n}_j = 0$ and $\widehat{\mathbf{C}}^{(j)} = [\mathbf{0}]_{n \times n}$. Next, we define the cumulative numbers $\hat{\theta}_0 = 0$, $\hat{\theta}_j = \hat{\eta}_1 + \cdots + \hat{\eta}_j$, $j \geq 1$ and, for j values with $\hat{\eta}_j > 0$, we extract the list

$$\hat{\alpha}^{(j)} = [\hat{\alpha}_1^j, \ldots, \hat{\alpha}_{\hat{\eta}_j}^j] = \overline{\alpha}(\hat{\theta}_{j-1} + 1 : \hat{\theta}_j) \tag{50}$$

and compute the number of interstory dampers $\hat{n}_j = sum(\hat{\alpha}^{(j)})$ allocated in $\mathcal{B}^{(j)}$. For j values with $\hat{n}_j = 0$, there are no dampers allocated in $\mathcal{B}^{(j)}$ and we set $\widehat{\mathbf{C}}^{(j)} = [\mathbf{0}]_{n \times n}$; for j values with $\hat{n}_j > 0$, we compute the list of indexes $\hat{\pi}^{(j)} = \pi(\hat{\alpha}^{(j)})$ that contains the positions of the $\hat{n}_j$ nonzero elements in $\hat{\alpha}^{(j)}$ and obtain the list of interstory positions $\hat{p}^{(j)} = \hat{\sigma}^{(j)}(\hat{\pi}^{(j)})$ where the $\hat{n}_j$ dampers are allocated. To compute the corresponding list of damping coefficients, we define the cumulative numbers $\widehat{N}_0 = 0$, $\widehat{N}_j = \hat{n}_1 + \cdots + \hat{n}_j$, for $j \geq 1$, and extract the coefficient sublist $\hat{c}^{(j)} = d(\widehat{N}_{j-1} + 1 : \widehat{N}_j)$. After obtaining the actuation scheme $\widehat{\mathcal{C}}^{(j)} = [\hat{p}^{(j)}, \hat{c}^{(j)}]$, the associated damping matrix $\widehat{\mathbf{C}}^{(j)}$ can be computed as indicated in Equation (9).

Analogously, to obtain the interbuilding damping matrices $\widetilde{\mathbf{C}}^{(j)}$, $j = 1, \ldots, m-1$, we consider the lists of allowed interbuilding damper positions $\tilde{\sigma}^{(j)} = [\tilde{\sigma}_1^j, \ldots, \tilde{\sigma}_{\tilde{\eta}_j}^j]$, $j = 1, \ldots, m-1$. For j values with $\tilde{\eta}_j = 0$, there can not be any interbuilding damper implemented between buildings $\mathcal{B}^{(j)}$ and $\mathcal{B}^{(j+1)}$, and we set $\tilde{n}_j = 0$ and $\widetilde{\mathbf{C}}^{(j)} = [\mathbf{0}]_{n \times n}$. For j values with $\tilde{\eta}_j > 0$, we consider the cumulative numbers $\tilde{\theta}_0 = 0$, $\tilde{\theta}_j = \tilde{\eta}_1 + \cdots + \tilde{\eta}_j$, $j \geq 1$, extract the sublist

$$\tilde{\alpha}^{(j)} = [\tilde{\alpha}_1^{(j)}, \ldots, \tilde{\alpha}_{\tilde{\eta}_j}^{(j)}] = \overline{\alpha}(\hat{\eta} + \tilde{\theta}_{j-1} + 1 : \hat{\eta} + \tilde{\theta}_j) \tag{51}$$

and compute the number of interbuilding dampers $\tilde{n}_j = sum(\tilde{\alpha}^{(j)})$ allocated between $\mathcal{B}^{(j)}$ and $\mathcal{B}^{(j+1)}$. For j values with $\tilde{n}_j = 0$, there are no dampers allocated between $\mathcal{B}^{(j)}$ and $\mathcal{B}^{(j+1)}$, and we set $\widetilde{\mathbf{C}}^{(j)} = [\mathbf{0}]_{n \times n}$; for j values with $\tilde{n}_j > 0$, we compute the list of indexes $\tilde{\pi}^{(j)} = \pi(\tilde{\alpha}^{(j)})$, which contains the positions of the $\tilde{n}_j$ nonzero elements in $\tilde{\alpha}^{(j)}$, and obtain the list of interbuilding positions $\tilde{p}^{(j)} = \tilde{\sigma}^{(j)}(\tilde{\pi}^{(j)})$ where the dampers are allocated. To compute the corresponding list of damping coefficients, we define the cumulative numbers $\widetilde{N}_0 = 0$, $\widetilde{N}_j = \tilde{n}_1 + \cdots + \tilde{n}_j$, for $j \geq 1$, and extract the sublist $\tilde{c}^{(j)} = d(\hat{n} + \widetilde{N}_{j-1} + 1 : \hat{n} + \widetilde{N}_j)$. From the interbuilding damping scheme $\widetilde{\mathcal{C}}^{(j)} = [\tilde{p}^{(j)}, \tilde{c}^{(j)}]$, the associated damping matrix $\widetilde{\mathbf{C}}^{(j)}$ can be computed as indicated in Equation (12).

To illustrate the described procedure, let us consider the scheme of allowed damper positions in Equation (45), the overall list of allocated dampers in Equation (49) and the list of damping coefficients $d = [d_1, \ldots, d_7]$. First, we observe that $\hat{\eta}_2 = 0$ and set $\hat{n}_2 = 0$ and $\widehat{\mathbf{C}}^{(2)} = [\mathbf{0}]_{5 \times 5}$. Next, we compute the cumulative numbers

$$\hat{\theta}_0 = 0,\ \hat{\theta}_1 = 3,\ \hat{\theta}_2 = 3,\ \hat{\theta}_3 = 6. \tag{52}$$

For $j = 1$, we have $\hat{\eta}_1 = 3$ and extract the sublist

$$\hat{\alpha}^{(1)} = \overline{\alpha}(\hat{\theta}_0 + 1 : \hat{\theta}_1) = \overline{\alpha}(1:3) = [1,1,0], \tag{53}$$

which indicates that there are $\hat{n}_1 = sum(\hat{\alpha}^{(1)}) = sum([1,1,0]) = 2$ interstory dampers allocated in building $\mathcal{B}^{(1)}$. The list of positions of the nonzero elements in $\hat{\alpha}^{(1)}$ is $\hat{\pi}^{(1)} = \pi([1,1,0]) = [1,2]$ and the list of interstory positions where the dampers are allocated is $\hat{p}^{(1)} = \hat{\sigma}^{(1)}(\hat{\pi}^{(1)}) = \hat{\sigma}^{(1)}([1,2]) = [1,2]$. To obtain the list of damping coefficients, we compute the cumulative numbers $\widehat{N}_0 = 0$, $\widehat{N}_1 = 2$, and extract the coefficient sublist $\hat{c}^{(1)} = d(\widehat{N}_0 + 1 : \widehat{N}_1) = d(1:2) = [d_1, d_2]$. After computing the interstory actuation scheme $\widehat{\mathcal{C}}^{(1)} = [\hat{p}^{(1)}, \hat{c}^{(1)}]$, we apply Equation (9) and obtain

$$\widehat{\mathbf{C}}^{(1)} = \begin{bmatrix} d_1 + d_2 & -d_2 & 0 & 0 & 0 \\ -d_2 & d_2 & 0 & 0 & 0 \\ 0 & 0 & 0 & 0 & 0 \\ 0 & 0 & 0 & 0 & 0 \\ 0 & 0 & 0 & 0 & 0 \end{bmatrix}. \tag{54}$$

For $j = 3$, we have $\hat{\eta}_3 = 3$ and obtain the sublist $\hat{\alpha}^{(3)} = \overline{\alpha}(\hat{\theta}_2 + 1 : \hat{\theta}_3) = \overline{\alpha}(4:6) = [1,0,1]$ and the number of dampers $\hat{n}_3 = sum(\hat{\alpha}^{(3)}) = 2$. The nonzero elements in $\hat{\alpha}^{(3)}$ are placed at positions $\hat{\pi}^{(3)} = \pi([1,0,1]) = [1,3]$, which produces the list of interstory damper positions $\hat{p}^{(3)} = \hat{\sigma}^{(3)}(\hat{\pi}^{(3)}) = \hat{\sigma}^{(3)}([1,3]) = [1,3]$. With the values $\hat{n}_2 = 0$ and $\hat{n}_3 = 2$, we can complete the cumulative numbers $\widehat{N}_2 = 2$, $\widehat{N}_3 = 4$, and extract the coefficient sublist $\hat{c}^{(3)} = d(\widehat{N}_2 + 1 : \widehat{N}_3) = d(3:4) = [d_3, d_4]$. According to Equation (9), the corresponding damping matrix is

$$\widehat{\mathbf{C}}^{(3)} = \begin{bmatrix} d_3 & 0 & 0 & 0 & 0 \\ 0 & d_4 & -d_4 & 0 & 0 \\ 0 & -d_4 & d_4 & 0 & 0 \\ 0 & 0 & 0 & 0 & 0 \\ 0 & 0 & 0 & 0 & 0 \end{bmatrix}. \tag{55}$$

To obtain the interbuilding actuation schemes $\widetilde{\mathcal{C}}^{(1)}$ and $\widetilde{\mathcal{C}}^{(2)}$, we consider the total number of allowed interstory positions $\hat{\eta} = 6$, the numbers of allowed interbuilding dampers $\tilde{\eta}_1 = 2$ and $\tilde{\eta}_2 = 2$, and the cumulative values $\tilde{\theta}_0 = 0$, $\tilde{\theta}_1 = 2$ and $\tilde{\theta}_2 = 4$. For $\widetilde{\mathcal{C}}^{(1)}$, we extract the sublist

$$\tilde{\alpha}^{(1)} = \overline{\alpha}(\hat{\eta} + \tilde{\theta}_0 + 1 : \hat{\eta} + \tilde{\theta}_1) = \overline{\alpha}(7:8) = [0,1] \tag{56}$$

and compute the number of interbuilding dampers $\tilde{n}_1 = sum(\tilde{\alpha}^{(1)}) = 1$ allocated between $\mathcal{B}^{(1)}$ and $\mathcal{B}^{(2)}$. The corresponding list of indexes $\tilde{\pi}^{(1)} = \pi(\tilde{\alpha}^{(1)}) = [2]$ produces the list of interbuilding positions $\tilde{p}^{(1)} = \tilde{\sigma}^{(1)}(\tilde{\pi}^{(1)}) = [5]$ where the dampers are allocated. To obtain the list of damping coefficients, we consider the total number of allocated interstory dampers $\hat{n} = 4$ and the cumulative values $\widetilde{N}_0 = 0$, $\widetilde{N}_1 = 1$, and extract the sublist $\tilde{c}^{(1)} = d(\hat{n} + \widetilde{N}_0 + 1 : \hat{n} + \widetilde{N}_1) = d(5:5) = [d_5]$. Next, by applying Equation (12), we obtain the corresponding interbuilding damping matrix

$$\widetilde{\mathbf{C}}^{(1)} = \begin{bmatrix} 0 & 0 & 0 & 0 & 0 \\ 0 & 0 & 0 & 0 & 0 \\ 0 & 0 & 0 & 0 & 0 \\ 0 & 0 & 0 & 0 & 0 \\ 0 & 0 & 0 & 0 & d_5 \end{bmatrix}. \tag{57}$$

Finally, for the interbuilding damping system $\widetilde{\mathcal{C}}^{(2)}$, we obtain the sublist of allocated dampers

$$\tilde{\alpha}^{(2)} = \overline{\alpha}(\hat{\eta} + \tilde{\theta}_1 + 1 : \hat{\eta} + \tilde{\theta}_2) = \overline{\alpha}(9 : 10) = [1, 1], \tag{58}$$

the number of dampers $\tilde{n}_2 = sum(\tilde{\alpha}^{(2)}) = 2$, the list of indexes $\widetilde{\pi}^{(2)} = \pi(\tilde{\alpha}^{(2)}) = [1, 2]$ and the cumulative value $\widetilde{N}_2 = 3$, which produces the list of positions $\tilde{p}^{(2)} = \tilde{\sigma}^{(2)}(\widetilde{\pi}^{(2)}) = [4, 5]$ and the list of damping coefficients $\tilde{c}^{(2)} = d(\hat{n} + \widetilde{N}_1 + 1 : \hat{n} + \widetilde{N}_2) = d(6 : 7) = [d_6, d_7]$. In this case, by applying Equation (12), we get the interbuilding damping matrix

$$\widetilde{\mathbf{C}}^{(2)} = \begin{bmatrix} 0 & 0 & 0 & 0 & 0 \\ 0 & 0 & 0 & 0 & 0 \\ 0 & 0 & 0 & 0 & 0 \\ 0 & 0 & 0 & d_6 & 0 \\ 0 & 0 & 0 & 0 & d_7 \end{bmatrix}. \tag{59}$$

From matrices $\widehat{\mathbf{C}}^{(j)}$, $j = 1, \ldots, m$ and $\widetilde{\mathbf{C}}^{(j)}$, $j = 1, \ldots, m-1$, we can obtain the overall damping matrices $\widehat{\mathbf{C}}$ and $\widetilde{\mathbf{C}}$ given in Equations (22) and (23), respectively, and the overall state-space model for the damped MBS can be computed as indicated in Equation (30).

3.3. Objective Function and Optimization Constraints

For a linear system $\mathcal{S}$ with the state-space model

$$\mathcal{S} : \begin{cases} \dot{\mathbf{x}}(t) = \mathbf{A}\mathbf{x}(t) + \mathbf{B}\mathbf{w}(t), \\ \mathbf{z}(t) = \mathbf{C}_z\mathbf{x}(t), \end{cases} \tag{60}$$

the H_∞ system-norm

$$\gamma(\mathcal{S}) = \gamma(\mathbf{A}, \mathbf{B}, \mathbf{C}_z) = \sup_{\|\mathbf{w}\|_2 \neq 0} \frac{\|\mathbf{z}\|_2}{\|\mathbf{w}\|_2} \tag{61}$$

indicates the maximum energy-gain from the external disturbance $\mathbf{w}(t)$ to the controlled output $\mathbf{z}(t)$ [26,33]. Broadly speaking, the H_∞ approach aims at minimizing the effect of the worst-case scenario by designing a system $\mathcal{S}$ with a minimum γ-value. In our case, we consider the seismic ground acceleration $\ddot{w}(t)$ as external disturbance and take the overall vector of interstory drifts $\hat{\mathbf{r}}(t)$ given in Equation (35) as controlled output, which can be computed from the state vector using the controlled-output matrix $\mathbf{C}_{\hat{r}}$ in Equation (36). As described in the previous section, a particular configuration of the DDS is specified by a list of optimization variables $v = [\overline{\alpha}, d]$, where $\overline{\alpha} = [\overline{\alpha}_1, \ldots, \overline{\alpha}_{\overline{\eta}}]$ is the list of dampers allocation and $d = [d_1, \ldots, d_{n_d}]$ is the list of damping coefficients. According to Equation (30), the corresponding damped system $\mathcal{S}(v)$ admits the state-space representation

$$\mathcal{S}(v) : \begin{cases} \dot{\mathbf{x}}(t) = \{\mathbf{A}_b + \mathbf{A}_d(v)\}\mathbf{x}(t) + \mathbf{B}\ddot{w}(t), \\ \hat{\mathbf{r}}(t) = \mathbf{C}_{\hat{r}}\mathbf{x}(t), \end{cases} \tag{62}$$

where $\mathbf{A}_b$, $\mathbf{B}$ and $\mathbf{C}_{\hat{r}}$ are constant matrices and $\mathbf{A}_d(v)$ has the form indicated in Equation (32). The associated H_∞-cost of that DDS configuration is

$$\gamma(v) = \gamma(\mathbf{A}_b + \mathbf{A}_d(v), \mathbf{B}, \mathbf{C}_{\hat{r}}). \tag{63}$$

The GA solver included in the Matlab Global Optimization Toolbox allows working simultaneously with discrete and continuous optimization variables. Taking advantage of that feature, the allocation variables $\overline{\alpha}_1, \ldots, \overline{\alpha}_{\overline{\eta}}$ are declared as discrete variables and the constraints

$$0 \leq \overline{\alpha}_i \leq 1, \; i = 1, \ldots, \overline{\eta} \tag{64}$$

are used to define them as binary variables. To specify the total number of dampers, we have to set the condition $\sum_{i=1}^{\overline{\eta}} \overline{\alpha}_i = n_d$. As the selected optimization solver does not admit equality constraints on the discrete variables, we use the following pair of linear constraints to impose that condition

$$n_d - \epsilon \leq \sum_{i=1}^{\overline{\eta}} \overline{\alpha}_i, \quad \sum_{i=1}^{\overline{\eta}} \overline{\alpha}_i \leq n_d + \epsilon, \tag{65}$$

where ϵ is a small positive number. We also consider the sublists of interbuilding dampers allocation $\tilde{\alpha}^{(j)} = [\tilde{\alpha}_1^{(j)}, \ldots, \tilde{\alpha}_{\tilde{\eta}_j}^{(j)}], j = 1, \ldots, m-1$, defined in Equation (51) and enforce a full-linked configuration by setting the linear constraints

$$\sum_{i=1}^{\tilde{\eta}_j} \tilde{\alpha}_i^{(j)} \geq 1, \; j = 1, \ldots, m-1. \tag{66}$$

As to the continuous optimization variables $d_1, \ldots, d_{n_d}$, in this work we assume that all the dampers are FVDs with a maximum damping capacity c_{max} and the DDS has a maximum overall damping c_{tot}. Accordingly, we set the bound constraints

$$0 \leq d_i \leq c_{\text{max}}, \; i = 1, \ldots, n_d \tag{67}$$

and the linear constraint

$$\sum_{i=1}^{n_d} d_i \leq c_{\text{tot}}. \tag{68}$$

In summary, the DDS optimal configuration can be obtained by solving the constrained optimization problem:

$$\mathcal{P}_{\Gamma} : \begin{cases} \text{minimize } \gamma(v) = \gamma(\mathbf{A}_b + \mathbf{A}_d(v), \mathbf{B}, \mathbf{C}_{\hat{z}}), \\ \text{subject to } v \in \Gamma, \end{cases} \tag{69}$$

where Γ denotes the set of all admissible v-lists that satisfy the system of constraints specified in Equations (64)–(68).

Remark 4. *Using the* `ss()` *function of the Matlab Control System Toolbox [34], a state-space representation of the linear time-invariant model in Equation (60) can be created with* `sys=ss(A,B,Cz,0)`. *After that, the corresponding H_∞ system-norm in Equation (61) can be readily computed with the function* `hinfnorm` *of the Matlab Robust Control Toolbox [30] in the form* `gamma=hinfnorm(sys)`.

Remark 5. *It should be observed that the GA solver stops after a certain number of generations, producing only a suboptimal solution to the optimization problem $\mathcal{P}_{\Gamma}$. Moreover, due to its stochastic character, distinct solutions are typically obtained in different runs of the solver. From a practical point of view, however, suboptimal solutions with small γ-values are frequently able to define DDSs with high-performance characteristics and, in that sense, can be taken as acceptable solutions to the considered design problem.*

4. DDS Designs

To illustrate the effectiveness and flexibility of the proposed design methodology, in this section we design three different DDSs for the seismic protection of a MBS formed by a row of $m = 5$ identical five-story buildings with story masses ($\times 10^5$ kg) $m_1 = 2.152$, $m_2 = 2.092$, $m_3 = 2.070$, $m_4 = 2.048$, $m_5 = 2.661$ and story stiffness coefficients ($\times 10^8$ N/m) $k_1 = 1.47$, $k_2 = 1.13$, $k_3 = 0.99$, $k_4 = 0.89$ and $k_5 = 0.84$ [35]. The building damping matrix (in Ns/m)

$$\mathbf{C}_b = 10^5 \times \begin{bmatrix} 2.602 & -0.924 & 0 & 0 & 0 \\ -0.924 & 2.196 & -0.810 & 0 & 0 \\ 0 & -0.810 & 1.995 & -0.728 & 0 \\ 0 & 0 & -0.728 & 1.867 & -0.687 \\ 0 & 0 & 0 & -0.687 & 1.274 \end{bmatrix} \tag{70}$$

has been computed as a Rayleigh damping matrix with 2% of relative damping in the first and fifth modes [32]. The DDSs damping devices are assumed to be linear FVDs with maximum damping capacity $c_{\max} = 3.0 \times 10^7$ Ns/m. The maximum DDS damping capacity has been set to $c_{\text{tot}} = 1.5 \times 10^8$ Ns/m and the total number of damping devices has been restricted to $n_d = 12$. Also, a full-linked configuration has been enforced on all the considered DDSs by forbidding empty interbuilding damping systems.

For the first damping configuration (DC1), we select the schemes of allowed interstory damper positions

$$\hat{\sigma}^{(1)} = [1,2,3,4,5],\ \hat{\sigma}^{(2)} = [\,],\ \hat{\sigma}^{(3)} = [1,2,3,4,5],\ \hat{\sigma}^{(4)} = [\,],\ \hat{\sigma}^{(5)} = [1,2,3,4,5] \tag{71}$$

and allowed interbuilding damper positions

$$\tilde{\sigma}^{(1)} = [1,2,3,4,5],\ \tilde{\sigma}^{(2)} = [1,2,3,4,5],\ \tilde{\sigma}^{(3)} = [1,2,3,4,5],\ \tilde{\sigma}^{(4)} = [1,2,3,4,5], \tag{72}$$

which are schematically represented in Figure 5 by blue dashed and red dotted rectangles, respectively. That configuration keeps $\mathcal{B}^{(2)}$ and $\mathcal{B}^{(4)}$ as nonactuated buildings and allows implementing interstory dampers at all the interstory levels of buildings $\mathcal{B}^{(1)}$, $\mathcal{B}^{(3)}$ and $\mathcal{B}^{(5)}$, and interbuilding dampers at all interbuilding positions. In the optimization variable $v = [\overline{\alpha}, d]$, the list of dampers allocations $\overline{\alpha}$ contains $\overline{\eta} = 35$ binary variables, and the list of damping coefficients d includes $n_d = 12$ continuous variables. For DC1, the solution provided by the GA solver to the optimization problem in Equation (69) includes the optimal allocation list

$$\overline{\alpha} = [\overbrace{1,1,1,0,0}^{\hat{\alpha}^{(1)}},\ \overbrace{1,1,1,0,0}^{\hat{\alpha}^{(3)}},\ \overbrace{0,1,1,0,0}^{\hat{\alpha}^{(5)}},\ \overbrace{0,0,0,1,0}^{\tilde{\alpha}^{(1)}},\ \overbrace{0,0,1,0,0}^{\tilde{\alpha}^{(2)}},\ \overbrace{0,0,1,0,0}^{\tilde{\alpha}^{(3)}},\ \overbrace{0,0,1,0,0}^{\tilde{\alpha}^{(4)}}], \tag{73}$$

which corresponds to the system of interstory and interbuilding dampers displayed as blue and red small dashpots in Figure 5, respectively, and the damping coefficients collected in the first row of Table 1.

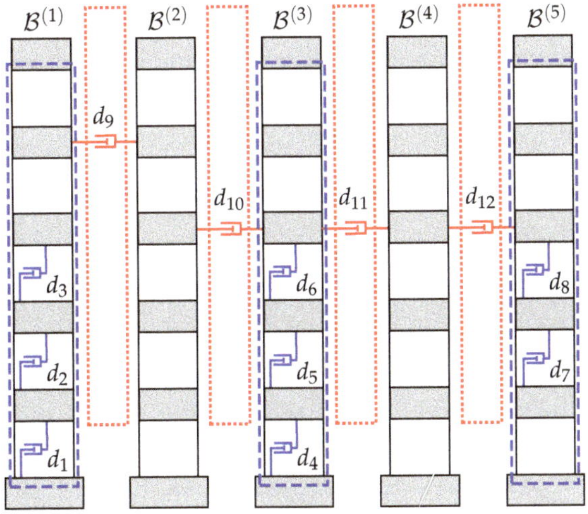

Figure 5. Damping configuration DC1. Full-linked distributed damping system (DDS) with three actuated buildings, $\hat{n} = 8$ interstory dampers and $\tilde{n} = 4$ interbuilding dampers.

Table 1. Values of the damping coefficients corresponding to the linked damping configurations DC1, DC2 and DC3 ($\times 10^7$ Ns/m).

Conf.	d_1	d_2	d_3	d_4	d_5	d_6	d_7	d_8	d_9	d_{10}	d_{11}	d_{12}
DC1	1.4480	1.5137	1.3414	2.0098	2.1014	1.9725	1.5460	1.3074	0.4334	0.3683	0.8023	0.1528
DC2	1.8626	1.9882	1.7986	1.6611	2.1895	1.9489	1.6484	0.6311	0.4362	0.1519	0.0006	0.6828
DC3	2.6792	2.7151	2.5843	2.1407	0.2790	0.9873	0.7835	0.0009	1.3963	0.9310	0.0007	0.5018

The second damping configuration (DC2), schematically displayed in Figure 6, allows placing dampers in the interstory positions

$$\hat{\sigma}^{(1)} = [\,],\ \hat{\sigma}^{(2)} = [1,2,3,4,5],\ \hat{\sigma}^{(3)} = [\,],\ \hat{\sigma}^{(4)} = [1,2,3,4,5],\ \hat{\sigma}^{(5)} = [\,] \tag{74}$$

and the interbuilding positions

$$\tilde{\sigma}^{(1)} = [3,4,5],\ \tilde{\sigma}^{(2)} = [3,4,5],\ \tilde{\sigma}^{(3)} = [3,4,5],\ \tilde{\sigma}^{(4)} = [3,4,5], \tag{75}$$

which constrains the placement of interstory dampers to buildings $\mathcal{B}^{(2)}$ and $\mathcal{B}^{(4)}$ and restricts the placements of interbuilding dampers to the upper three interbuilding levels. In this second case, the total number of optimization variables is reduced to 34, with $\overline{\eta} = 22$ binary allocation variables and $n_d = 12$ continuous damping-coefficient variables. The optimal solution attained by the GA solver for DC2 includes the allocation list

$$\alpha = [\,\overbrace{1,1,1,1,0}^{\hat{\alpha}^{(2)}},\ \overbrace{0,1,1,1,0}^{\hat{\alpha}^{(4)}},\ \overbrace{0,0,1}^{\tilde{\alpha}^{(1)}},\ \overbrace{0,0,1}^{\tilde{\alpha}^{(2)}},\ \overbrace{0,0,1}^{\tilde{\alpha}^{(3)}},\ \overbrace{1,1,0}^{\tilde{\alpha}^{(4)}}\,], \tag{76}$$

and the damping coefficients presented in the second row of Table 1.

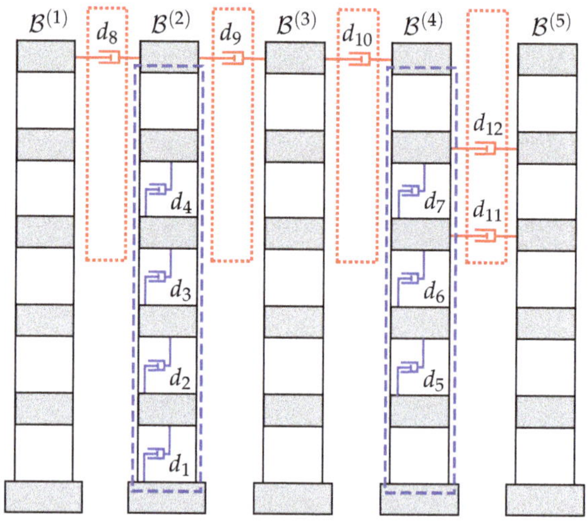

Figure 6. Damping configuration DC2. Full-linked DDS with two actuated buildings, $\hat{n} = 7$ interstory dampers and $\tilde{n} = 5$ interbuilding dampers.

Finally, we consider the third damping configuration (DC3) displayed in Figure 7, which allows placing interbuilding dampers at all interbuilding positions and restricts interstory damper implementation to building $\mathcal{B}^{(1)}$. For that case, the schemes of allowed interstory and interbuilding damping positions are

$$\hat{\sigma}^{(1)} = [1,2,3,4,5],\ \hat{\sigma}^{(2)} = [\,],\ \hat{\sigma}^{(3)} = [\,],\ \hat{\sigma}^{(4)} = [\,],\ \hat{\sigma}^{(5)} = [\,] \tag{77}$$

and

$$\tilde{\sigma}^{(1)} = [1,2,3,4,5],\ \tilde{\sigma}^{(2)} = [1,2,3,4,5],\ \tilde{\sigma}^{(3)} = [1,2,3,4,5],\ \tilde{\sigma}^{(4)} = [1,2,3,4,5], \tag{78}$$

respectively, and the total number of optimization variables is 37 with $\overline{\eta} = 25$ binary damper-allocation variables and $n_d = 12$ continuous damping-coefficient variables. The optimal allocation list obtained for DC3 is

$$\overline{\alpha} = [\overbrace{1,1,1,1,0}^{\hat{\alpha}^{(1)}},\ \overbrace{0,0,1,1,1}^{\tilde{\alpha}^{(1)}},\ \overbrace{0,0,1,0,1}^{\tilde{\alpha}^{(2)}},\ \overbrace{0,0,0,0,1}^{\tilde{\alpha}^{(3)}},\ \overbrace{0,1,0,0,1}^{\tilde{\alpha}^{(4)}}], \tag{79}$$

and the corresponding damping coefficients are collected in the third row of Table 1.

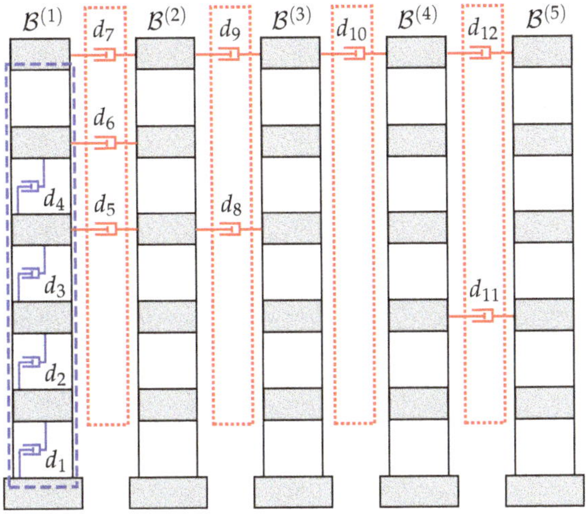

Figure 7. Damping configuration DC3. Full-linked DDS with a single actuated building, $\hat{n} = 4$ interstory dampers and $\tilde{n} = 8$ interbuilding dampers.

Looking at the damping coefficient values in Table 1 and the structure of the optimal damper placements in Figures 5–7, the following facts can be observed: (i) damping coefficients of interbuilding dampers are about one order of magnitude lower than those obtained for interstory dampers, (ii) interstory dampers tend to be placed in low building levels, and (iii) interbuilding dampers are preferably allocated at upper interbuilding positions. These facts are consistent with the results obtained in preliminary works on DDSs for MBSs [22,23]. It can also be observed in Table 1 that there are some dampers with particularly small damping coefficients. Specifically, damping coefficients d_{11} in DC2 and d_8 and d_{11} in DC3 are remarkably small when compared with the coefficient values of all the other interstory and interbuilding dampers. It should be noted that those small coefficients correspond to dampers placed at particularly low interbuilding positions and can be interpreted as a numerical side-effect of constraining the overall number of dampers to exactly $n_d = 12$ elements. From a practical point of view, those residual dampers can be removed without any significant loss of performance and, consequently, the DDSs corresponding to the optimal configurations DC2 and DC3 could be implemented with a set of 11 and 10 dampers, respectively. Regarding the optimal γ-values, the data in Table 2 indicate that the considered damping configurations are all able to produce a significant reduction of the system H_∞-norm when compared with the nonactuated MBS. In particular, an H_∞-norm decrease around 82% is attained by DC3 and larger reductions of about 88% are achieved by DC1 and DC2. The better results obtained by DC1 and DC2 suggest a superior performance of widely distributed interstory damping schemes. However, it is worth highlighting the potential implementation advantages of DC3, which would only require internal modifications of building $\mathcal{B}^{(1)}$. As to the computational aspects, the selected GA solver has shown to be very effective in dealing with the mixture of discrete and continuous variables, the variety of optimization constraints and the relatively large number of optimization variables. All the presented damping configurations have been obtained with a common random seed and using the standard parameter setting for large-scale GA optimization problems suggested in the Matlab Global Optimization Toolbox. Considering the dimension and complexity of the problem and the modest computing resources (see Remark 8), the computation times are notably short, specially when the GA solver is run in parallel mode. Finally, the large number of objective-function evaluations required to obtain the different optimal configurations indicates that the computational cost of evaluating the objective function can certainly be a critical bottleneck for the overall computational effectiveness of the proposed design methodology. In that sense, the presented matrix formulation for the damped multibuilding model has proved to be a relevant contribution.

Table 2. Computational design characteristics of the linked DDS configurations DC1, DC2 and DC3.

Conf.	Act. Build.	H_∞ Norm	opt. vars.	Generations	funct. aval.	Time (s)	Parallel Time (s)
Plain conf.	—	0.8090	—	—	—	—	—
DC1	1, 3, 5	0.0897	47	206	41,400	273.5	67.9
DC2	2, 4	0.0970	34	161	32,400	242.8	57.5
DC3	1	0.1457	37	288	57,800	409.0	103.8

Remark 6. *As suggested in the GA solver documentation for problems with a large number of optimization variables [31], we have introduced some modifications in the default parameter setting of the solver. Specifically, we have set the values 200 for the* `PopulationSize`*, 0.9 for the* `CrossoverFraction`*, 20 for the* `EliteCount` *and 500 for the* `MaxGenerations` *parameters. To take advantage of the CPU multi-core architecture, the GA solver has been enforced to run in parallel mode by enabling the option* `UseParallel`*. Also, to improve the relative accuracy in the computation of the H_∞-norm, the tolerance in the function* `hinfnorm` *has been decreased to 10^{-3} [30].*

Remark 7. *As indicated in Remark 5, the stochastic character of the GA solver typically produces distinct suboptimal solutions in different runs of the solver. For simplicity, in this work the Matlab order* `rng(125)` *has been used to set a common random seed for all the computed DDS configurations. That random seed has been arbitrarily chosen, which confirms the effectiveness of the proposed design methodology and indicates that improved results could be possibly obtained by exploring a wider set of random seeds [29].*

Remark 8. *The computation time values presented in Table 2 should only be taken as approximate references, in the sense that small variations can be observed in the computation time of different runs of the GA solver. Moreover, the computation time in parallel mode can be significantly affected by the available number of CPU cores. In this work, all the computations have been carried out with Matlab 2019a on a regular desktop computer equipped with an Intel Core i7-8700 CPU at 3.20 GHz, 16 GB RAM and a 480GB SSD hard drive.*

5. Seismic Responses

To illustrate the behavior of the different DDSs designed in Section 4, we have carried out a proper set of numerical simulations using the full-scale 180-component of El Centro 1940 seismic record as ground acceleration disturbance (see Figure 8). Specifically, for the nonactuated MBS and the damping configurations DC1, DC2 and DC3, we have computed the maximum absolute interstory drifts

$$|\hat{r}_i^j|_{\max} = \max_{0\le t\le T_w} |\hat{r}_i^j(t)|, i = 1,\dots,5,\ j = 1,\dots,5, \tag{80}$$

the maximum absolute story total-accelerations

$$|a_i^j|_{\max} = \max_{0\le t\le T_w} |a_i^j(t)|, i = 1,\dots,5,\ j = 1,\dots,5 \tag{81}$$

and the maximum interbuilding approachings

$$(\tilde{r}_i^j)_{\max} = \max_{0 \leq t \leq T_w} \tilde{r}_i^j(t), i = 1, \ldots, 5, j = 1, \ldots, 4, \quad (82)$$

where $\hat{r}_i^j(t)$, $a_i^j(t)$ and $\tilde{r}_i^j(t)$ are the output variables defined in Equations (33), (37) and (40), respectively, and T_w denotes the total duration of the seismic disturbance.

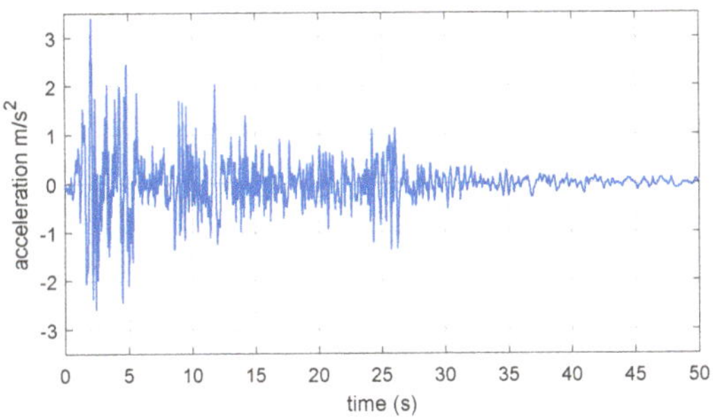

Figure 8. Full-scale 180-component of El Centro 1940 ground-acceleration seismic record with an absolute acceleration-peak of 3.417 m/s^2. Data available at Strong-Motion Virtual Data Center (VDC) (ftp://strongmotioncenter.org/vdc/smdb/1940/c/139u37el.c0a).

The obtained interstory-drift peak-values are presented in Figure 9. A global view of the plots in that figure indicates that the three designed damping configurations are all able to produce an overall reduction of the interstory-drift peak-values when compared with the response of the nonactuated MBS (black solid lines with rectangles). In a more detailed inspection of the DC1 response (red solid lines with circles), it can be appreciated that the interstory-drift peak-value reduction is particularly effective in the actuated buildings $\mathcal{B}^{(1)}$, $\mathcal{B}^{(3)}$ and $\mathcal{B}^{(5)}$. A slightly poorer performance can be observed in the nonactuated buildings $\mathcal{B}^{(2)}$ and $\mathcal{B}^{(4)}$, whose seismic protection is provided through the linking interbuilding dampers (see Figure 5). A similar behavior can be observed in the response of the damping configuration DC2 (blue dashed lines with asterisks), where $\mathcal{B}^{(2)}$ and $\mathcal{B}^{(4)}$ are actuated buildings and $\mathcal{B}^{(1)}$, $\mathcal{B}^{(3)}$ and $\mathcal{B}^{(5)}$ are nonactuated (see Figure 6). In this case, it is worth noting the loss of performance in the upper level of building $\mathcal{B}^{(2)}$, which is an effect that has been observed in previous works [22] and can be associated to the action of the interbuilding links. Finally, for the single-actuated-building configuration DC3 (green dotted lines with triangles), the best results are also attained in the actuated building $\mathcal{B}^{(1)}$, and a moderate but gradual increase of the interstory-drift peak-values can be observed in the nonactuated buildings as we move away from $\mathcal{B}^{(1)}$. Also in this case, a loss of performance associated to the interbuilding links can be appreciated in the upper level of the actuated building $\mathcal{B}^{(1)}$. The obtained story total-acceleration peak-values and maximum interbuilding approachings are displayed in Figures 10 and 11, respectively, using the same colors, line styles and symbols. The plots in Figure 10 confirm that, despite having only included the interstory drifts in the optimization index, the considered H_∞ design approach can produce a notable reduction of the acceleration peak-values. That reduction is more relevant in the actuated buildings and smaller but yet significant in the nonactuated ones. Also in this case, it can be appreciated the progressive loss of performance of the configuration DC3 as we move away from the only actuated building $\mathcal{B}^{(1)}$. Regarding the approaching peak-values, the plots in Figure 11 show that the three damping configurations are able to keep the interbuilding approachings within remarkably low levels. In fact, for the considered seismic disturbance, an interbuilding gap of 3.5 cm can be considered a safe interbuilding separation in all the cases. It is worth noting that the best approaching results are attained by the single-actuated-building configuration DC3, which can be explained by the stronger interbuilding damping system of that configuration.

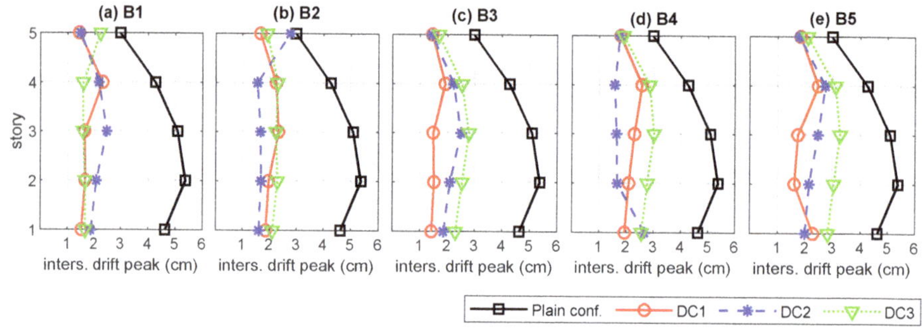

Figure 9. Maximum absolute interstory drifts corresponding to the nonactuated multibuilding system (plain configuration) and the damping configurations DC1, DC2 and DC3.

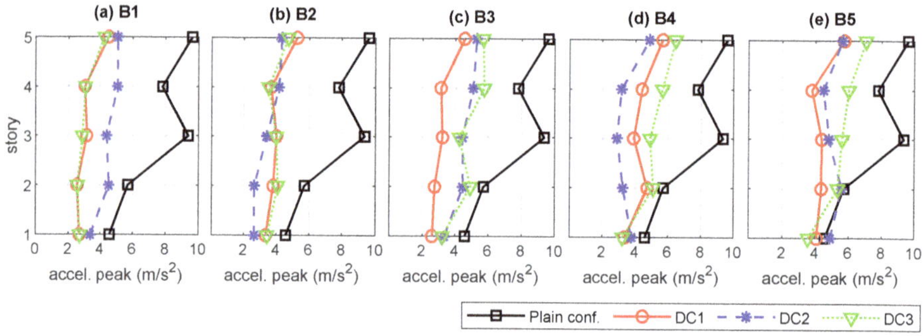

Figure 10. Maximum absolute story total-accelerations attained by the nonactuated multibuilding system (plain configuration) and the damping configurations DC1, DC2 and DC3.

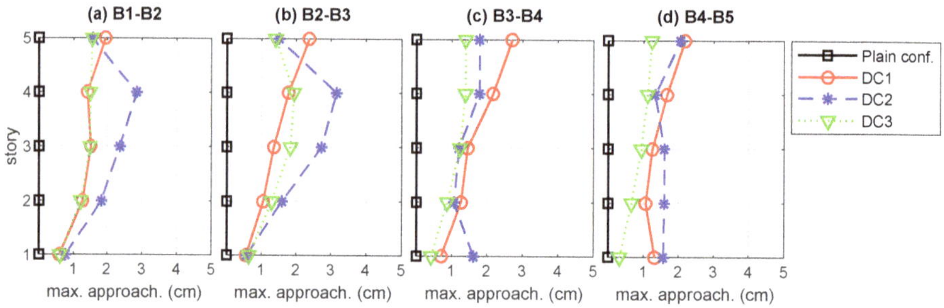

Figure 11. Maximum interbuilding approachings produced by the nonactuated multibuilding system (plain configuration) and the damping configurations DC1, DC2 and DC3.

To gain a better understanding of the relevance of linked configurations in the seismic protection of the overall MBS, we have considered an additional unlinked damping configuration DC4 (see Figure 12), which has been obtained by removing the interbuilding dampers in the linked configuration DC2. The interstory-drift and acceleration peak-values produced by DC4 are displayed in Figures 13 and 14, respectively, and the corresponding maximum approachings are presented in Figure 15. The peak-values produced by the nonactuated MBS and the linked configuration DC2 are also included in those figures as a reference. Looking at the plots of interstory-drift peak-values in Figure 13, it can be appreciated that the unlinked configuration DC4 (red dotted lines with triangles) produces better results than the linked configuration DC2 in the actuated buildings $\mathcal{B}^{(2)}$ and $\mathcal{B}^{(4)}$, but it provides null protection to the nonactuated buildings $\mathcal{B}^{(1)}$, $\mathcal{B}^{(3)}$ and $\mathcal{B}^{(5)}$. The plots of acceleration peak-values in Figure 14 indicate that, in addition of providing null protection to the nonactuated buildings, the unlinked configuration DC4 attains worse results than the linked configuration DC2 in building $\mathcal{B}^{(4)}$ and the first story of building $\mathcal{B}^{(2)}$. As to the interbuilding approachings, the plots in Figure 15 show that large approaching peak-values are produced at the top level of all the buildings by the unlinked configuration DC4, which would require interbuilding separations of about 20 cm to avoid pounding.

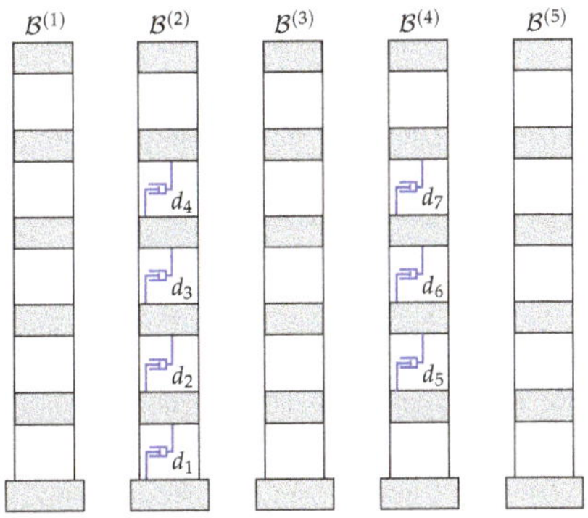

Figure 12. Fully unlinked damping configuration DC4 obtained by suppressing the interbuilding dampers in the linked configuration DC2.

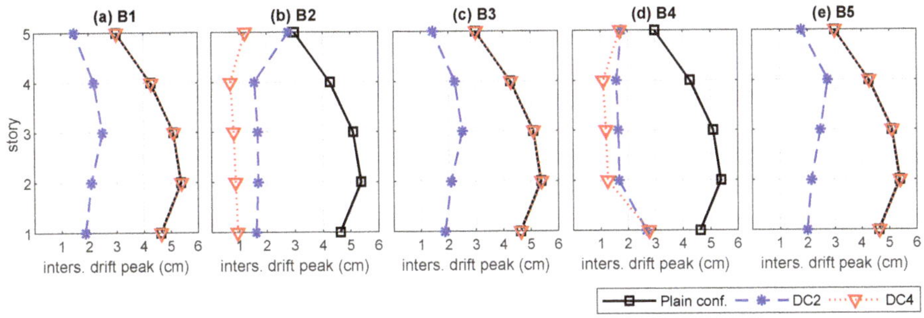

Figure 13. Maximum absolute interstory drifts corresponding to the nonactuated multibuilding system (plain configuration), the linked configuration DC2 and the unlinked configuration DC4.

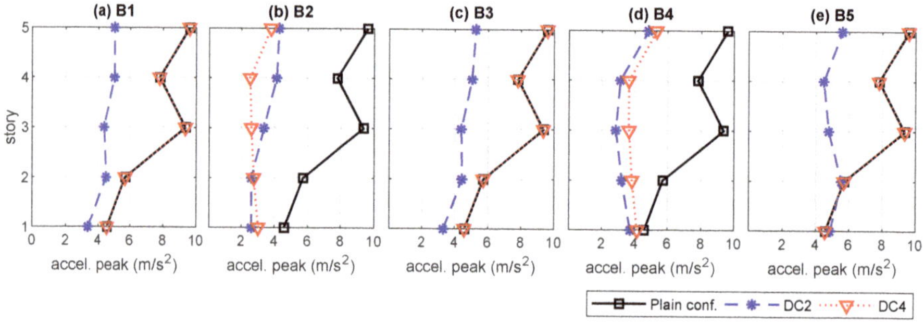

Figure 14. Maximum absolute story total-accelerations attained by the nonactuated multibuilding system (plain configuration), the linked configuration DC2 and the unlinked configuration DC4.

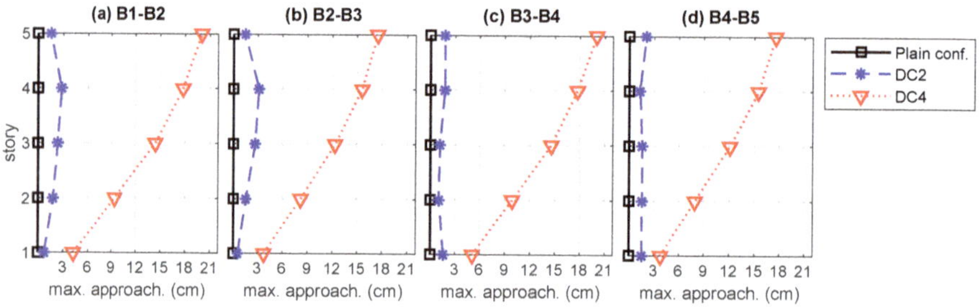

Figure 15. Maximum interbuilding approachings produced by the nonactuated multibuilding system (plain configuration), the linked configuration DC2 and the unlinked configuration DC4.

Finally, to summarize the global behavior of the nonactuated MBS and the discussed damping configurations, we consider the overall peak-values of absolute interstory drifts

$$|\hat{r}|_{\max} = \max_{1\leq j\leq m}\left(\max_{1\leq i\leq n}|\hat{r}_i^j|_{\max}\right), \tag{83}$$

absolute story total-accelerations

$$|a|_{\max} = \max_{1\leq j\leq m}\left(\max_{1\leq i\leq n}|a_i^j|_{\max}\right) \tag{84}$$

and interbuilding approachings

$$(\tilde{r})_{\max} = \max_{1\leq j\leq m-1}\left(\max_{1\leq i\leq n}(\tilde{r}_i^j)_{\max}\right). \tag{85}$$

The obtained overall peak-values and the corresponding H_∞-norms are collected in Table 3. The data in the table indicate that, with respect to the nonactuated MBS, the DDS configurations DC1 and DC2 produce reductions of about 50% in the overall interstory-drift peak-value and around 40% in the overall acceleration peak-value. The overall reductions attained by the single-actuated-building configuration DC3 are around 40% in the interstory-drift and slightly below 30% in the acceleration responses. The overall maximum approaches are around 3 cm for DC1 and DC2 and inferior to 2 cm for DC3. Those values indicate that the linked DDSs obtained with the proposed design strategy can provide an overall seismic protection for the MBS. At the same time, the values corresponding to DC4 illustrate the inefficacy of unlinked configurations in mitigating the interstory-drift and acceleration seismic response of the overall MBS and clearly demonstrate their possible detrimental effects on pounding risk.

Table 3. H_∞ norm and overall maximum peak-values corresponding to the nonactuated multibuilding system (plain configuration), the linked configurations DC1, DC2 and DC3, and the unlinked configuration DC4.

Conf.	Act. Build.	H_∞ Norm	Max. drift (cm)	Max. accel. (m/s^2)	Max. Approach. (cm)
Plain	—	0.8090	5.38	9.62	0.00
DC1	1, 3, 5	0.0897	2.56	5.71	2.70
DC2	2, 4	0.0970	2.76	5.62	3.15
DC3	1	0.1457	3.25	7.02	1.92
DC4	2, 4	0.6272	5.38	9.62	20.05

Remark 9. *The plots in Figure 15 indicate that null interbuilding approachings are produced by the nonactuated MBS, which can be explained by the synchronized response of the identical buildings subjected to the same seismic excitation. From a practical point of view, however, it should be observed that there are a number of factors, such as the differential variation of the building structural parameters over time, the effect of live loads (weight of persons, furniture, equipment, movable partitions, etc.) or the soil-structure interaction, that can brake the ideal synchrony of the buildings response and, consequently, increase the risk of pounding.*

6. Conclusions

In this work, we have investigated the design of distributed damping systems (DDSs) for the overall seismic protection of multiple adjacent buildings. The considered DDSs include two different kinds of damping devices: interstory dampers, which are implemented inside the buildings, and external interbuilding damping links. To keep the problem complexity within reasonable limits, we have assumed that the damping elements are linear fluid viscous dampers and the buildings have been considered as linear planar frames with identical dynamic characteristics. The main objective of the study is designing suitable DDS configurations that are able to mitigate the buildings seismic response by reducing the interstory-drift and story-acceleration peak-values and, at the same time, are capable of cutting down the risk of interbuilding collisions (pounding) by producing small interbuilding approachings. Typically, designing high-performance DDSs involves solving a mixed allocation-tuning optimization problem, which includes both determining convenient damper positions and computing proper values for the damper parameters. The proposed design methodology is based on an effective matrix formulation of the damped multibuilding system, follows an H_∞ approach that permits avoiding costly numerical simulations of seismic time-responses, exploits the computational advantages of state-of-the-art genetic algorithm (GA) solvers, and allows setting actuation schemes of particular interest such as full-linked configurations or nonactuated buildings. To illustrate the main features of the presented design strategy, three different DDS configurations have been computed for a system of five adjacent multistory buildings. Also, to explore the performance characteristics of the obtained DDS configurations, a convenient set of numerical simulations of the corresponding seismic responses have been carried out using the full-scale 180-component of El Centro 1940 seismic record as ground acceleration input. Considering the obtained results, the following points can be highlighted: (i) properly designed DDSs can provide an overall seismic protection to systems of multiple adjacent buildings, being able to mitigate the buildings seismic response and reduce the pounding risk; (ii) full-linked DDS configurations should be used to attain the seismic protection of nonactuated buildings and to produce low levels of pounding risk; (iii) a simultaneous reduction of the buildings interstory-drift and story-accelerations peak values can be attained with the considered H_∞ approach; (iv) the proposed design methodology is highly flexible, being able to produce high-performance DDS configurations for a wide variety of actuation schemes; and (v) the proposed approach is computationally effective in dealing with large-scale problems. Regarding that last point, it should be observed that computational efficiency is a critical factor in DDS design of multibuilding problems. In this context, a fast evaluation of the objective function and the possibility of running the GA solver in parallel mode are elements of singular relevance.

After the positive results obtained in the present work, we believe that further research effort should be invested in obtaining a deeper understanding of the problem and removing some of the model simplifications introduced in this paper. In that sense, some lines of particular interest include the usage of inerter-based vibration absorbers [36,37], the study of the effects of interstory and interbuilding velocities on the multibuilding damper allocation problem [38,39], the analysis of the effects produced by soil-structure interaction [40] and seismic-wave propagation [41] on large multibulding problems, and the formulation of extended design strategies for elastic-plastic structures [42] and/or nonlinear damping devices [43].

Author Contributions: Conceptualization, formal analysis, investigation, methodology and writing—review & editing were conducted collaboratively by all the authors; software, F.P.-Q. and J.R.-M.; visualization, J.R.-M. and J.M.R.; writing—original draft, F.P.-Q. and J.R.-M. All authors have read and agreed to the published version of the manuscript.

Funding: This research was partially supported by the Spanish Ministry of Economy and Competitiveness under Grant DPI2015-64170-R (MINECO/FEDER) and by the Italian Ministry of Education, University and Research under the Project "Department of Excellence LIS4.0—Lightweight and Smart Structures for Industry 4.0".

Conflicts of Interest: The authors declare no conflict of interest.

Abbreviations

The following abbreviations are used in this manuscript:

ATOP	allocation-tuning optimization problem
DDS	distributed damping system
FVD	fluid viscous damper
GA	genetic algorithm
MBS	multibuilding system

References

1. Soong, T.T.; Spencer, B.F., Jr. Supplemental energy dissipation: state-of-the-art and state-of-the-practice. *Eng. Struct.* **2002**, *24*, 243–259. [CrossRef]
2. Ali, M.M.; Moon, K.S. Structural developments in tall buildings: Current trends and future prospects. *Archit. Sci. Rev.* **2007**, *50*, 205–223. [CrossRef]
3. Takewaki, I.; Fujita, K.; Yamamoto, K.; Takabatake, H. Smart passive damper control for greater building earthquake resilience in sustainable cities. *Sustain. Cities Soc.* **2011**, *1*, 3–15. [CrossRef]
4. Wang, S.; Mahin, S.A. Seismic retrofit of a high-rise steel moment-resisting frame using fluid viscous dampers. *Struct. Des. Tall Special Build.* **2017**, *26*, 1–11. [CrossRef]
5. Aydin, E.; Ozturk, B.; Dutkiewicz, M. Analysis of efficiency of passive dampers in multistorey buildings. *J. Sound Vib.* **2019**, *439*, 17–28. [CrossRef]
6. Li, Z.; Shu, G.; Huang, Z. Proper configuration of metallic energy dissipation system in shear-type building structures subject to seismic excitation. *J. Constr. Steel Res.* **2019**, *154*, 177–189. [CrossRef]

7. De Domenico, D.; Ricciardi, G.; Takewaki, I. Design strategies of viscous dampers for seismic protection of building structures: A review. *Soil Dyn. Earthq. Eng.* **2019**, *118*, 144–165. [CrossRef]
8. López García, D.; Soong, T.T. Efficiency of a simple approach to damper allocation in MDOF structures. *J. Struct. Control* **2002**, *9*, 19–30. [CrossRef]
9. Liu, W.; Tong, M.; Lee, G.C. Optimization methodology for damper configuration based on building performance indices. *J. Struct. Eng.* **2005**, *131*, 1746–1756. [CrossRef]
10. Main, J.A.; Krenk, S. Efficiency and tuning of viscous dampers on discrete systems. *J. Sound Vib.* **2005**, *286*, 97–122. [CrossRef]
11. Fujita, K.; Yamamoto, K.; Takewaki, I. An evolutionary algorithm for optimal damper placement to minimize interstorey-drift transfer function in shear building. *Earthq. Struct.* **2010**, *1*, 289–306. [CrossRef]
12. Del Gobbo, G.M.; Williams, M.S.; Blakeborough, A. Comparing fluid viscous damper placement methods considering total-building seismic performance. *Earthq. Eng. Struct. Dyn.* **2018**, *47*, 2864–2886. [CrossRef]
13. Singh, M.P.; Moreschi, L.M. Optimal seismic response control with dampers. *Earthq. Eng. Struct. Dyn.* **2001**, *30*, 553–572. [CrossRef]
14. Huang, X. Evaluation of genetic algorithms for the optimum distribution of viscous dampers in steel frames under strong earthquakes. *Earthq. Struct.* **2018**, *14*, 215–227.
15. Cetin, H.; Aydin, E.; Ozturk, B. Optimal design and distribution of viscous dampers for shear building structures under seismic excitations. *Front. Built Environ.* **2019**, *5*, 1–13. [CrossRef]
16. Del Gobbo, G.M. Placement of fluid viscous dampers to improve total-building seismic performance. In Proceedings of the CSCE Annual Conference, Laval, Montreal, QC, Canada, 12–15 June 2019; pp. 1–9.
17. Tubaldi, E. Dynamic behavior of adjacent buildings connected by linear viscous/viscoelastic dampers. *Struct. Control Health Monit.* **2015**, *22*, 1086–1102. [CrossRef]
18. Kandemir-Mazanoglu, E.C.; Mazanoglu, K. An optimization study for viscous dampers between adjacent buildings. *Mech. Syst. Signal Process.* **2017**, *89*, 88–96. [CrossRef]
19. Liu, M.Y.; Wang, A.P.; Chiu, M.Y. Optimal vibration control of adjacent building structures interconnected by viscoelastic dampers. *J. Civ. Eng. Arch.* **2017**, *11*, 468–476.
20. Patel, C.C. Dynamic response of viscous damper connected similar multi-degree of freedom structures. *Int. J. Earth Sci. Eng.* **2011**, *4*, 1068–1071.
21. Patel, C.C.; Jangid, R.S. Dynamic response of identical adjacent structures connected by viscous damper. *Struct. Control Health Monit.* **2014**, *21*, 205–224. [CrossRef]
22. Palacios-Quiñonero, F.; Rubió-Massegú, J.; Rossell, J.M.; Karimi, H.R. Integrated design of hybrid interstory-interbuilding multi-actuation schemes for vibration control of adjacent buildings under seismic excitations. *Appl. Sci.* **2017**, *7*, 323. [CrossRef]
23. Palacios-Quiñonero, F.; Rubió-Massegú, J.; Rossell, J.; Rodellar, J. Interstory-interbuilding actuation schemes for seismic protection of adjacent identical buildings. *Smart Struct. Syst.* **2019**, *24*, 67–81.
24. Del Gobbo, G.M.; Blakeborough, A.; Williams, M.S. Improving total-building seismic performance using linear fluid viscous dampers. *Bull. Earthq. Eng.* **2018**, *16*, 4249–4272. [CrossRef]
25. Yang, J.N.; Lin, S.; Kim, J.-H.; Agrawal, A.K. Optimal design of passive energy dissipation systems based on H_∞ and H_2 performances. *Earthq. Eng. Struct. Dyn.* **2002**, *31*, 921–936. [CrossRef]
26. Palacios-Quiñonero, F.; Rubió-Massegú, J.; Rossell, J.M.; Karimi, H.R. Optimal passive-damping design using a decentralized velocity-feedback H_∞ approach. *Model. Identif. Control* **2012**, *33*, 87–97. [CrossRef]
27. Abdeddaim, M.; Ounis, A.; Djedoui, N.; Shrimali, M.K. Pounding hazard mitigation between adjacent planar buildings using coupling strategy. *J. Civ. Struct. Health Monit.* **2016**, *6*, 603–617. [CrossRef]
28. Jankowski, R.; Mahmoud, S. Linking of adjacent three-storey buildings for mitigation of structural pounding during earthquakes. *Bull. Earthq. Eng.* **2016**, *14*, 3075–3097. [CrossRef]
29. Palacios-Quiñonero, F.; Rubió-Massegú, J.; Rossell, J.M.; Karimi, H.R. Design of inerter-based multi-actuator systems for vibration control of adjacent structures. *J. Frankl. Inst.* **2019**, *356*, 7785–7809.
30. Balas, G.J.; Chiang, R.Y.; Packard, A.K.; Safonov, M.G. *MATLAB Robust Control Toolbox User's Guide, 2019a*; The MathWorks, Inc.: Natick, MA, USA, 2019.

31. MathWorks. *MATLAB Global Optimization Toolbox User's Guide, 2019a*; The MathWorks, Inc.: Natick, MA, USA, 2019.
32. Chopra, A. *Dynamics of Structures. Theory and Applications to Earthquake Engineering*, 3rd ed.; Prentice Hall: Upper Saddle River, NJ, USA, 2007.
33. Wang, Y.; Lynch, J.P.; Law, K.H. Decentralized H_∞ controller design for large-scale civil structures. *Earthq. Eng. Struct. Dyn.* **2009**, *38*, 377–401. [CrossRef]
34. MathWorks. *MATLAB Control System Toolbox User's Guide, 2019a*; The MathsWorks, Inc.: Natick, MA, USA, 2019.
35. Kurata, N.; Kobori, T.; Takahashi, M.; Niwa, N.; Midorikawa, H. Actual seismic response controlled building with semi-active damper system. *Earthq. Eng. Struct. Dyn.* **1999**, *28*, 1427–1447. [CrossRef]
36. Wang, Q.; Qiao, H.; De Domenico, D.; Zhu, Z.; Xie, Z. Wind-induced response control of high-rise buildings using inerter-based vibration absorbers. *Appl. Sci.* **2019**, *9*, 5045. [CrossRef]
37. Petrini, F.; Giaralis, A.; Wang, Z. Optimal tuned mass-damper-inerter (TMDI) design in wind-excited tall buildings for occupants' comfort serviceability performance and energy harvesting. *Eng. Struct.* **2020**, *204*, 1–16. [CrossRef]
38. Adachi, F.; Fujita, K.; Tsuji, M.; Takewaki, I. Importance of interstory velocity on optimal along-height allocation of viscous oil dampers in super high-rise buildings. *Eng. Struct.* **2013**, *56*, 489–500. [CrossRef]
39. Logotheti, V.E.; Kafetzi, T.C.; Papagiannopoulos, G.A.; Karabalis, L.K. On the use of interstorey velocity for the seismic retrofit of steel frames with viscous dampers. *Soil Dyn. Earthq. Eng.* **2020**, *129*, 1–14. [CrossRef]
40. Sarcheshmehpour, M.; Estekanchi, H.E.; Ghannad, M.A. Optimum placement of supplementary viscous dampers for seismic rehabilitation of steel frames considering soil–structure interaction. *Struct. Des. Tall Special Build.* **2020**, *29*, 1–17. [CrossRef]
41. Savor Novak, M.; Lazarevic, D.; Atalic, J.; Uros, M. Influence of multiple-support excitation on seismic response of reinforced concrete arch bridges. *Appl. Sci.* **2020**, *10*, 17. [CrossRef]
42. Akehashi, H.; Takewaki, I. Optimal viscous damper placement for elastic-plastic MDOF structures under critical double impulse. *Front. Built Environ.* **2019**, *5*, 1–17. [CrossRef]
43. De Domenico, D.; Ricciardi, G. Earthquake protection of structures with nonlinear viscous dampers optimized through an energy-based stochastic approach. *Eng. Struct.* **2019**, *179*, 523–539. [CrossRef]

MDPI
St. Alban-Anlage 66
4052 Basel
Switzerland
Tel. +41 61 683 77 34
Fax +41 61 302 89 18
www.mdpi.com

Applied Sciences Editorial Office
E-mail: applsci@mdpi.com
www.mdpi.com/journal/applsci